多模态大模型
技术与开发实战

薛栋 孙中奇 孙挥 编著

人民邮电出版社

北京

图书在版编目（CIP）数据

多模态大模型技术与开发实战 / 薛栋，孙中奇，孙挥编著. -- 北京：人民邮电出版社，2025. -- ISBN 978-7-115-65538-7

Ⅰ．TP18

中国国家版本馆 CIP 数据核字第 2024RJ2903 号

内 容 提 要

本书循序渐进地阐述了多模态大模型的核心开发技术与应用实战的知识。全书共 10 章，分别讲解了多模态技术概述、多模态模型与框架、多模态数据处理、多模态表示学习、多模态嵌入表示、多模态大模型的训练、多模态大模型的评估与验证、基于多模态大模型的翻译系统、基于多模态大模型的音视频广义零样本学习系统、基于 Diffusion Transformer 的文生图系统。全书简洁而不失技术深度，内容丰富全面，案例翔实，以通俗易懂的文字介绍了复杂的知识体系，易于阅读，是学习多模态大模型开发的实用教程。

本书适用于已经了解了 Python 语言基础语法，想进一步学习大模型开发、自然语言处理、计算机视觉处理、机器学习、深度学习技术的读者，还可以作为各类院校相关专业师生的学习用书和培训学校的教材。

◆ 编　著　薛　栋　孙中奇　孙　挥
  责任编辑　张　涛
  责任印制　王　郁　焦志炜

◆ 人民邮电出版社出版发行　北京市丰台区成寿寺路 11 号
  邮编　100164　电子邮件　315@ptpress.com.cn
  网址　https://www.ptpress.com.cn
  大厂回族自治县聚鑫印刷有限责任公司印刷

◆ 开本：787×1092　1/16
  印张：23
  字数：637 千字
  2025 年 2 月第 1 版
  2025 年 2 月河北第 1 次印刷

定价：109.80 元

读者服务热线：(010)81055410　印装质量热线：(010)81055316
反盗版热线：(010)81055315

前　　言

　　多模态技术是指通过融合不同类型的数据（如文本、图像、音频、视频等）来增强系统的信息处理和理解能力的技术。这项技术近年来发展迅速，得益于人工智能领域的重大进展。多模态技术在许多领域展现了巨大的潜力，如自然语言处理、计算机视觉、语音识别和生成对抗网络等。

　　多模态技术的应用场景广泛，涵盖了从消费级应用到工业级应用的各个方面。

- **智能助手与聊天机器人**：通过结合语音、文本和图像数据，多模态技术可以提高人与智能助手和聊天机器人的交互体验。
- **医疗健康**：结合影像、电子病历和基因数据，多模态技术能够帮助医务人员更准确地诊断疾病。
- **自动驾驶**：整合视觉和雷达数据，多模态技术可以显著提高自动驾驶系统的感知和决策能力。
- **内容生成与推荐**：在社交媒体和电商平台上，多模态技术结合用户的文本、图像和行为数据，可以实现更精准的内容生成和推荐。
- **安防监控**：多模态技术结合视频监控和音频数据，可以提升异常行为检测和安全监控的效果。

　　多模态大模型不仅是技术进步的结果，更是企业创新和发展的重要驱动力。随着企业和开发者对智能化需求的不断增加，如何高效地开发多模态大模型成为一个迫切的课题。本书通过系统的知识讲解、丰富的实战案例和实用的应用技巧，帮助读者掌握多模态大模型开发技术的精髓，为广大企业和开发者提供一条通往智能化应用的高效路径。

本书的特色

- **全面覆盖多模态技术**

  本书从多模态技术的基本概念出发，系统介绍了多模态数据类型、模型框架、数据处理、表示学习、嵌入表示、模型训练、评估与验证等内容，涵盖了多模态技术的各个方面，提供了全面的知识体系。读者可以通过本书掌握多模态技术的核心理论和实战技能，为深入研究和应用多模态技术打下坚实基础。

- **深度剖析经典模型**

  书中详细解析了多模态领域的经典模型，通过具体案例展示经典模型的原理、架构和应用方法，帮助读者深入理解模型内部机制，学会在实际项目中应用这些模型。

- **实战项目指导**

  本书不仅注重理论讲解，更注重实战应用。每章都配有具体的项目案例，如基于 S3D MIL-NCE 的多模态文本到视频检索、多模态图像搜索引擎、CLIP 模型的训练与微调、多模态翻译系统等。这些项目案例结合了新的研究成果和实际应用场景，可以帮助读者在动手实践过程中掌握多模态技术。

- **实例多、源代码完整**

  本书讲解了多模态大模型开发的多个知识点，每个知识点都用典型的例子进行了演示，并且提供了完整的源代码。

- **前沿技术追踪**

  本书紧跟多模态技术的发展前沿，介绍了新的研究进展和应用趋势，如多模态对抗生成网

前言

络、零样本学习、多模态嵌入表示等，帮助读者了解当前多模态技术的发展动态，拓宽视野，跟上技术潮流。

- **注重数据处理与增强**
  数据是多模态技术的核心，本书专门设立章节详细讲解多模态数据的预处理、数据增强和特征提取方法，涵盖文本、图像、音频、视频等多种数据类型。通过这些内容，读者可以学会如何高效地处理多模态数据，提高模型的训练效果和性能。

- **模型评估与验证**
  本书深入探讨了多模态模型的评估与验证方法，包括模型性能评估、数据准备与分割、交叉验证、模态间一致性验证等，帮助读者全面掌握模型评估技术。

- **实用工具和框架**
  书中介绍了多种常用的深度学习框架和工具，如 TensorFlow、PyTorch、Keras 等，并结合这些框架和工具讲解如何构建和训练多模态大模型。这些内容将帮助读者熟练掌握各类深度学习工具的使用，提升实际操作能力。

- **丰富的示例代码**
  本书大部分章节配有详细的示例代码，涵盖数据处理、模型构建、训练、评估等多个环节，读者可以直接运行和修改这些代码，进行实验和项目开发。通过动手实践，读者可以更好地理解书中的内容，提高编程和实战能力。

总之，本书通过理论与实践相结合的方式，全面系统地介绍了多模态技术的核心开发技术和实际应用方法。阅读本书，读者不仅可以掌握多模态技术的基本原理和前沿进展，还能通过实战项目提高实际操作能力，为从事多模态技术研究和应用打下坚实的基础。

致谢

本书在编写过程中，得到了人民邮电出版社编辑的大力支持，正是编辑的求实、耐心和高效，才使得本书能够在短时间内出版。在此，编者还要特别感谢 X-D Lab（心动实验室，https://github.com/X-D-Lab）的支持与帮助。X-D Lab 长期致力于人工智能领域的探索与研究，其核心成员包括高校教授、企业高管及大模型专家等。X-D Lab 成员的专业知识和宝贵建议对本书内容的完善起到了至关重要的作用，编者欢迎读者与 X-D Lab 交流沟通（邮箱：xd.lab2023@gmail.com。QQ 群号：362742959）。读者加入本书提供的 QQ 群，可以下载本书配套的源程序。

另外，编者十分感谢家人给予的巨大支持。编者水平有限，书中纰漏之处在所难免，恳请读者提出宝贵的意见或建议，以便修订并使之更加完善。本书编辑联系邮箱：zhangtao@ptpress.com.cn。

感谢读者购买本书，希望本书能成为读者编程路上的领航者，祝读者阅读快乐！

编者

目 录

第1章 多模态技术概述 ······1
1.1 多模态技术简介 ······1
- 1.1.1 什么是多模态 ······1
- 1.1.2 多模态技术的发展历史 ······2
- 1.1.3 多模态技术的应用场景 ······3

1.2 多模态数据类型 ······4
- 1.2.1 常见的多模态类型 ······4
- 1.2.2 图像数据 ······4
- 1.2.3 文本数据 ······5
- 1.2.4 音频数据 ······5
- 1.2.5 视频数据 ······6

1.3 多模态技术的基本原理 ······6
- 1.3.1 数据融合与对齐 ······6
- 1.3.2 多模态表示学习 ······7
- 1.3.3 多模态推理 ······8

第2章 多模态模型与框架 ······10
2.1 常用的深度学习框架 ······10
- 2.1.1 TensorFlow ······10
- 2.1.2 PyTorch ······11
- 2.1.3 Keras ······11
- 2.1.4 Transformer ······12

2.2 多模态模型技术 ······13
- 2.2.1 ViLBERT 模型 ······13
- 2.2.2 VisualBERT 模型 ······14
- 2.2.3 OpenAI CLIP 模型 ······15
- 2.2.4 UNITER 模型 ······16
- 2.2.5 LXMERT 模型 ······17
- 2.2.6 SigLIP 模型 ······18
- 2.2.7 LoRA 微调技术 ······18
- 2.2.8 LLaVA 模型 ······18

2.3 预训练模型 ······19
- 2.3.1 预训练模型简介 ······19
- 2.3.2 使用预训练模型 ······20
- 2.3.3 预训练模型的微调 ······24

第3章 多模态数据处理 ······28
3.1 数据预处理 ······28
- 3.1.1 文本预处理 ······28
- 3.1.2 图像预处理 ······39
- 3.1.3 音频预处理 ······43
- 3.1.4 视频预处理 ······50

3.2 数据增强 ······51
- 3.2.1 文本数据增强 ······51
- 3.2.2 图像数据增强 ······54
- 3.2.3 音频数据增强 ······55
- 3.2.4 视频数据增强 ······57

3.3 特征提取 ······59
- 3.3.1 特征在大模型中的关键作用 ······60
- 3.3.2 文本特征提取 ······60
- 3.3.3 图像特征提取 ······64
- 3.3.4 音频特征提取 ······67
- 3.3.5 视频特征提取 ······68

第4章 多模态表示学习 ······71
4.1 多模态表示学习介绍 ······71
- 4.1.1 多模态表示学习简介 ······71
- 4.1.2 多模态表示学习的主要方法 ······71

4.2 多模态表示学习方法 ······72
- 4.2.1 表示融合 ······72
- 4.2.2 联合学习 ······76
- 4.2.3 表示对齐 ······79
- 4.2.4 模态间关系建模 ······83

4.3 基于 S3D MIL-NCE 的多模态文本到视频检索 ······86
- 4.3.1 项目介绍 ······86
- 4.3.2 准备工作 ······87

4.3.3 视频加载和可视化 …… 87
4.3.4 加载视频并定义文本查询 …… 89
4.3.5 预处理视频和查询 …… 90
4.3.6 展示结果 …… 90

第5章 多模态嵌入表示 …… 92

5.1 多模态嵌入基础 …… 92
5.1.1 多模态嵌入介绍 …… 92
5.1.2 多模态嵌入的应用 …… 92
5.2 图像嵌入 …… 93
5.2.1 图像嵌入介绍 …… 93
5.2.2 图像特征提取 …… 94
5.2.3 模态对齐 …… 97
5.2.4 CLIP 模型 …… 100
5.3 文本嵌入 …… 101
5.3.1 多模态模型中的文本嵌入 …… 101
5.3.2 基于 CLIP 模型的文本嵌入 …… 103
5.4 音频嵌入 …… 105
5.4.1 音频特征提取 …… 105
5.4.2 常用音频嵌入模型 …… 106
5.5 多模态图像搜索引擎 …… 109
5.5.1 项目介绍 …… 109
5.5.2 CLIP 模型的配置参数 …… 110
5.5.3 数据集处理 …… 111
5.5.4 实现 Bangla CLIP 模型 …… 115
5.5.5 基于文本的图像搜索 …… 117
5.5.6 基于 Streamlit 的 Web 客户端 …… 118

第6章 多模态大模型的训练 …… 121

6.1 模型训练的过程 …… 121
6.2 训练策略 …… 122
6.2.1 预训练与微调 …… 122
6.2.2 多任务学习 …… 125
6.2.3 全量微调 …… 127
6.2.4 对比学习 …… 129
6.2.5 参数高效微调 …… 131
6.2.6 迁移学习 …… 133

6.2.7 人类反馈强化学习 …… 135
6.2.8 动态学习率调整 …… 137
6.2.9 SFT 微调 …… 138
6.3 CLIP 模型训练与微调 …… 141
6.3.1 项目介绍 …… 141
6.3.2 创建文本和图像配对数据集 …… 142
6.3.3 构建多模态模型 …… 145
6.3.4 训练模型 …… 161
6.3.5 模型微调 …… 162
6.3.6 调试运行 …… 162

第7章 多模态大模型的评估与验证 …… 164

7.1 模型评估 …… 164
7.1.1 模型评估的必要性 …… 164
7.1.2 评估指标 …… 165
7.1.3 单模态性能评估 …… 168
7.1.4 多模态融合性能评估 …… 170
7.1.5 效率与资源使用 …… 171
7.1.6 定性评估和复杂场景评估 …… 172
7.1.7 语音命令识别系统 …… 173
7.2 模型验证 …… 183
7.2.1 模型验证的必要性 …… 183
7.2.2 数据准备和分割 …… 184
7.2.3 交叉验证 …… 185
7.2.4 嵌套交叉验证 …… 188
7.2.5 模态间一致性验证 …… 189
7.2.6 模型鲁棒性验证 …… 190
7.2.7 验证指标 …… 192
7.3 多模态大模型评估基准 …… 195
7.3.1 MM-Vet …… 195
7.3.2 MMEvalPro …… 196
7.3.3 MMT-Bench …… 197
7.4 CLIP 模型的增强训练与评估 …… 197
7.4.1 项目介绍 …… 197
7.4.2 定义数据集 …… 198
7.4.3 创建模型 …… 201
7.4.4 分词器 …… 207

  7.4.5 损失函数 ·················· 209
  7.4.6 模型训练 ·················· 210
  7.4.7 模型评估 ·················· 218
  7.4.8 文本重写 ·················· 221

第8章 基于多模态大模型的翻译系统 ·················· 225

 8.1 背景介绍 ······················ 225
 8.2 系统分析 ······················ 225
  8.2.1 系统需求分析 ············ 225
  8.2.2 技术架构分析 ············ 226
  8.2.3 项目介绍 ·················· 226
 8.3 准备数据集 ··················· 227
  8.3.1 Multi30k 数据集介绍 ···· 227
  8.3.2 下载 Multi30k 数据集 ··· 227
  8.3.3 下载 WIT 数据集中的图像数据 ·················· 229
 8.4 数据集处理 ··················· 230
  8.4.1 PyTorch 数据集类 ········ 230
  8.4.2 数据处理和后处理 ······ 233
  8.4.3 数据集填充 ··············· 236
  8.4.4 获取 Multi30k 数据集的数据 ·························· 237
  8.4.5 获取 WIT 数据集的数据 ···· 239
  8.4.6 获取 WMT 数据集的文本数据 ······················ 241
 8.5 多模态大模型 ················ 243
  8.5.1 功能函数 ·················· 243
  8.5.2 适配器模型 ··············· 244
  8.5.3 获取文本输入的嵌入表示 ·· 245
  8.5.4 多模态模型类 ············ 245
  8.5.5 多模态文本生成任务模型 ·· 247
  8.5.6 分布式训练 ··············· 248
  8.5.7 模型训练和测试 ········· 249
  8.5.8 主程序 ····················· 253

第9章 基于多模态大模型的音视频广义零样本学习系统 ·············· 256

 9.1 背景介绍 ······················ 256
 9.2 系统分析 ······················ 256
  9.2.1 系统需求分析 ············ 256
  9.2.2 功能分析 ·················· 257
 9.3 系统配置 ······················ 258
  9.3.1 命令行接口 ··············· 258
  9.3.2 数据集处理 ··············· 260
  9.3.3 辅助函数 ·················· 267
 9.4 特征提取 ······················ 275
  9.4.1 从 ActivityNet 数据集提取特征 ·························· 275
  9.4.2 从 UCF101 数据集提取特征 ·························· 278
  9.4.3 从 VGGSound 数据集提取特征 ·························· 280
 9.5 多模态模型 ··················· 283
  9.5.1 多模态数据学习模型 ··· 283
  9.5.2 性能评估指标 ············ 289
  9.5.3 模型优化器 ··············· 295
  9.5.4 模型训练和验证 ········· 296
  9.5.5 模型的评估 ··············· 300
  9.5.6 主文件 ····················· 302
 9.6 调试运行 ······················ 307

第10章 基于 Diffusion Transformer 的文生图系统 ·················· 310

 10.1 Diffusion Transformer 介绍 ···· 310
  10.1.1 Diffusion Transformer 的特点 ······················· 310
  10.1.2 Stable Diffusion 和 Diffusion Transformer 的区别和联系 ····················· 310
 10.2 项目介绍 ····················· 311
 10.3 准备预训练模型 ············ 312
 10.4 扩散模型核心模块 ········· 313
  10.4.1 计算高斯分布概率 ···· 313
  10.4.2 实现扩散模型 ·········· 314
  10.4.3 模型扩展 ················ 328
  10.4.4 采样器调度 ············ 330
 10.5 训练模型 ····················· 333
  10.5.1 定义不同配置的 DiT 模型 ······················· 333

10.5.2 最小训练脚本……341
10.5.3 实现 DiT 模型……343
10.5.4 DiT 模型的标准训练……344
10.5.5 DiT 模型的全精度训练……347
10.5.6 DiT 模型的特征预训练……352
10.5.7 DiT 模型的原始训练……352
10.5.8 DiT 模型的禁用 TF32 模式训练……352

10.6 生成图像……353
10.6.1 预训练生成……353
10.6.2 基于 DDP 的图像生成……354

10.7 调试运行……357

第1章 多模态技术概述

多模态技术是一种结合和处理多种类型数据（如文本、图像、音频和视频）的技术，旨在通过融合这些不同模态的信息来提高模型的理解和生成能力。这种开发方法广泛应用于人工智能和机器学习领域，能够实现更复杂的任务，如文生图、视频分析和多模态对话系统。多模态技术的关键在于设计和训练能够有效融合和处理多源数据的模型，从而使系统具备更高的智能和更好的适应性。本章将详细讲解多模态开发的基础知识，介绍多模态开发的历史、数据类型和原理，为读者对后面知识的学习打下基础。

1.1 多模态技术简介

本节将详细讲解多模态技术的基本知识。

1.1.1 什么是多模态

多模态（Multimodal）涉及多种不同类型数据或信号的处理和融合，每种数据类型或信号被称为一种模态。常见的模态包括文本、图像、音频、视频等。多模态技术旨在同时利用这些不同模态的数据，以实现更全面、更准确的理解和决策。

1. 核心概念
- 模态：一种特定类型的数据或信号。例如，文本是一种模态，图像是一种模态，音频也是一种模态。
- 多模态融合：将来自不同模态的数据进行结合和综合，以利用各模态的优势，从而提升系统的整体性能。例如，通过结合视觉和听觉信息，系统可以更准确地识别和理解环境。

2. 技术和方法
- 深度学习：尤其是卷积神经网络（CNN, Convolutional Neural Network）和循环神经网络（RNN, Recurrent Neural Network）在处理图像和序列数据方面表现出色。
- 注意力机制：用于选择和加权不同模态的信息，提升模型的性能。
- 模态预训练模型：如 OpenAI 的 CLIP 和 DALL-E，能够通过大规模预训练，在多种模态间实现优秀的泛化能力。

3. 技术挑战
- 数据对齐和同步：不同模态的数据可能具有不同的时间和空间特性，需要进行有效的对齐和同步。
- 信息融合：设计算法以有效地融合不同模态的信息，避免信息丢失或冲突。
- 模型复杂性：多模态模型往往比单模态模型更复杂，需要更多的计算资源和更大的数据集来训练。

4. 未来展望

多模态技术有望在更多领域实现突破，如智能家居、自动驾驶、教育和娱乐等。随着计算能力和数据获取手段的不断提升，多模态技术将变得更加普及和强大，为人工智能的发展带来新的机遇和挑战。

1.1.2 多模态技术的发展历史

多模态技术的发展历史充满了创新和突破，涉及多个学科的交叉融合。下面将简要介绍多模态技术发展的几个重要阶段。

1. 初期探索阶段（20世纪80年代至90年代）

（1）背景与基础研究。
- 在20世纪80年代，计算机视觉和自然语言处理作为独立的研究领域开始发展。
- 在20世纪90年代，研究者们开始探索将不同模态的数据结合起来，以提高系统的性能。例如，早期的语音识别系统尝试结合口型识别，提高语音识别的准确率。

（2）关键技术与成果。
- 视觉和语音信号处理技术的初步发展。
- 计算机视觉和自然语言处理领域的基础算法和模型。

2. 融合与协同阶段（21世纪初期）

（1）背景与基础研究。
- 21世纪初期，随着计算能力的提升和互联网数据的爆炸式增长，多模态数据的获取变得更加容易。
- 在这一时期，研究者们开始致力于融合来自不同模态的数据，以实现更复杂和智能的系统。

（2）关键技术与成果。
- 图像和文本结合的初步应用，如图像标注和图文搜索。
- 多模态传感器融合技术在机器人和自动驾驶领域开始应用。

3. 深度学习时代（21世纪10年代）

（1）背景与基础研究。
- 在21世纪10年代，深度学习的崛起为多模态技术带来了新的机遇。深度神经网络在图像识别、语音识别和自然语言处理等单模态任务中表现出色，激发了研究者将这些技术应用于多模态任务。

（2）关键技术与成果。
- 卷积神经网络在图像处理方面取得重大突破。
- 循环神经网络和长短期记忆网络：在处理序列数据（如文本和语音）方面表现优异。
- 生成对抗网络（GAN, Generative Adversarial Network）：在图像生成和风格迁移方面取得显著进展。
- 多模态模型：如Show and Tell、Visual Question Answering（VQA），以及DeepMind的AlphaGo（融合视觉和棋局信息）。

4. 多模态预训练模型的兴起（21世纪20年代）

（1）背景与基础研究。
- 在21世纪20年代，预训练和微调（fine-tuning）策略在自然语言处理和计算机视觉领域大获成功，进一步推动了多模态预训练模型的发展。

（2）关键技术与成果。
- BERT和GPT系列：二者在自然语言处理领域的成功，引发了多模态预训练模型的研究。
- CLIP：由OpenAI开发，能够从文本描述中理解图像，并实现图文匹配和搜索。
- DALL-E：由OpenAI开发，能够根据文本描述生成高质量的图像。
- Flamingo：由DeepMind开发，实现了强大的多模态理解和生成能力。

5. 未来阶段
- 跨模态学习：进一步提高不同模态之间的信息互通和共享。
- 实时多模态处理：在实时应用中处理和融合多模态数据，如智能驾驶和增强现实。
- 多模态交互：开发更加自然和高效的人机交互系统，如虚拟助手和机器人。
- 伦理与隐私：在多模态技术应用中确保数据隐私和伦理规范。

多模态技术的发展历程显示出其广阔的应用前景和持续的创新潜力。随着技术的不断进步，多模态技术将会在更多领域实现突破，为人工智能的发展注入新的动力。

1.1.3 多模态技术的应用场景

多模态技术因能够融合和处理不同类型的数据而具备广泛的应用场景。多模态技术的主要应用领域如下。

1. 图像与视频分析
- 文生图：通过文本描述生成对应的图像，应用于艺术创作、广告设计等。
- 视觉问答（VQA）：根据图像内容回答问题，应用于辅助盲人、教育等领域。
- 视频字幕生成：结合视频和音频信息生成自动字幕，应用于影视制作、视频会议等。

2. 自然语言处理与理解
- 多模态聊天机器人：结合文本、语音和图像进行自然互动，提升用户体验，应用于客服、智能助手等领域。
- 情感分析：结合文本和音频分析用户情感，应用于社交媒体分析、客户服务等。

3. 医疗健康
- 辅助诊断：结合医疗图像（如 X 光片、CT 图像）和文本记录，辅助医生进行诊断，应用于医疗影像分析、电子病历分析等。
- 远程医疗：通过视频和语音进行远程诊疗，应用于医疗资源匮乏地区。

4. 自动驾驶与智能交通
- 环境感知：结合车载摄像头和雷达数据，感知和理解周围环境，应用于自动驾驶汽车、智能交通管理等。
- 驾驶员监控：通过视频和语音监控驾驶员状态，提高行车安全。

5. 安防监控
- 行为识别：结合视频和音频监控识别异常行为，应用于公共安全、智能监控系统等。
- 身份验证：结合面部识别和语音识别进行身份验证，应用于安防系统、门禁系统等。

6. 教育与培训
- 虚拟教师：结合文本、图像和语音进行互动教学，应用于在线教育、培训等。
- 沉浸式学习：通过多模态技术提供沉浸式学习体验，应用于语言学习、职业培训等。

7. 娱乐与媒体
- 内容生成：通过文本生成图像或视频内容，应用于游戏开发、影视制作等。
- 虚拟现实（VR）与增强现实（AR）：结合多模态数据提供更为沉浸的体验，应用于游戏、电影、广告等领域。

8. 人机交互
- 智能助手：结合语音、图像和文本进行自然的人机交互，应用于智能家居、智能手机等。
- 增强现实（AR）应用：结合图像和语音提供增强现实体验，应用于购物、旅游等。

9. 电子商务与推荐系统
- 视觉搜索：用户上传图像进行商品搜索，应用于电商平台。

- 多模态推荐：结合用户的浏览记录、文本评论和图像信息进行商品推荐，应用于个性化推荐系统。

总之，多模态技术通过整合和分析来自不同模态的数据，能够达到对数据更全面和深入的理解，从而在各个应用场景中发挥重要作用。这不仅提高了系统的智能化水平，也为用户带来了更加丰富和便捷的体验。

1.2 多模态数据类型

常见的多模态数据类型包括文本、图像、音频和视频等，这些数据类型可以相互补充。通过整合不同模态的信息，系统可以更好地理解和生成复杂的内容。例如，音频可以补充视频内容，图像和视频可以提供直观的视觉信息，文本可以描述图像，也可以提供详细的解释和上下文。多模态技术在自然语言处理、计算机视觉、语音识别等领域具有广泛的应用前景，能够提升人工智能系统的智能化和交互性。

1.2.1 常见的多模态类型

多模态技术通过融合和处理多种类型的数据，能够提供更全面和智能的解决方案，极大地拓展了人工智能的应用范围和能力。在目前的技术条件下，常见的多模态类型如下。

- 文生图（Text-to-Image Generation）：根据文本描述生成相应的图像，例如输入描述"一个红色的苹果在桌子上"后生成相应的图像。
- 文生视频（Text-to-Video Generation）：根据文本描述生成相应的视频片段，例如输入描述"一个人在海边散步"后生成对应的视频。
- 视觉问答（Visual Question Answering，VQA）：根据图像内容回答自然语言问题，例如给定一张图片和问题"图片中有几只猫？"后，系统回答"两只猫"。
- 图像描述生成（Image Captioning）：根据图像生成相应的文本描述，例如输入一张小狗的图片后生成描述"一只小狗在草地上玩耍"。
- 语音转文本（Speech-to-Text）：将语音信号转换成文本，例如将录音内容转写成文本文档。
- 文本生成语音（Text-to-Speech，TTS）：将文本内容转换成语音，例如输入文本"你好，今天天气很好"后生成对应的语音输出。
- 视频描述生成（Video Captioning）：结合视频和音频信息为视频生成描述性字幕，例如为无声视频添加字幕，描述视频中的主要事件和动作。
- 多模态情感分析（Multimodal Sentiment Analysis）：结合文本、语音和面部表情分析情感，例如在视频通话中分析对方的语音语调和面部表情，判断其情感状态。
- 多模态翻译（Multimodal Translation）：结合多种模态信息进行语言翻译，例如结合文本和图像进行文档翻译。

1.2.2 图像数据

图像数据是由像素（Pixel）组成的二维数组，每个像素包含了图像中的颜色和亮度信息。颜色信息通常以红、绿、蓝（RGB）三种基本颜色的组合来表示，每种颜色的取值范围在 0~255。除 RGB 之外，还有其他表示颜色的方式，如灰度图像只有单通道表示亮度信息。

在现实应用中，图像数据的获取方式多种多样，可以通过数字摄像头、扫描仪、卫星遥感设备等获取，也可以通过计算机图形学技术生成。在获取图像数据后，还需要对图像数据进行预处理，如调整大小、裁剪、去噪等，以便于后续的分析和应用。

在多模态数据中，图像数据经常与其他类型的数据结合使用，例如以下几种。
- 文本数据与图像数据结合：在图像描述生成、图像分类、图像检索等任务中，图像数据与文本描述相结合，帮助系统理解图像内容。
- 音频数据与图像数据结合：在视频处理任务中，图像数据与音频数据结合，进行视频内容分析、行为识别等。
- 传感器数据与图像数据结合：在智能交通、智能家居等领域，图像数据与传感器数据结合，进行环境感知、行为监测等。

图像数据的处理和分析涉及许多技术，包括图像特征提取、图像分割、目标检测、图像分类、图像生成等。随着深度学习技术的发展，特别是卷积神经网络的兴起，图像数据的处理和分析能力不断提升，使得图像识别、视频分析、医学影像分析等领域产生了革命性的进步。

1.2.3　文本数据

文本数据是多模态数据中的一种数据类型，是由字符和单词组成的语言信息的表示形式。文本数据可以包含在许多不同的上下文中，包括书面文档、网络页面、社交媒体帖子、电子邮件等。在多模态数据中，文本数据通常与其他类型的数据（如图像、音频、视频）结合使用，以提供更丰富和全面的信息。

在多模态应用中，文本数据的特点如下。
- 符号性质：文本数据由字符和单词组成，每个字符和单词都有其特定的符号含义。
- 结构化和非结构化：文本数据可以是结构化的（如表格数据、标记语言数据）或非结构化的（如自然语言文本），处理方式有所不同。
- 语义丰富：文本数据通常包含丰富的语义信息，能够表达复杂的概念和思想。
- 多样性：文本数据的形式和内容多种多样，可以是长篇大论、简短评论、表格数据等。

在多模态数据中，文本数据常常与其他模态的数据结合使用，例如下面两种。
- 图像数据与文本数据结合：在图像标注、图像搜索、图像描述生成等任务中，图像数据与文本数据相结合，帮助系统理解图像内容。
- 音频数据与文本数据结合：在语音识别、语音转写等任务中，音频数据转换成文本数据，便于进一步处理和分析。

文本数据的处理和分析通常涉及自然语言处理（NLP，Natural Language Processing）技术，包括文本清洗、分词、词性标注、句法分析、命名实体识别、情感分析等。随着深度学习技术的发展，特别是循环神经网络（RNN）和 Transformer 模型的出现，文本数据的处理和分析能力不断提升，为文本理解、信息检索、机器翻译等领域带来了新的突破。

1.2.4　音频数据

音频数据是多模态数据中的一种数据类型，它代表了声音信号的数字化表示。音频数据通常以数字形式存储，可以通过麦克风、录音设备等采集，也可以通过数字化处理从模拟声音信号转换而来。在多模态数据中，音频数据常常与其他类型的数据（如图像、文本、视频）结合使用，以提供更丰富和全面的信息。

在多模态应用中，音频数据的主要特点如下。
- 时域表示：音频数据通常以时间序列的形式表示，每个时间点上的采样值代表了声音信号的振幅。
- 频域表示：音频数据可以通过傅里叶变换等方法转换成频域表示，用于分析声音信号的频谱特征。

- 声学特性：音频数据反映了声音信号的声学特性，如频率、振幅、时长、音调等。
- 语音信息：音频数据中包含了说话者的语音信息，可以包括语言、语调、情感等内容。

在多模态数据中，音频数据通常与其他模态的数据结合使用，例如，文本数据与音频数据结合：在语音识别、语音合成等任务中，将音频数据转换成文本数据，方便进行自然语言处理。

音频数据的处理和分析涉及信号处理、语音处理等技术，包括音频特征提取、语音识别、语音合成、音乐信息检索等。随着深度学习技术的发展，特别是循环神经网络和卷积神经网络的出现，音频数据的处理和分析能力不断提升，使语音识别、音乐生成、情感分析等领域取得了新的进展。

1.2.5 视频数据

视频数据是多模态数据中的一种数据类型，它是由一系列连续的图像帧组成的，以时间序列的方式记录了连续的视觉信息。每个图像帧都是图像数据的一部分，具有图像数据的特性，同时又涵盖了更丰富的时间维度信息。在多模态数据中，视频数据通常与其他类型的数据（如音频、文本）结合使用，以提供更全面和丰富的信息。

视频数据的特点如下。

- 时序性：视频数据是由一系列连续的图像帧组成的，每一帧都代表了视频在某个时刻的图像内容。
- 时空相关性：视频数据不仅包含了图像数据的空间信息，还包含了时间维度的变化信息，反映了视频中物体的运动和变化。
- 动态性：视频数据能够捕捉到真实世界中的动态场景和行为，具有丰富的视觉信息。
- 复杂性：视频数据通常包含大量的图像帧，处理和分析视频数据需要消耗大量的计算资源。

在多模态数据中，视频数据通常与其他模态的数据结合使用，例如下面两种。

- 文本数据与视频数据结合：在视频字幕生成、视频内容理解等任务中，文本数据与视频数据相结合，帮助系统理解视频内容。
- 音频数据与视频数据结合：在视频处理任务中，音频数据与视频数据相结合，进行语音识别、音乐分析等。

视频数据的处理和分析涉及视频编解码、视频特征提取、视频内容理解、视频内容生成等技术。随着深度学习技术的发展，特别是卷积神经网络和循环神经网络的应用，视频数据的处理和分析能力不断提升，使视频内容理解、视频检索、视频生成等领域取得了新的进展。

1.3 多模态技术的基本原理

多模态技术的基本原理是通过将多种模态的数据进行融合，并结合相应的模型和算法进行处理和分析，以提取数据之间的关联性和信息融合的特征，从而实现对复杂真实世界的综合感知和智能化应用。

1.3.1 数据融合与对齐

数据融合与对齐是多模态技术中的重要步骤，旨在将来自不同感知模态的数据整合在一起，并确保它们具有相似的表示形式和语义内容，以便后续的处理和分析。下面将详细介绍数据融合与对齐的过程和方法。

1. 数据预处理

在进行数据融合与对齐之前，首先需要对不同模态的原始数据进行预处理，以确保数据的质量和一致性。数据预处理的步骤包括数据清洗、去噪、归一化、采样率调整等，旨在消除数据中的噪声和不一致性，使数据具有可比性和可融合性。

2. 特征提取

针对不同模态的数据，需要设计相应的特征提取方法，将原始数据转换成特征表示形式。特征提取的目的是从数据中提取出具有代表性和区分性的特征，以便于后续的融合和对齐。常用的特征提取方法包括传统的特征提取算法（如 SIFT、HOG 等）以及基于深度学习的特征提取方法（如卷积神经网络）。

3. 数据对齐

数据对齐是指将来自不同模态的数据映射到一个统一的表示空间中，使它们具有相似的语义内容和表示形式。数据对齐的过程通常涉及将不同模态的特征进行匹配和对齐，以确保它们在统一表示空间中具有一致的语义信息。常用的数据对齐方法包括基于距离度量的方法、基于相关性分析的方法、基于学习的方法等。

4. 融合与整合

在数据对齐之后，不同模态的数据已经具有了相似的表示形式和语义内容，可以进行进一步的融合和整合。融合与整合的目的是将不同模态的信息进行有机结合，以获得更加全面和丰富的信息。常用的融合与整合方法包括加权平均法、特征级融合、决策级融合等。

5. 增强与补全

在数据融合与对齐的过程中，可能会出现一些模态间的信息缺失或不完整的情况。针对这些问题，需要设计相应的增强与补全方法，以补充缺失的信息，提高数据的完整性和一致性。常用的增强与补全方法包括插值法、生成对抗网络等。

通过数据融合与对齐，可以将来自不同感知模态的数据整合在一起，为后续的多模态分析和应用提供基础。这一过程涉及数据预处理、特征提取、数据对齐、融合与整合以及增强与补全等多个步骤，需要综合考虑不同模态数据的特点和要求，设计合适的方法和算法。

1.3.2 多模态表示学习

多模态表示学习是指学习如何将多模态数据映射到一个统一的表示空间，以便于后续的处理和分析。多模态表示学习的核心目标是挖掘不同模态之间的相关性和共享信息，将其转化为一个统一的特征表示，从而实现对多模态数据的整合和统一处理。下面将详细介绍多模态表示学习的方法和技术。

1. 共享表示学习

共享表示学习是一种常用的多模态表示学习方法，其思想是通过学习一个共享的表示空间，将不同模态的数据映射到这个共享空间，使不同模态的数据具有相似的表示形式。常用的共享表示学习方法包括联合主成分分析（Joint Principal Component Analysis，JPCA）、联合稀疏编码（Joint Sparse Coding，JSC）、联合独立成分分析（Joint Independent Component Analysis，JICA）等。

2. 异构网络嵌入

异构网络嵌入是一种基于图嵌入的多模态表示学习方法，其思想是将不同模态的数据表示为一个图结构，然后通过图嵌入算法将这个图结构映射到一个低维的向量空间中，从而实现对多模态数据的表示学习。常用的异构网络嵌入方法包括图卷积网络（Graph Convolutional Network，GCN）、异构信息网络嵌入（Heterogeneous Information Network Embedding，HIN-Embedding）等。

3. 深度神经网络方法

近年来，深度神经网络方法在多模态表示学习中取得了很大的成功。这类方法通过设计深度神经网络结构，将不同模态的数据输入网络，并通过网络的层次结构学习到一个统一的特征表示。常用的深度神经网络方法包括多模态卷积神经网络（Multimodal Convolutional Neural Network，MCNN）、多模态递归神经网络（Multimodal Recurrent Neural Network，MRNN）等。

4. 生成对抗网络方法

生成对抗网络（Generative Adversarial Network，GAN）是一种通过对抗训练的方式学习多模态表示的方法。在这种方法中，一个生成器网络负责生成多模态数据，而一个判别器网络则负责判断生成的数据是否真实。通过不断地迭代训练，生成器网络学习到一个统一的多模态表示，使生成的数据具有较高的真实性和一致性。

5. 弱监督学习方法

弱监督学习方法是一种利用大量无标注数据和少量有标注数据进行多模态表示学习的方法。这类方法通过设计合适的损失函数和训练策略，使模型能够从无标注数据中学习到一个统一的多模态表示，从而提高模型的泛化能力和鲁棒性。

通过上述方法，可以实现对多模态数据的统一表示学习，从而为后续的多模态分析和应用打下基础。这些方法涵盖了共享表示学习、异构网络嵌入、深度神经网络方法、生成对抗网络方法和弱监督学习方法等多种技术手段，能够有效地挖掘不同模态之间的相关性和共享信息，实现对多模态数据的综合分析和应用。

1.3.3 多模态推理

多模态推理是指在多模态数据的基础上进行推理和决策，以实现更加综合和准确的推断结果。多模态推理的核心目标是利用多模态数据之间的关联性和信息融合的特点，提高推理和决策的准确性和鲁棒性。下面将详细介绍多模态推理的方法和技术。

1. 跨模态信息融合

多模态推理的关键在于如何有效地融合来自不同模态的信息，包括对文本、图像、音频等不同模态的数据进行融合和整合，以提取出更丰富和全面的信息。常用的信息融合方法包括特征级融合、决策级融合和模型级融合等。

2. 跨模态语义理解

多模态推理需要对不同模态的数据进行语义理解，以确保推理过程中考虑到了不同模态数据之间的语义关联。这包括将文本数据转换为语义表示、对图像数据进行物体识别和场景理解、对音频数据进行语音识别和情感分析等。

3. 跨模态推理模型

为了实现多模态推理，需要设计相应的推理模型，以处理来自不同模态的数据并产生推断结果。常用的跨模态推理模型包括多模态神经网络、跨模态图网络、跨模态注意力网络等。这些模型能够有效地整合不同模态的信息，实现更加准确和全面的推理。

4. 跨模态推理任务

多模态推理涉及多种任务和应用场景，包括图像标注、视频内容理解、情感分析、语音转写等。在这些任务和应用场景中，需要将来自不同模态的数据进行综合推理（即跨模态推理），从而实现对复杂真实世界的全面理解和应用。

5. 强化学习方法

近年来，强化学习方法在多模态推理中得到了广泛应用。通过设计合适的奖励函数和策略，强化学习方法能够实现从多模态数据中学习到一个有效的决策策略，从而提高推理的准确性和鲁棒性。

通过上述方法，可以实现对多模态数据的综合推理，从而实现对复杂真实世界的全面理解和应用。这些方法涵盖了跨模态信息融合、跨模态语义理解、跨模态推理模型、跨模态推理任务和强化学习方法等多种技术手段，能够有效地挖掘不同模态之间的相关性和信息融合的特点，提高推理和决策的准确性和鲁棒性。

第2章 多模态模型与框架

本章将详细讲解多模态模型与相关框架的知识。首先介绍 TensorFlow、PyTorch 和 Keras 这三大深度学习框架，然后详细分析 ViLBERT、VisualBERT 和 OpenAI CLIP 等多模态模型，最后阐述预训练模型的概念及其使用和微调的方法。

2.1 常用的深度学习框架

常用的深度学习框架包括 TensorFlow、PyTorch 和 Keras。TensorFlow 由 Google 开发，具有高度灵活性和可扩展性，适用于从研究到生产的各个阶段；PyTorch 由 Facebook 开发，以动态计算图和直观的编程体验著称，广受研究界和工业界欢迎；Keras 则是一个简洁易用的高层 API，适合快速构建和训练深度学习模型。

2.1.1 TensorFlow

TensorFlow 是由 Google Brain 团队开发并于 2015 年发布的一个开源深度学习框架，旨在简化机器学习模型的开发和部署，支持从研究实验到生产环境的各种应用。

1. 主要特点

（1）多平台支持：TensorFlow 可以在多个硬件平台上运行，包括 CPU、GPU 和 TPU（Tensor Processing Unit）。TensorFlow 还支持跨平台部署，能够在服务器、桌面、移动设备和嵌入式系统上运行。

（2）灵活性和可扩展性：TensorFlow 使用计算图（Computational Graph）来表示计算任务，支持静态图（Static Graph）和动态图（Eager Execution）模式。用户可以灵活地构建复杂的模型，并对其进行优化和调整。

（3）丰富的生态系统：TensorFlow 拥有一个庞大且活跃的开发者社区，其生态系统包括多个子项目。

- TensorFlow Hub：一个用于发布、发现和重用机器学习模型组件的平台。
- TensorFlow Lite：一个轻量级解决方案，用于在移动和嵌入式设备上部署模型。
- TensorFlow.js：一个在浏览器和 Node.js 环境中运行的深度学习库。
- TensorFlow Extended（TFX）：一个用于构建和管理生产级机器学习管道的端到端平台。

（4）高级 API 和工具：TensorFlow 提供了多种高级 API，如 Keras，用于简化模型的构建和训练。它还提供了 TensorBoard 和 TensorFlow Datasets，前者是一个用于可视化和调试的工具，后者可用于简化数据加载和预处理。

2. 应用场景

TensorFlow 被广泛应用于各种机器学习和深度学习任务，包括但不限于图像和视频识别、自然语言处理、强化学习、时间序列分析和生成对抗网络。

3. 社区和支持

TensorFlow 拥有一个活跃的开源社区，用户可以通过 GitHub、Stack Overflow、TensorFlow

论坛等渠道获取支持和资源。Google 还提供了详尽的文档和教程,帮助用户快速上手并解决实际问题。

总之,TensorFlow 凭借其灵活性、可扩展性和强大的生态系统,成为深度学习领域的重要工具,广泛应用于学术研究和工业界。

2.1.2 PyTorch

PyTorch 是由 Facebook 的人工智能研究实验室(FAIR)开发并于 2016 年发布的开源深度学习框架,其以动态计算图和易用性迅速赢得了研究界和工业界的广泛关注和使用。

1. 主要特点

(1)动态计算图:PyTorch 采用动态计算图(Dynamic Computational Graph),允许用户在运行时定义和修改模型。这使得调试更加直观,代码更加简洁易读,非常适合在研究和开发阶段使用。

(2)灵活性和易用性:PyTorch 的 API 设计简洁直观,符合 Pythonic 风格,使用体验与 NumPy 类似。这种设计使新手和有经验的开发者都能快速上手并高效地进行模型开发。

(3)自动微分:PyTorch 的自动微分引擎(Autograd)支持任意复杂度的梯度计算。通过 Autograd,用户可以轻松实现反向传播算法,自动计算梯度,从而加速模型的训练过程。

(4)强大的生态系统:PyTorch 拥有丰富的生态系统和工具支持。

- TorchVision:专门用于计算机视觉的工具包,包含常用的数据集、模型和图像变换工具。
- TorchText:用于自然语言处理的工具包,提供了文本数据处理和模型构建的工具。
- TorchAudio:用于音频处理和分析的工具包,支持音频数据的加载、变换和特征提取。
- PyTorch Lightning:一个轻量级的框架,旨在简化 PyTorch 代码,促进可重复性和模块化。

(5)与深度学习社区的紧密结合:PyTorch 在学术界和工业界的使用非常广泛,其开发者和用户社区非常活跃。许多最新的深度学习研究论文都基于 PyTorch 实现,这也促进了技术的快速迭代和应用。

(6)支持分布式计算:PyTorch 提供了强大的分布式计算能力,支持在多个 GPU 和多个节点上进行训练,从而加速大规模模型的训练过程。

2. 应用场景

PyTorch 被广泛应用于各种深度学习任务,包括但不限于以下几种。

- 计算机视觉:图像分类、目标检测、图像分割等。
- 自然语言处理:文本分类、情感分析、机器翻译等。
- 强化学习:智能体训练、策略优化等。
- 生成模型:生成对抗网络、变分自编码器等。

3. 社区和支持

PyTorch 拥有一个庞大且活跃的社区,用户可以通过 GitHub、Stack Overflow、PyTorch 论坛等渠道获取支持和资源。Facebook 和其他贡献者还提供了详尽的文档、教程和示例代码,可以帮助用户快速上手并解决实际问题。

总之,PyTorch 凭借其动态计算图、灵活性、易用性和强大的生态系统,成为深度学习领域的重要工具。无论是在学术研究还是工业应用中,PyTorch 都展示了其强大的功能和广泛的适用性。

2.1.3 Keras

Keras 是一个用于构建和训练深度学习模型的高层神经网络 API,最初由 François Chollet 开发,并于 2015 年发布。Keras 以其简洁、模块化和可扩展的设计,迅速成为深度学习社区中的重要工具。Keras 现在是 TensorFlow 的高级 API。

1. 主要特点

- **简洁易用**：Keras 的 API 设计简洁直观，旨在简化深度学习模型的构建、训练和评估过程。用户可以用少量的代码快速构建复杂的神经网络模型。
- **模块化设计**：Keras 采用模块化设计，各个组件（如神经网络层、损失函数、优化器等）都是独立的、可插拔的模块，用户可以自由组合和扩展，构建自定义模型。
- **多后端支持**：尽管 Keras 现在主要作为 TensorFlow 的高级 API 使用，但它最初支持多个后端，包括 TensorFlow、Theano 和 Microsoft CNTK。多后端支持使得 Keras 能够灵活适应不同的计算需求和环境。
- **高级功能**：Keras 提供了许多高级功能，如内置的支持卷积神经网络、循环神经网络等常用模型，简化了处理图像、文本、序列数据的操作。同时，Keras 还支持多输入、多输出模型，以及模型的并行训练。
- **强大的社区和生态系统**：Keras 拥有活跃的社区和丰富的生态系统，包括大量的开源项目、第三方库和工具。Keras 的文档详尽，提供了大量的教程和示例，可以帮助用户快速上手。
- **与 TensorFlow 深度集成**：Keras 自从成为 TensorFlow 的高级 API，便得到了 TensorFlow 的全面支持，能够利用 TensorFlow 强大的计算能力和工具生态系统，如 TensorFlow Lite、TensorFlow.js 等。

2. 应用场景

Keras 被广泛应用于各种深度学习任务，特别适合快速原型开发和实验。常见的应用场景如下。

- **计算机视觉**：图像分类、目标检测、图像分割等。
- **自然语言处理**：文本分类、情感分析、机器翻译等。
- **时间序列分析**：预测、分类、异常检测等。
- **生成模型**：自动编码器、生成对抗网络等。

总之，Keras 以其简洁、模块化和易用性，成为深度学习领域的重要工具。它适用于快速构建和训练深度学习模型，特别是在研究和原型开发阶段。与 TensorFlow 的深度集成使得 Keras 能够充分利用 TensorFlow 的强大功能和生态系统，进一步提升了其应用价值。

2.1.4 Transformer

Transformer 是一种深度学习模型架构，它的出现彻底优化了自然语言处理领域的许多处理任务，如机器翻译、文本生成和语义理解等。Transformer 的核心机制是自注意力（Self-Attention），它使得模型能够并行处理序列数据，并捕捉序列中元素之间的依赖关系。

1. 主要组成

（1）自注意力机制。

自注意力是 Transformer 的关键创新之一，它允许模型为序列中的每个元素计算一个权重分布，这些权重表示该元素与序列中其他元素的关系。这使模型可以在计算每个元素的表示时，考虑到序列中所有其他元素的信息。

自注意力的计算是通过查询（Query）、键（Key）和值（Value）三个向量完成的，通过这些向量，模型计算每个元素与其他元素的相似度，并根据这个相似度来加权组合元素的信息。

（2）多头注意力（Multi-Head Attention）。

为了捕捉不同类型的关系，Transformer 使用多头注意力机制，将输入的数据分成多组，并在每组上独立地执行自注意力操作。然后将这些不同注意力头的结果拼接起来，通过线性变换得到最终的输出。这使得模型可以同时关注到不同的特征或关系。

（3）前馈神经网络（Feed-Forward Neural Network，FFN）。

在每个自注意力层之后，Transformer 包含一个前馈神经网络，它由两个线性变换和一个非线性激活函数组成。FFN 在每个序列位置上独立地应用，有助于模型捕捉复杂的模式。

（4）残差连接和层归一化（Residual Connections and Layer Normalization）。

Transformer 在每一层中使用残差连接和层归一化，以帮助训练深层网络，防止梯度消失，并提高模型的稳定性和收敛速度。

（5）编码器-解码器架构（Encoder-Decoder Architecture）。

原始的 Transformer 采用了"编码器-解码器"架构，适用于序列到序列的任务，如机器翻译。

- 编码器（Encoder）：负责处理输入序列，将其转化为一个固定长度的上下文表示。
- 解码器（Decoder）：接收编码器的输出，并生成目标序列，如翻译后的文本。

（6）位置编码（Positional Encoding）。

由于 Transformer 的自注意力机制没有固有的顺序信息，模型通过添加位置编码来表示序列中每个位置的信息。这些编码通常是通过正弦和余弦函数生成的，并被加到输入嵌入中。

2. Transformer 的应用

- 自然语言处理：Transformer 被广泛应用于机器翻译、文本生成、问答系统、文本分类等任务中。
- 计算机视觉：通过 Vision Transformer（ViT）等模型，Transformer 也开始应用于图像分类、目标检测等视觉任务。
- 多模态任务：Transformer 被用于处理包含多种模态（如图像、文本）的任务，如视觉问答、图像字幕生成等。

总之，Transformer 是一种灵活而强大的架构，因其并行计算能力和捕捉长程依赖关系的能力，在深度学习领域得到了广泛应用。

2.2 多模态模型技术

多模态模型是一类能够处理和融合来自不同模态（如文本、图像、音频等）数据的机器学习模型，它们通过联合学习不同模态的特征，实现更丰富和准确的理解和生成任务。多模态模型在诸如视觉问答（VQA）、图文生成和跨模态检索等应用中表现出色，它们能够同时理解和关联图像与文本信息，从而提高任务的效果和性能。

2.2.1 ViLBERT 模型

ViLBERT（Vision-and-Language BERT）是一种多模态模型，专门用于处理视觉和语言任务。ViLBERT 扩展了 BERT（Bidirectional Encoder Representations from Transformers）模型，通过并行处理图像和文本特征，来增强对多模态数据的理解能力。

1. 架构

ViLBERT 的架构基于 BERT，但进行了显著的扩展，以处理图像和文本两种模态的信息。ViLBERT 架构的主要特点如下。

（1）双流处理器：ViLBERT 使用两个独立的 Transformer 流，一个用于处理文本特征，另一个用于处理视觉特征，每个流都有自己的嵌入层和多层 Transformer 编码器。

（2）交互层：在两个独立的流之间，ViLBERT 引入了多层跨模态交互（Co-attention）机制。这些交互层允许视觉和语言特征在多个层次上进行信息交换和融合，增强了模型对多模态信息的综合理解。

（3）输入特征。
- 文本输入：文本输入首先通过 BERT 的嵌入层进行编码，生成词向量序列。
- 图像输入：图像输入通过预训练的卷积神经网络（如 ResNet）提取视觉特征，并进一步通过线性变换映射到适合 Transformer 输入的特征空间。

2. 训练
ViLBERT 的训练过程包括如下两个阶段。
- 预训练：在大规模的图文配对数据集上进行预训练，使用多任务学习策略。常见的预训练任务包括图文匹配、遮掩语言建模（MLM，Masked Language Model）和遮掩对象预测。
- 微调：在具体的下游任务上进行微调，如视觉问答、图文检索、图文生成等。这一阶段的训练通常在较小的、任务特定的数据集上进行，以提高模型在特定应用场景中的性能。

3. 应用场景
ViLBERT 可以应用于各种多模态任务，常见的应用场景如下。
- 视觉问答：通过理解图像内容和文本问题，生成准确的答案。
- 图文匹配：判断给定的图像和文本描述是否匹配，适用于图文检索和推荐系统。
- 图文生成：根据图像生成相应的文本描述，或根据文本生成对应的图像。

4. 优势
ViLBERT 的提出和发展，推动了多模态模型的研究和应用。随着计算资源和数据规模的不断增加，多模态模型将继续在更广泛的领域中发挥重要作用，促进跨模态理解和生成技术的发展。ViLBERT 的主要优势如下。
- 增强的多模态理解：通过双流处理器和跨模态交互层，ViLBERT 能够在更深层次上融合和理解视觉和语言特征。
- 灵活的架构：ViLBERT 的架构可以灵活地扩展和适应不同的多模态任务，通过预训练和微调策略，实现高效的模型训练和性能提升。
- 广泛的应用：ViLBERT 在多种多模态任务上表现优异，展示了其在实际应用中的广泛适用性。

综上所述，ViLBERT 通过创新的双流处理器和跨模态交互层设计，显著提升了多模态任务中的表现，为视觉和语言的联合理解提供了强大的支持。

2.2.2 VisualBERT 模型

VisualBERT 是一种多模态模型，旨在处理图像和文本之间的联合理解任务。VisualBERT 将图像和文本输入融合在一起，采用类似 BERT 的架构，通过预训练和微调，在各种视觉和语言任务中取得了显著的性能提升。

1. 架构
VisualBERT 架构的主要特点如下。
- 双向编码器：VisualBERT 采用了双向 Transformer 编码器的架构，同时处理图像和文本输入。它包含了图像编码器和文本编码器两个部分。
- 图像编码器：用于处理图像输入的编码器，通常采用预训练的卷积神经网络（如 ResNet）提取图像特征，然后通过自注意力机制将图像特征与文本特征进行交互。
- 文本编码器：用于处理文本输入的编码器，与传统的 BERT 相似，将输入文本进行分词、嵌入、位置编码等处理。
- 跨模态融合：在编码器的多层中，引入了跨模态融合机制，允许图像和文本特征在不同层次上进行信息交换和融合。

2. 训练

VisualBERT 模型的训练过程包括如下两个阶段。
- 预训练：在大规模的图文配对数据上进行预训练，通过图文匹配、遮掩语言建模等任务进行多任务学习，从而学习图像和文本之间的语义关联。
- 微调：在具体的下游任务上进行微调，例如视觉问答、图像标注、图文检索等任务。微调阶段通常在任务特定的数据集上进行，以提高模型在具体任务上的性能。

3. 应用场景

VisualBERT 模型可以被应用于多种视觉和语言任务，常见的应用场景如下。
- 视觉问答：根据图像和问题生成答案。
- 图像标注：给图像添加语义描述。
- 图像检索：根据文本描述检索相关图像。
- 图像生成：根据文本描述生成图像等。

4. 优势

VisualBERT 代表了一类成功的多模态模型，推动了视觉和语言的联合理解技术的发展。随着对跨模态理解需求的增加，类似的模型将继续受到关注，并在各种应用领域中发挥重要作用。VisualBERT 模型的主要优势如下。
- 跨模态理解：VisualBERT 能够有效地将图像和文本信息融合在一起，实现跨模态的深度理解。
- 适用性广泛：VisualBERT 在各种视觉和语言任务上表现出色，具有很强的通用性和适用性。
- 预训练与微调：采用预训练与微调策略，使 VisualBERT 能够在特定任务上快速收敛并取得良好性能。

综上所述，VisualBERT 通过融合图像和文本信息，采用类似 BERT 的架构，其性能在多种视觉和语言任务中得到显著提升，成为多模态理解领域的重要里程碑之一。

2.2.3 OpenAI CLIP 模型

OpenAI CLIP（Contrastive Language-Image Pre-training）是由 OpenAI 开发的一种多模态模型，旨在实现图像和文本之间的强大关联。CLIP 通过在大规模图像和文本对上进行预训练，使图像和文本能够在同一特征空间进行对比，从而实现了出色的图像和文本理解能力。

1. 架构

OpenAI CLIP 架构的主要特点如下。
- 双向编码器：CLIP 采用了两个独立的编码器，一个用于处理图像，另一个用于处理文本。这两个编码器共享参数，以便在同一特征空间中对图像和文本进行编码。
- 图像编码器：图像编码器通常采用预训练的卷积神经网络（如 ResNet）提取图像特征，然后通过线性投影将图像特征映射到与文本编码器相同的特征空间。
- 文本编码器：文本编码器将文本输入进行编码，通常采用 Transformer 架构进行编码，以产生与图像特征相对应的文本特征。
- 对比损失：CLIP 使用对比学习的方法，在训练过程中，通过最大化正确图像-文本对的相似性、最小化错误图像-文本对的相似性来训练模型。

2. 训练

OpenAI CLIP 的训练过程包括如下两个阶段。
- 预训练：在大规模图像和文本对数据集上进行预训练。CLIP 使用对比学习的方法，通过最大化正确对的相似性来学习图像和文本的嵌入表示。
- 微调：在特定任务上进行微调，如图像分类、图像检索、零样本学习等。微调阶段通常在任务特定的数据集上进行。

3. 应用场景

OpenAI CLIP 可以应用于多种视觉和语言任务，常见的应用场景如下。
- 图像分类：根据文本描述对图像进行分类。
- 图像检索：根据文本描述检索相关图像。
- 零样本学习：在没有任何标签的情况下学习对图像进行分类。
- 视觉问答：根据图像和问题生成答案。

4. 优势

OpenAI CLIP 代表了无监督学习在多模态领域的重要进展之一，它推动了图像和文本之间关联性学习的研究，并在多种任务上展示了令人印象深刻的性能。
- 无监督预训练：CLIP 使用无监督的对比学习方法进行预训练，无须标注图像-文本对，从而可以利用大规模未标记的数据进行训练。
- 泛化能力强：CLIP 在零样本学习等任务上表现出色，具有很强的泛化能力，可以处理在训练时未见过的类别和概念。
- 通用性：CLIP 是一个通用的多模态模型，适用于各种视觉和语言任务。

综上所述，OpenAI CLIP 通过对比学习实现了强大的图像和文本理解能力，成为多模态理解领域的重要里程碑之一。

2.2.4 UNITER 模型

UNITER（Unified Vision-Language Pre-training）是一种多模态预训练模型，旨在统一处理视觉和语言任务。它通过在大规模图像和文本对数据上进行预训练，实现了对图像和文本之间语义关联的学习，从而在各种视觉和语言任务中取得了优异表现。

1. 架构

UNITER 架构的主要特点如下。
- 双向编码器：UNITER 采用了双向 Transformer 编码器的架构，同时处理图像和文本输入。它包含了图像编码器和文本编码器两个部分。
- 图像编码器：用于处理图像输入的编码器，通常采用预训练的卷积神经网络（如 ResNet）提取图像特征，然后通过自注意力机制将图像特征与文本特征进行交互。
- 文本编码器：用于处理文本输入的编码器，与传统的 BERT 相似，将输入文本进行分词、嵌入、位置编码等处理。
- 跨模态融合：在编码器的多层中，引入了跨模态融合机制，允许图像和文本特征在不同层次上进行信息交换和融合。

2. 训练

UNITER 的训练过程包括如下两个阶段。
- 预训练：在大规模的图像和文本对数据上进行预训练。预训练任务通常包括图文匹配、遮掩语言建模等，以学习图像和文本之间的语义关联。
- 微调：在具体的下游任务上进行微调，例如视觉问答、图像标注、图文检索等任务。微调阶段通常在任务特定的数据集上进行，以提高模型在具体任务上的性能。

3. 优势

UNITER 代表了一类成功的多模态模型，推动了视觉和语言的联合理解技术的发展。随着对跨模态理解需求的增加，类似的模型将继续受到关注，并在各种应用领域中发挥重要作用。UNITER 模型的主要优势如下。
- 跨模态理解：UNITER 能够有效地将图像和文本信息融合在一起，实现跨模态的深度理解。
- 适用性广泛：UNITER 在各种视觉和语言任务上表现出色，具有很强的通用性和适用性。
- 预训练与微调：采用预训练与微调策略，使 UNITER 能够在特定任务上快速收敛并取得良好性能。

综上所述，UNITER 通过融合图像和文本信息，取得了在多种视觉和语言任务中的显著性能提升，为视觉和语言的联合理解提供了强大的工具。

2.2.5 LXMERT 模型

LXMERT（Learning Cross-Modality Encoder Representations from Transformers）是一种用于视觉和语言任务的多模态预训练模型，它通过在大规模图像和文本数据上进行预训练，实现了对图像和文本之间丰富的语义理解，并在各种视觉和语言任务中取得了显著的性能提升。

1. 架构

LXMERT 架构的主要特点如下。
- 双向编码器：LXMERT 采用了双向 Transformer 编码器的架构，同时处理图像和文本输入。它包含了图像编码器和文本编码器两个部分。
- 图像编码器：用于处理图像输入的编码器，通常采用预训练的卷积神经网络（如 ResNet）提取图像特征，然后通过自注意力机制将图像特征与文本特征进行交互。
- 文本编码器：用于处理文本输入的编码器，与传统的 BERT 相似，将输入文本进行分词、嵌入、位置编码等处理。
- 跨模态融合：在编码器的多层中，引入了跨模态融合机制，允许图像和文本特征在不同层次上进行信息交换和融合。

2. 训练

LXMERT 的训练过程包括如下两个阶段。
- 预训练：在大规模的图像和文本数据上进行预训练。预训练任务通常包括图文匹配、遮掩语言建模等，以学习图像和文本之间的语义关联。
- 微调：在具体的下游任务上进行微调，例如视觉问答、图像标注、图文检索等任务。微调阶段通常在任务特定的数据集上进行，以提高模型在具体任务上的性能。

3. 优势

LXMERT 代表了一类成功的多模态模型，推动了视觉和语言的联合理解技术的发展。随着对跨模态理解需求的增加，类似的模型将继续受到关注，并在各种应用领域中发挥重要作用。LXMERT 模型的主要优势如下。
- 跨模态理解：LXMERT 能够有效地将图像和文本信息融合在一起，实现跨模态的深度理解。
- 适用性广泛：LXMERT 在各种视觉和语言任务上表现出色，具有很强的通用性和适用性。
- 预训练与微调：采用预训练与微调策略，使 LXMERT 能够在特定任务上快速收敛并取得良好性能。

LXMERT 通过融合图像和文本信息，取得了在多种视觉和语言任务中的显著性能提升，为视觉和语言的联合理解提供了强大的工具。

2.2.6　SigLIP 模型

SigLIP 是一种多模态预训练模型，旨在通过结合视觉和语言信息来提升在多模态任务中的表现。SigLIP 是基于 CLIP 模型设计的，重点在于学习图像和文本的对齐表示，使得模型能够在多种任务中展现出强大的泛化能力。利用 SigLIP 的嵌入表示，用户可以轻松地在大规模图像库中找到与给定文本描述最匹配的图像，或反过来可找到与图像相关的文本描述。

SigLIP 的核心特点如下。

- 多模态对齐：SigLIP 通过对比学习的方法，将图像和文本嵌入同一个表示空间。在这个空间里，相似的图像和文本对会被映射到相近的位置，而不相关的图像和文本对则被拉开距离。这种对齐过程使得模型能够更好地理解和处理图像与文本之间的关系。
- 大规模预训练：SigLIP 在一个大规模的图像-文本对数据集上进行预训练，这种预训练使得模型能够学习到通用的视觉和语言表示，进而在下游任务中表现出色。
- 多样化应用场景：SigLIP 可以应用于多个多模态任务，如图像检索、图像字幕生成、视觉问答等。通过在这些任务上进行微调，SigLIP 能够很好地适应不同的应用场景。
- 高效模型设计：SigLIP 的架构设计注重计算效率和模型性能的平衡，使其在处理大规模数据时仍能保持较高的速度和准确度。
- 灵活性和扩展性：SigLIP 的多模态对齐表示方便与其他下游任务结合，提供了极大的灵活性。此外，它也可以与其他模型架构（如 Transformers）集成，进一步提升任务表现。

SigLIP 代表了多模态预训练模型的发展方向之一，通过有效地结合视觉和语言信息，推动了多模态任务的进步。

2.2.7　LoRA 微调技术

LoRA（Low-Rank Adaptation，低秩适应）是一种轻量级的模型微调技术，旨在高效地微调大型预训练模型，尤其是在资源有限的情况下。LoRA 微调技术通过引入低秩矩阵来减少参数的数量，从而在保持模型性能的同时，降低微调的计算和存储成本。LoRA 在自然语言处理、计算机视觉和跨模态任务中得到了广泛应用，其主要特点如下所示。

- 低秩矩阵表示：LoRA 的关键思想是将模型的权重矩阵分解为低秩矩阵的乘积。在微调过程中，LoRA 不直接调整原始的权重矩阵，而是引入两个低秩矩阵，通过它们的乘积来表示权重的变化。这种表示方式大大减少了微调时需要调整的参数数量。
- 高参数效率：通过使用低秩矩阵，LoRA 在微调过程中仅需要更新少量的参数，而不需要对整个模型的所有参数进行调整。这使 LoRA 成为一种非常高效的方法，特别适合在内存和计算资源有限的环境中进行模型微调。
- 与现有架构兼容：LoRA 可以与现有的模型架构（如 Transformer、BERT 等）无缝集成。它通过在模型的线性层或其他特定层插入低秩矩阵来实现微调，因此不需要对原始模型结构进行大幅修改。
- 减少过拟合：LoRA 只调整少量参数，减少了微调时的自由度，这在一定程度上有助于减少过拟合的风险，尤其是在小规模数据集上进行微调时。

总之，LoRA 为大型预训练模型的微调提供了一种有效且节约资源的解决方案，使在低资源环境中应用大型预训练模型成为可能。

2.2.8　LLaVA 模型

LlaVA（Large Language and Vision Assistant）是一种结合了大语言模型和视觉模型（Vision

Model）的多模态 AI 模型，它能够同时处理文本和图像输入，为用户提供更为丰富的对话与问答体验。

LLaVA 通过结合大语言模型和视觉模型，突破了传统单一模态（仅文本或仅图像）AI 模型的限制，能够处理更加复杂的任务。具体来说，LLaVA 模型的核心特点如下。

- 多模态输入：LLaVA 能够处理文本和图像的组合输入，通过对视觉数据和语言数据的理解，LLaVA 可以对图像进行详细描述、回答有关图像内容的问题，或基于图像生成相关的文本。
- 大语言模型：因为 LLaVA 利用当前先进的大语言模型（如 GPT-4 或其变体）作为其语言理解和内容生成的核心，所以，它在处理复杂语言任务上表现优异，如自然语言生成、问答、翻译等。
- 视觉模型集成：视觉功能通常基于强大的图像识别和理解模型，如 CLIP（Contrastive Language–Image Pretraining）或类似的模型。这些模型在大规模图像数据上预训练，能够很好地理解图像内容，并将其与语言模型的输出相结合。
- 应用场景广泛：LLaVA 模型可应用于多种场景，如智能助理、图像标注、教育工具、图像问答系统等。特别是在需要跨越文本和图像信息的任务中，LLaVA 展现了强大的能力。
- 高互动性：LLaVA 能够实时与用户交互，基于用户输入的图像和文本数据生成自然、连贯的回答。这种能力使其在需要高互动性的应用中非常有用，如虚拟助手、内容创作等。

LLaVA 是一种强大的多模态 AI 模型，结合了文本和图像的理解与生成能力，在多个应用场景中展现了优异的表现。随着深度学习和多模态模型的不断进步，LLaVA 代表了智能系统向更加自然和智能化方向发展的一个重要里程碑。

2.3 预训练模型

预训练模型是通过在大规模未标记数据上进行学习而生成的模型，它们能够捕捉数据中的统计特性和语义信息。这些模型通常在通用任务上进行预训练，如语言模型的掩码语言建模或图像模型的自监督学习，然后在特定任务上进行微调，以提高性能和泛化能力。例如本章前面介绍的多模态大模型 ViLBERT、VisualBERT、OpenAI CLIP、UNITER 和 LXMERT 都可以进行预训练，它们都是基于大规模未标记数据进行预训练的多模态模型，这些模型在预训练之后，通常可以在各种视觉和语言任务上进行微调，以适应特定的应用场景，提高性能。

2.3.1 预训练模型简介

预训练模型是指在大规模未标记数据上进行预训练的深度学习模型。这些模型通常通过无监督或半监督学习的方式，在大规模数据上学习数据的表示和语义信息，而无须人工标注标签。预训练模型的目的是捕捉数据中的统计结构和语义信息，以便在特定任务上进行微调，提高性能和泛化能力。

在实际应用中，预训练模型通常包括如下两个主要组成部分。

（1）模型架构：预训练模型的架构通常基于深度神经网络，如 Transformer、CNN 等。这些模型可以是单模态的（如仅处理文本或仅处理图像），也可以是多模态的（同时处理多种类型的数据，如图像和文本）。

（2）预训练任务：预训练模型在训练过程中执行的任务，通常是无监督或半监督的。常见的预训练任务如下。

- 语言模型预训练：如掩码语言建模（Masked Language Modeling，MLM），模型尝试从部分文本中预测被遮盖的单词。
- 图像自监督学习：模型尝试从图像中学习有用的表示，如图像补全、颜色化等任务。
- 多模态预训练：模型同时处理多模态数据，如图像和文本，学习它们之间的关联性。

预训练模型在预训练任务上学习到的参数和表示可以被迁移或微调到特定的下游任务上，如文本分类、图像分类、目标检测、文本生成等，以提高模型在这些任务上的性能。

常见的预训练模型包括 BERT、GPT、ResNet、Vision Transformers（ViT）等，这些预训练模型已经成为许多自然语言处理和计算机视觉任务的重要基础。

2.3.2 使用预训练模型

使用预训练模型是利用已经在大规模数据上进行预训练的模型参数，在特定任务上进行微调或者特征提取的一种常见策略。使用预训练模型，可以帮助提高模型性能、加速模型训练，并且降低在大规模数据上训练深度神经网络的成本，因此在实践中被广泛采用。使用预训练模型的一般流程如下。

1. **选择预训练模型**
   - 根据任务需求和数据集特点选择合适的预训练模型，例如 BERT、GPT、ResNet 等。
   - 可以根据任务如文本分类、目标检测、语言生成等来选择相应的预训练模型。
2. **获取预训练模型权重**

   从预训练模型的官方发布源、Hugging Face 模型中心或其他可靠来源获取模型的权重文件。
3. **加载预训练模型**
   - 使用深度学习框架（如 PyTorch、TensorFlow 等）加载预训练模型及其权重。
   - 通常可以使用相应框架提供的库来加载，如 Hugging Face Transformers 库。
4. **准备数据**
   - 根据任务准备相应的数据集，包括训练集、验证集和测试集。
   - 对数据进行预处理，如标记化、归一化、分批等操作。
5. **微调模型**
   - 如果任务需要，可以在特定任务的数据集上对预训练模型进行微调。
   - 设置损失函数、优化器等参数，并在训练集上进行训练。
6. **评估模型性能**
   - 使用验证集评估微调后模型的性能，调整超参数以提高性能。
   - 可以使用准确率、精确率、召回率等指标进行评估。
7. **模型应用**
   - 在测试集上评估最终模型的性能，获取模型在真实数据上的表现。
   - 根据需求将模型部署到生产环境中进行应用。

实例 2-1 演示了使用预训练的 UNITER 模型（基于 BERT）处理图像和文本的多模态任务的过程。首先加载了预训练的 UNITER 模型和分词器，对给定的图像和文本进行预处理，然后将图像和文本输入模型获取模型的表示，接着通过自定义的头部网络进行分类预测，最后输出预测结果的概率。

> **实例 2-1** 使用预训练的 UNITER 模型处理任务
> 源码路径：codes/2/xia.py 和 Unmo.py

（1）实例文件 xia.py 用于从指定 URL 下载预训练 UNITER 模型的权重文件（包括 pytorch_model.bin、config.json 和 vocab.txt），并将它们保存到指定目录。如果文件已经存在于目标目录中，则不会重复下载。文件 xia.py 的具体实现代码如下所示。

```python
import os
import requests

# 下载路径和保存路径
base_url= 'https://github.com/visualjoyce/transformers4vl-uniter-base/raw/main/'
output_dir = r'C:\qinghua\多模态\codes\2'
model_name = 'uniter-base'

# 创建目录（如果不存在）
os.makedirs(output_dir, exist_ok=True)

# 下载模型权重文件
files_to_download = ['pytorch_model.bin', 'config.json', 'vocab.txt']
for file_name in files_to_download:
    url = base_url + file_name
    output_path = os.path.join(output_dir, f'{model_name}_{file_name}')

    # 下载文件
    if not os.path.exists(output_path):
        response = requests.get(url, stream=True)
        with open(output_path, 'wb') as f:
            for chunk in response.iter_content(chunk_size=8192):
                if chunk:
                    f.write(chunk)
        print(f'{file_name} downloaded to {output_path}')
    else:
        print(f'{file_name} already exists at {output_path}')
```

（2）编写文件 Unmo.py，功能是使用预训练的 UNITER 模型处理图像和文本，并定制网络头部进行预测。具体实现代码如下所示。

```python
From tensorflow_datasets.image_classification.colorectal_histology_test import num_classes
from transformers import BertModel, BertTokenizer
from PIL import Image
from torchvision import transforms
import torch.nn as nn
import torch

# 加载模型和分词器
model_name = "visualjoyce/transformers4vl-uniter-base"
model = BertModel.from_pretrained(model_name)
tokenizer = BertTokenizer.from_pretrained(model_name)

# 定义自定义头部网络
class CustomHead(nn.Module):
    def __init__(self, input_size, output_size):
        super(CustomHead, self).__init__().__init__()
        self.fc = nn.Linear(input_size, output_size)
```

```python
    def forward(self, x):
        x = self.fc(x)
        return x

# 定义一个图像预处理函数
def preprocess_image(image_path):
    image = Image.open(image_path).convert("RGB")
    transform = transforms.Compose([
        transforms.Resize((224, 224)),
        transforms.ToTensor(),
        transforms.Normalize(mean=[0.485, 0.456, 0.406], std=[0.229, 0.224, 0.225]),
    ])
    image = transform(image).unsqueeze(0)   # 增加 batch 维度
    return image

# 定义一个文本预处理函数
def preprocess_text(question):
    inputs=tokenizer(question,return_tensors="pt", padding=True, truncation=True)
    return inputs

# 加载图像和文本
image_path = "path_to_image.png"   # 修改为图像的实际路径
question = "What is the color of the cat?"

# 预处理图像和文本
image_tensor = preprocess_image(image_path)
text_inputs = preprocess_text(question)

# 将图像和文本输入模型
outputs = model(**text_inputs)

# 获取模型的最后一层表示
last_hidden_states = outputs.last_hidden_state

# 定义自定义头部,并传入模型表示的大小和自定义头部输出的大小
custom_head=CustomHead(input_size=last_hidden_states.shape[-1],output_size=num_classes)
logits = custom_head(last_hidden_states)

# 示例: 在 logits 上添加 softmax 层并进行预测
softmax = nn.Softmax(dim=1)
probs = softmax(logits)

# 输出预测结果
print(probs)
```

上述代码的实现流程如下。

- 加载预训练模型和分词器:使用 visualjoyce/transformers4vl-uniter-base 预训练模型的权重

初始化 BERT 模型。使用相同的预训练模型初始化分词器。
- 定义自定义头部网络：创建了一个自定义头部网络 CustomHead，该网络接收模型表示作为输入，并输出预测结果。在初始化函数中定义了一个全连接层，输入大小为模型表示的大小，输出大小为数据集的类别数量。
- 定义图像预处理和文本预处理函数：通过 preprocess_image 函数加载、调整大小和标准化图像。通过 preprocess_text 函数将文本转换为模型可以接收的格式。
- 加载图像和文本数据：设置图像和文本的路径和问题，然后对图像和文本进行预处理。
- 将图像和文本输入模型：将预处理后的文本输入 BERT 模型中，获取模型输出。
- 获取模型的最后一层表示：从模型输出中获取最后一层的表示。
- 定义自定义头部并进行预测：使用自定义头部网络对模型表示进行分类预测，将预测结果通过 softmax 函数得到概率。
- 输出预测结果：打印预测结果概率，执行后会输出：

```
You are using a model of type uniter to instantiate a model of type bert. This is not
supported for all configurations of models and can yield errors.
Some weights of BertModel were not initialized from the model checkpoint at visualjo
yce/transformers4vl-uniter-base and are newly initialized: ['encoder.layer. 6.output
.LayerNorm.weight','encoder.layer.2.attention.self.value.bias', 'encoder. layer.11.o
utput.dense.weight',
###省略部分输出
You should probably TRAIN this model on a down-stream task to be able to use it for
predictions and inference.
Asking to truncate to max_length but no maximum length is provided and the model has
 no predefined maximum length. Default to no truncation.
tensor([[[0.1159, 0.0673, 0.1049, 0.0776, 0.1197, 0.0731, 0.0883, 0.0689],
         [0.0823, 0.1458, 0.0731, 0.0669, 0.0518, 0.1224, 0.1127, 0.1070],
         [0.1064, 0.0826, 0.0909, 0.0919, 0.1011, 0.0750, 0.0639, 0.1121],
         [0.0924, 0.0629, 0.1253, 0.0675, 0.0825, 0.1019, 0.0811, 0.0725],
         [0.1722, 0.0981, 0.1051, 0.1325, 0.0635, 0.1115, 0.1225, 0.1473],
         [0.0919, 0.1118, 0.1306, 0.2269, 0.0681, 0.1108, 0.1476, 0.0910],
         [0.0796, 0.0823, 0.1016, 0.0819, 0.1281, 0.0842, 0.0816, 0.0920],
         [0.1339, 0.1364, 0.0656, 0.0850, 0.1175, 0.1316, 0.1187, 0.1114],
         [0.0567, 0.0842, 0.0508, 0.0561, 0.1148, 0.0971, 0.0791, 0.0889],
         [0.0687, 0.1285, 0.1521, 0.1138, 0.1529, 0.0924, 0.1046, 0.1090]]],
       grad_fn=<SoftmaxBackward0>)
```

在上面的输出结果中，请先暂时忽略前面的警告，只看后面的输出结果。这个输出结果是经过 softmax 处理后的预测概率，这是一个三维的张量，具体含义如下。
- 第一维：表示 batch 中每个样本的索引。
- 第二维：表示每个样本的预测结果，这里每个样本有 8 个类别的预测概率。
- 第三维：表示每个类别的预测概率。

在这个例子中，输出的张量是一个 [1，10，8] 的张量，具体含义如下。
- 1 表示只有一个样本。
- 10 表示这个样本有 10 个预测结果。
- 8 表示每个预测结果有 8 个类别的预测概率。

所以，例如第一个样本的第一个预测结果，预测各个类别的概率分别为[0.1159, 0.0673, 0.1049,

0.0776, 0.1197, 0.0731, 0.0883, 0.0689]。

上面的输出结果表明了模型对每个类别的预测概率，开发者可以根据具体的需求选择最高概率的类别作为模型的预测结果。

2.3.3 预训练模型的微调

在前面实例 2-1 的输出结果中输出如下所示的警告信息：

```
You are using a model of type uniter to instantiate a model of type bert. This is not supported for all configurations of models and can yield errors.
####省略后面的警告信息
```

这是因为在实例 2-1 中加载了一个 visualjoyce/transformers4vl-uniter-base 的模型，但是由于该模型是未经过微调的，因此输出的结果可能是随机的。要解决这个问题，可以对这个模型进行微调，或者使用一个经过预训练并且在你的数据上微调过的模型。

预训练模型的微调是指将预先训练好的模型在特定任务上进行进一步训练，以适应特定任务的数据和要求。

1. **预训练模型微调的原理**
- 迁移学习：微调利用了迁移学习的思想，预训练模型在大规模数据上进行了预训练，学习到了通用的语言或图像表示。微调则在特定任务的数据集上进一步调整这些表示，使其适应特定任务。
- 参数调整：在微调过程中，模型的参数会根据特定任务的数据进行调整。通常是在预训练模型的基础上，通过反向传播和优化算法（如随机梯度下降）来微调参数。

2. **预训练模型微调的架构**

（1）预训练模型：常见的预训练模型包括 BERT、GPT、RoBERTa、UniLM、Vision Transformers 等，它们在大规模文本或图像数据上进行了预训练。

（2）微调层：在预训练模型之上添加了用于特定任务的额外层，通常包括自定义的分类头部、池化层等。这些层用于从预训练模型的输出中提取特征，并进行特定任务的预测。

（3）损失函数：根据任务的性质选择合适的损失函数，如交叉熵损失函数用于分类任务，均方误差用于回归任务等。

（4）微调过程。
- 前向传播：输入数据经过预训练模型和微调层，得到预测结果。
- 计算损失：预测结果与真实标签计算损失。
- 反向传播：通过反向传播计算梯度，并更新模型参数以最小化损失。
- 优化：使用优化算法（如随机梯度下降、Adam 等）来更新模型的参数。
- 重复迭代：在训练数据上多次迭代，直到模型收敛或达到设定的停止条件。

3. **微调步骤**

（1）加载预训练模型：加载预训练模型及其权重，可以使用像 Hugging Face Transformers 这样的库从预训练模型的 URL 或本地文件加载模型。

（2）修改模型结构：根据任务的特点，可能需要修改模型的结构。例如，添加自定义的分类头部或其他层来适应特定的任务。

（3）准备数据：准备用于微调的数据集。数据集通常需要进行预处理，并转换为模型所需的格式。

（4）定义训练参数：设置微调的超参数，如学习率、批量大小、训练轮数等。

（5）微调模型：使用准备好的数据集对模型进行训练。在训练过程中，模型的权重会根据数据集进行调整，以更好地适应特定任务。

（6）评估模型：使用验证集或测试集评估微调后的模型性能。这可以帮助确定模型是否已经适当地微调，并且是否需要进一步调整。

（7）调整和优化：根据评估结果，可能需要调整模型结构或超参数，并重新进行微调，直到达到满意的性能。

（8）应用和部署：在完成微调后，可以将模型应用于实际任务中，并进行部署以供使用。

总之，微调可以使预训练模型适应各种任务，例如文本分类、命名实体识别、图像分类等，而无须从头开始训练模型。

请看实例 2-2，功能是使用预训练的 BERT 模型对图像和文本进行处理，并在自定义的分类任务上进行微调。本实例的实现步骤包括加载预训练的 BERT 模型和分词器、预处理图像和文本数据、定义自定义头部和微调模型、定义损失函数和优化器、进行模型微调、在微调完成后进行预测。代码中展示了 3 个训练周期（epoch）的损失变化，并最终输出预测的类别。

实例 2-2　对 UNITER 模型进行微调
源码路径：codes/2/tiao.py

文件 tiao.py 的具体实现代码如下所示。

```python
import torch
from transformers import BertModel, BertTokenizer, BertConfig
from PIL import Image
from torchvision import transforms
import torch.nn as nn
import torch.optim as optim

# 加载预训练的 BERT 模型和 tokenizer
model_name = "visualjoyce/transformers4vl-uniter-base"  # 或者你使用的其他预训练 BERT 模型
config = BertConfig.from_pretrained(model_name)
tokenizer = BertTokenizer.from_pretrained(model_name)
model = BertModel.from_pretrained(model_name, config=config)

# 自定义头部
class CustomHead(nn.Module):
    def __init__(self, input_size, output_size):
        super(CustomHead, self).__init__()
        self.fc = nn.Linear(input_size, output_size)

    def forward(self, x):
        x = self.fc(x)
        return x

# 图像预处理
def preprocess_image(image_path):
    image = Image.open(image_path).convert("RGB")
    transform = transforms.Compose([
```

```python
        transforms.Resize((224, 224)),
        transforms.ToTensor(),
        transforms.Normalize(mean=[0.485,0.456,0.406], std=[0.229, 0.224, 0.225]),
    ])
    image = transform(image).unsqueeze(0)   # 添加 batch 维度
    return image

# 文本预处理
def preprocess_text(question):
    inputs=tokenizer(question,return_tensors="pt", padding=True, truncation=True)
    return inputs

# 加载图像和文本
image_path = "path_to_image.png"   # 修改为图像的实际路径
question = "What is the color of the cat?"

# 预处理图像和文本
image_tensor = preprocess_image(image_path)
text_inputs = preprocess_text(question)

# 准备微调任务的标签(示例)
labels = torch.tensor([1])   # 示例标签

# 自定义微调部分
class MyModel(nn.Module):
    def __init__(self, bert_model, num_classes):
        super(MyModel, self).__init__()
        self.bert = bert_model
        self.custom_head = CustomHead(input_size=self.bert.config.hidden_size, output_size=num_classes)

    def forward(self, input_ids, attention_mask, image):
        outputs = self.bert(input_ids=input_ids, attention_mask=attention_mask)
        pooled_output = outputs.pooler_output
        logits = self.custom_head(pooled_output)
        return logits

# 初始化模型
num_classes = 2   # 假设输出类别数为 2
model = MyModel(model, num_classes)

# 定义损失函数和优化器
loss_fn = nn.CrossEntropyLoss()
optimizer = optim.AdamW(model.parameters(), lr=1e-5)

# 模型微调
model.train()
num_epochs = 3
```

```
for epoch in range(num_epochs):
    optimizer.zero_grad()
    logits=model(input_ids=text_inputs.input_ids, attention_mask=text_inputs.attention_mask, image=image_tensor)
    loss = loss_fn(logits, labels)
    loss.backward()
    optimizer.step()
    print(f"Epoch [{epoch + 1}/{num_epochs}], Loss: {loss.item()}")

# 在微调完成后进行预测
model.eval()
with torch.no_grad():
    logits=model(input_ids=text_inputs.input_ids, attention_mask=text_inputs.attention_mask, image=image_tensor)
    predicted_class = torch.argmax(logits, dim=1)

print("Predicted class:", predicted_class.item())
```

上述代码的实现流程如下。

（1）加载预训练的 BERT 模型和分词器，定义自定义的头部和预处理函数。加载配置文件和预训练模型；定义自定义头部，用于最终的分类任务；定义图像和文本的预处理函数以适应模型的输入需求。

（2）加载图像和文本数据，并进行预处理。图像通过 PIL 库打开并转换为 RGB 格式，然后通过 torchvision 库进行标准化和转换；文本通过 tokenizer 进行分词和编码。

（3）准备微调任务的标签，并定义包含 BERT 模型和自定义头部的模型结构。标签是用于监督学习的目标变量，模型结构包括预训练的 BERT 模型和自定义的分类头部。

（4）定义损失函数和优化器，进行模型微调。使用交叉熵损失函数和 AdamW 优化器，针对预处理后的图像和文本输入，迭代优化模型参数，以最小化训练过程中的损失函数值。

（5）在模型微调完成后进行预测，并输出预测结果。切换模型到评估模式，使用微调后的模型对新的输入数据进行预测，输出预测的类别标签。这样就完成了对预训练 UNITER 模型的微调和应用。执行代码后会输出：

```
Epoch [1/3], Loss: 0.6956343054771423
Epoch [2/3], Loss: 0.28804779052734375
Epoch [3/3], Loss: 0.07029898464679718
Predicted class: 1
```

第 3 章 多模态数据处理

多模态数据处理是对图像、文本、音频和视频等不同类型的数据进行统一处理的过程。首先，数据预处理针对不同模态的数据进行清理和标准化。然后，通过数据增强技术，如图像翻转、文本同义词替换、音频变速和视频剪辑等，增加数据的多样性。接着，使用专门的标注工具为不同模态的数据添加标签。最后，通过数据融合与对齐，将不同模态的数据整合在一起，以便进行综合分析和建模，从而提升模型的性能。本章将详细讲解多模态数据处理的知识，帮助读者为后面的学习打下基础。

3.1 数据预处理

数据预处理是数据分析和建模的关键步骤，旨在提高数据质量，使其适合后续处理和分析。针对不同的数据模态有不同的预处理方法，具体的数据预处理包括文本预处理、图像预处理、音频预处理和视频预处理。

3.1.1 文本预处理

在多模态大模型应用中，文本数据的预处理包括数据清洗与处理、数据转换与整合、数据标准化与归一化等操作。下面的内容将详细讲解上述有关操作。

1. 数据清洗与处理

数据清洗和处理是数据预处理过程的一部分，它涉及对原始数据进行修复、填充、删除和转换，以使其适合用于训练和测试机器学习模型。

（1）处理缺失值。

假设有一个 CSV 文件 room.csv，其中包含有关房屋的信息，如下所示：

```
area,rooms,price
1200,3,250000
1000,,200000
1500,4,300000
,,180000
```

在这个 CSV 文件中，数据存在缺失值，例如某些行的 rooms 列为空。此时可以使用 TensorFlow Transform（TFT）来处理这些缺失值，同时对数据进行标准化，实例 3-1 演示了这一用法。

实例 3-1　使用 TFT 处理 CSV 文件中的缺失值
源码路径：codes/3/que.py

实例文件 que.py 的具体实现代码如下所示。

```python
import apache_beam as beam    # 导入 apache_beam 模块
import tensorflow as tf
import tensorflow_transform as tft
import tensorflow_transform.beam as tft_beam
import tempfile
import csv

# 定义 CSV 文件读取和解析函数
def parse_csv(csv_row):
    columns = tf.io.decode_csv(csv_row, record_defaults=[[0], [0.0], [0]])
    return {
        'area': columns[0],
        'rooms': columns[1],
        'price': columns[2]
    }

# 定义读取 CSV 文件并应用预处理的函数
def preprocess_data(csv_file):
    raw_data = (
            pipeline
            | 'ReadCSV' >> beam.io.ReadFromText(csv_file)
            | 'ParseCSV' >> beam.Map(parse_csv)
    )

    with tft_beam.Context(temp_dir=tempfile.mkdtemp()):
        transformed_data, transformed_metadata = (
                (raw_data, feature_spec)
                | tft_beam.AnalyzeAndTransformDataset(preprocessing_fn)
        )

    return transformed_data, transformed_metadata

# 定义特征元数据
feature_spec = {
    'area': tf.io.FixedLenFeature([], tf.int64),
    'rooms': tf.io.FixedLenFeature([], tf.float32),
    'price': tf.io.FixedLenFeature([], tf.int64),
}

# 定义数据预处理函数,处理缺失值和标准化
def preprocessing_fn(inputs):
    processed_features = {
        'area':tft.scale_to_z_score(inputs['area']),
        'rooms':tft.scale_to_0_1(tft.impute(inputs['rooms'], tft.constants.FLOAT_MIN)),
        'price': inputs['price']
    }
    return processed_features

# 读取 CSV 文件并应用预处理
with beam.Pipeline() as pipeline:
    transformed_data, transformed_metadata = preprocess_data('room.csv')

# 显示处理后的数据和元数据
for example in transformed_data:
    print(example)
print('Transformed Metadata:', transformed_metadata.schema)
```

在上述代码中,首先定义了 CSV 文件读取和解析函数(parse_csv),然后定义了特征元数据

（feature_spec）。接着，定义了数据预处理函数（preprocessing_fn），该函数使用 tft.impute 填充了 rooms 列中的缺失值，同时对 area 列进行了标准化。随后，使用 Beam 管道读取 CSV 文件并应用预处理，然后输出处理后的数据和元数据。运行代码后，将看到填充了缺失值并进行了标准化的数据，以及相应的元数据信息。执行代码后会输出：

```
{'area': 1.0, 'rooms': 0.0, 'price': 250000}
{'area': -1.0, 'rooms': -0.5, 'price': 200000}
{'area': 0.0, 'rooms': 0.5, 'price': 300000}
{'area': 0.0, 'rooms': 0.0, 'price': 180000}
Transformed Metadata: feature {
  name: "area"
  type: INT
  presence {
    min_fraction: 1.0
  }
  shape {
  }
}
feature {
  name: "rooms"
  type: FLOAT
  presence {
    min_fraction: 1.0
  }
  shape {
  }
}
feature {
  name: "price"
  type: INT
  presence {
    min_fraction: 1.0
  }
  shape {
  }
}
```

对上述输出结果的说明如下。

- 每一行都是预处理后的数据样本，其中 area 和 rooms 列经过缩放或填充处理，price 列保持不变。
- area 列经过缩放处理，例如 1200 经过标准化后为 1.0。
- rooms 列经过填充和缩放处理，例如 1000 填充为-1.0 并标准化为-0.5。
- price 列保持不变，例如 250000。
- 最后，输出了转换后的元数据模式，显示了每个特征的类型和存在性信息。

（2）异常值检测与处理。

在人工智能和数据分析中，异常值（Outliers）是指与大部分数据点在统计上显著不同的数据点。异常值可能是由于错误、噪声、测量问题或其他异常情况引起的，它们可能会对模型的训练和性能产生负面影响。因此，异常值检测和处理是数据预处理的重要步骤之一。

例如，实例 3-2 是一个使用 PyTorch 进行异常值检测与处理的例子，将使用孤立森林（Isolation Forest）算法进行异常值检测，并对异常值进行处理。

实例 3-2 使用 PyTorch 进行异常值检测与处理
源码路径：codes/3/yi.py

实例文件 yi.py 的具体实现代码如下所示。

3.1 数据预处理

```python
import torch
from sklearn.ensemble import IsolationForest
from torch.utils.data import Dataset, DataLoader
import numpy as np

# 生成一些带有异常值的随机数据
data = np.random.randn(100, 2)
data[10] = [10, 10]   # 添加一个异常值
data[20] = [-8, -8]   # 添加一个异常值

# 使用孤立森林算法进行异常值检测
clf = IsolationForest(contamination=0.1)   # 设置异常值比例
pred = clf.fit_predict(data)
anomalies = np.where(pred == -1)[0]   # 异常值索引

# 打印异常值索引
print("异常值索引:", anomalies)

# 自定义数据集类
class CustomDataset(Dataset):
    def __init__(self, data, anomalies):
        self.data = data
        self.anomalies = anomalies

    def __len__(self):
        return len(self.data)

    def __getitem__(self, idx):
        sample = self.data[idx]
        label = 1 if idx in self.anomalies else 0   # 标记异常值为1,正常值为0
        return torch.tensor(sample, dtype=torch.float32), label

# 创建数据集实例
dataset = CustomDataset(data, anomalies)

# 创建数据加载器
dataloader = DataLoader(dataset, batch_size=10, shuffle=True)

# 遍历数据加载器并输出样本及其标签
for batch in dataloader:
    samples, labels = batch
    print("样本:", samples)
    print("标签:", labels)
```

在上述代码中,首先生成了一些带有异常值的随机数据。然后,使用孤立森林(Isolation Forest)算法对数据进行异常值检测,通过指定 contamination 参数来设置异常值比例。接着,定义了一个自定义数据集类 CustomDataset,其中异常值的索引被标记为1,正常值的索引被标记为0。最后,我们创建了数据集实例和数据加载器,遍历数据加载器并输出样本及其标签,从而演示了如何使用 PyTorch 进行异常值检测与处理。

执行代码后,输出的内容是每个批次的样本和标签。每个批次的样本是一个张量,包含了一批数据样本,而对应的标签是一个张量,指示了每个样本是正常值(标签为 0)还是异常值(标签为 1)。例如,输出中的第一个批次的样本如下所示。

```
样本: tensor([[ 0.3008,  1.6835],
        [ 0.9125,  1.5915],
        [-0.3871, -0.0249],
        [-0.2126, -0.2027],
```

```
            [-0.5890,  1.2867],
            [ 1.9692, -1.6272],
            [ 0.4465,  0.9076],
            [ 0.1764, -0.2811],
            [ 0.9241, -0.3346],
            [ 0.5370,  0.2201]])
标签: tensor([0, 0, 0, 0, 0, 1, 0, 0, 0, 0])
```

在这个例子中,正常值样本的标签为 0,异常值样本的标签为 1。这个标签信息可以用于训练机器学习模型来进行异常值检测任务。

(3)处理重复数据。

处理数据集中的重复数据涉及具体的数据集和问题场景。通常,数据集中的重复数据可能会影响模型的性能和训练结果,因此需要进行适当的处理。在实际应用中,通常使用 Python 库 Pandas 来处理重复数据。例如下面是一个使用 Pandas 来处理重复数据的例子,假设有一个简单的文件 dataset.csv,其内容如下所示。

```
feature1,feature2,label
1.2,2.3,0
0.5,1.8,1
1.2,2.3,0
2.0,3.0,1
0.5,1.8,1
```

这个 CSV 文件包含 3 列内容:feature1 列、feature2 列和 label 列。其中,前两列是特征,最后一列是标签。注意,第 1 行和第 3 行以及第 2 行和第 5 行存在重复数据。在处理重复数据时,我们需要根据特定的情况来决定是否删除这些重复数据。实例 3-3 使用 Pandas 来处理重复数据。

实例 3-3　使用 Pandas 来处理重复数据
源码路径:codes/3/chong.py

实例文件 chong.py 用于处理文件 dataset.csv 中的重复数据,具体实现代码如下所示。

```python
import pandas as pd
# 读取数据集
data = pd.read_csv('dataset.csv')

# 检测重复数据
duplicates = data[data.duplicated()]

# 删除重复数据
data_no_duplicates = data.drop_duplicates()

# 打印处理后的数据集大小
print("原始数据集大小:", data.shape)
print("处理后数据集大小:", data_no_duplicates.shape)
```

执行代码后会输出:

```
原始数据集大小: (5, 3)
处理后数据集大小: (3, 3)
```

上述输出结果显示,原始数据集包含 5 行和 3 列,处理后的数据集包含 3 行和 3 列。这表明你成功地处理了数据集中的重复数据,将重复的样本行删除,从而得到了一个不包含重复数据的数据集。

2. 数据转换与整合

数据集的数据转换和整合是指在处理数据集时，对数据进行一系列的变换、处理和整合，以便于后续的分析、建模或其他任务的执行。这些操作包括特征选择与抽取、特征变换与降维等。

（1）特征选择与抽取。

特征选择和特征抽取是在机器学习和数据分析中常用的技术，旨在从原始数据中选择最有价值的特征或从特征中抽取出更有意义的信息。它们可以帮助减少特征的维度，提高模型的性能，降低过拟合的风险。下面是一个使用 PyTorch 实现数据集特征选择和抽取的例子（见实例 3-4）。在这个例子中，我们将使用 PCA 进行特征抽取，并使用相关性系数进行特征选择。

> **实例 3-4** 使用 PyTorch 实现数据集特征选择和抽取
> 源码路径：codes/3/te.py

实例文件 te.py 的具体实现代码如下所示。

```python
import torch
import torch.nn as nn
from sklearn.decomposition import PCA
from sklearn.feature_selection import SelectKBest, f_regression

# 创建一个示例数据集
data = torch.rand((100, 5))   # 100 个样本,每个样本有 5 个特征

# 特征抽取：PCA
pca = PCA(n_components=2)   # 将数据映射到 2 维空间
pca_data = pca.fit_transform(data.numpy())

print("PCA 抽取后的数据大小:", pca_data.shape)

# 特征选择：相关性选择
target = torch.rand((100,))   # 目标变量
select_k_best = SelectKBest(score_func=f_regression, k=3)   # 选择与目标变量相关性较高的 3 个特征
selected_data = select_k_best.fit_transform(data.numpy(), target.numpy())

print("相关性选择后的数据大小:", selected_data.shape)
```

在这个例子中，首先创建一个随机的数据集，其中有 100 个样本和 5 个特征。然后，使用 PCA 将特征抽取到 2 维空间，并使用相关性系数进行特征选择，选择与目标变量相关性较高的 3 个特征。最后，打印出抽取和选择后的数据集大小。执行代码后会输出：

```
PCA 抽取后的数据大小: (100, 2)
相关性选择后的数据大小: (100, 3)
```

上述输出结果与预期一致，PCA 抽取后的数据大小为(100，2)，相关性选择后的数据大小为(100，3)，说明特征抽取和特征选择操作已经成功执行。

（2）特征变换与降维。

数据集的特征变换和降维是在保留尽可能多信息的前提下，减少特征的维度。这有助于降低计算成本、加速训练和提高模型的泛化能力。当涉及使用 TensorFlow 进行数据集的特征变换和降维操作时，常见的方法是结合 TensorFlow 的数据处理功能和 PCA（主成分分析）技术进行处理。例如下面是一个完整的 TensorFlow 实现（见实例 3-5），展示了加载数据集、进行特征标准化、应用 PCA 进行降维的过程。

实例 3-5　使用 TensorFlow 实现特征变换与降维
源码路径： codes/3/jiang.py

实例文件 jiang.py 的具体实现代码如下所示。

```python
import numpy as np
import tensorflow as tf
from sklearn.datasets import load_iris
from sklearn.preprocessing import StandardScaler
from sklearn.decomposition import PCA

# 加载数据集
iris = load_iris()
data = iris.data

# 特征标准化
scaler = StandardScaler()
scaled_data = scaler.fit_transform(data)

# TensorFlow 数据集转换
features_tensor = tf.convert_to_tensor(scaled_data, dtype=tf.float32)
dataset = tf.data.Dataset.from_tensor_slices(features_tensor)

# 定义 PCA 模型
pca = PCA(n_components=2)

# 训练 PCA 模型并降维
pca_result = pca.fit_transform(scaled_data)

print("PCA 抽取后的数据大小:", pca_result.shape)
print("PCA 主成分:", pca.components_)
print("PCA 主成分方差解释率:", pca.explained_variance_ratio_)
```

在这个例子中，首先加载了一个名为 iris 的经典数据集。然后使用 Scikit-learn 的 StandardScaler 进行特征标准化，将数据转换为均值为 0、方差为 1 的分布。接下来，使用 TensorFlow 将数据集转换为张量，并构建了一个只包含特征的数据集。然后，定义了一个 PCA 模型，设置降维后的特征数为 2。将标准化后的数据输入 PCA 模型，训练 PCA 模型并进行降维。最后，输出降维后的数据大小、主成分以及主成分的方差解释率。

执行代码后会输出：

```
PCA 抽取后的数据大小: (150, 2)
PCA 主成分: [[ 0.52106591 -0.26934744  0.5804131   0.56485654]
 [ 0.37741762  0.92329566  0.02449161  0.06694199]]
PCA 主成分方差解释率: [0.72962445 0.22850762]
```

上述输出显示了经过 PCA 抽取后的数据大小、主成分矩阵以及主成分的方差解释率。下面是对输出结果的解释。

- PCA 抽取后的数据大小：表示经过 PCA 抽取后的数据集形状为（150，2），其中 150 是样本数，2 是降维后的特征数。
- PCA 主成分：二维数组表示主成分矩阵，每行代表一个主成分，每列代表一个原始特征。这里有两个主成分，每个主成分都是原始特征的线性组合。
- PCA 主成分方差解释率：这是一个数组，表示每个主成分对数据方差的解释程度。在这个例子中，第一个主成分解释了总方差的约 72.96%，第二个主成分解释了总方差的约 22.85%。

这个输出结果展示了 PCA 在降维过程中对数据进行的有效的特征提取，同时也提供了每个主成分的相对重要性。

注意：PCA 是一种无监督的降维技术，其目标是通过找到数据中的主成分来实现数据的降维。这个例子仅仅是 TensorFlow 在数据处理和特征变换方面的一个示范，在实际应用中可能需要根据具体问题进行调整。

3. 数据标准化与归一化

数据标准化和归一化是常用的数据预处理技术，用于将数据调整为特定范围或分布，以便于模型的训练和优化。这是数据预处理中的一项重要技术，它对于训练和优化机器学习模型具有重要作用。

（1）特征缩放和归一化。

特征缩放和归一化是数据预处理中常用的技术，用于将特征的值调整到一定范围内，以便提高模型的训练效果。在 TensorFlow Transform 中，可以使用 tft.scale_to_z_score 和 tft.scale_to_0_1 等函数来实现特征缩放和归一化。

- 特征缩放。

特征缩放是将特征的值缩放到一定的范围，通常是将特征值映射到均值为 0、标准差为 1 的正态分布。这有助于减少特征值之间的差异，使模型更稳定地进行训练。在 TensorFlow Transform 中，函数 tft.scale_to_z_score 用于将特征缩放到标准正态分布，使用格式如下：

```
processed_features = {
    'feature1': tft.scale_to_z_score(inputs['feature1']),
    # 其他特征处理
}
```

- 归一化。

归一化是将特征的值映射到 0~1 的范围内，通常使用特征值减去最小值，再用所得差值除以最大值与最小值之差。这有助于保持特征值之间的相对关系，并且适用于一些模型（例如神经网络）的输入。在 TensorFlow Transform 中，函数 tft.scale_to_0_1 用于将特征归一化到 0~1 的范围，使用格式如下：

```
processed_features = {
    'feature2': tft.scale_to_0_1(inputs['feature2']),
    # 其他特征处理
}
```

在实际应用中，可以根据数据的特点和模型的需求，选择适当的特征缩放或归一化方法。这些预处理技术有助于提高模型的收敛速度和稳定性，从而改善模型的性能。在使用 TensorFlow Transform 进行数据预处理时，可以将这些函数嵌入到 preprocessing_fn 中，以便对特征进行合适的处理。

实例 3-6 演示了 TensorFlow Transform 使用函数 scale_to_z_score 和函数 scale_to_0_1 进行特征缩放和归一化处理的过程。

实例 3-6 使用 TensorFlow 选择鸢尾花分类的最佳模型

源码路径：codes/3/suogui.py

实例文件 suogui.py 的具体实现代码如下所示。

```
# 定义 CSV 文件读取和解析函数
def parse_csv(csv_row):
    columns = tf.io.decode_csv(csv_row, record_defaults=[[0], [0.0], [0]])
```

```python
    return {
        'area': columns[0],
        'rooms': columns[1],
        'price': columns[2]
    }

# 定义特征元数据
feature_spec = {
    'area': tf.io.FixedLenFeature([], tf.int64),
    'rooms': tf.io.FixedLenFeature([], tf.float32),
    'price': tf.io.FixedLenFeature([], tf.int64),
}

# 定义数据预处理函数,处理特征缩放和归一化
def preprocessing_fn(inputs):
    processed_features = {
        'area_scaled': tft.scale_to_z_score(inputs['area']),
        'rooms_normalized': tft.scale_to_0_1(inputs['rooms']),
        'price': inputs['price']
    }
    return processed_features

# 读取 CSV 文件并应用预处理
def preprocess_data(csv_file):
    raw_data = (
            pipeline
            | 'ReadCSV' >> beam.io.ReadFromText(csv_file)
            | 'ParseCSV' >> beam.Map(parse_csv)
    )

    with tft_beam.Context(temp_dir=tempfile.mkdtemp()):
        transformed_data, transformed_metadata = (
                (raw_data, feature_spec)
                | tft_beam.AnalyzeAndTransformDataset(preprocessing_fn)
        )

    return transformed_data, transformed_metadata

# 定义数据管道
with beam.Pipeline() as pipeline:
    transformed_data, transformed_metadata = preprocess_data('data.csv')

# 显示处理后的数据和元数据
for example in transformed_data:
    print(example)
print('Transformed Metadata:', transformed_metadata.schema)
```

在这个例子中,'area'特征被缩放到标准正态分布,'rooms'特征被归一化到 0~1 的范围内。其余的'price'特征保持不变。执行代码后,将看到处理后的数据样本以及转换后的元数据模式:

```
{'area_scaled': -0.331662684, 'rooms_normalized': 0.6, 'price': 250000}
{'area_scaled': -0.957780719, 'rooms_normalized': 0.0, 'price': 200000}
{'area_scaled': 0.294811219, 'rooms_normalized': 0.8, 'price': 300000}
{'area_scaled': 1.378632128, 'rooms_normalized': 0.0, 'price': 180000}
Transformed Metadata: (schema definition)
```

在上面的输出结果中,'area_scaled'特征已经被缩放到标准正态分布范围内,'rooms_normalized'特征已经归一化到 0~1 的范围内,'price'特征保持不变。同时,读者还会看到转换后的元数据模

式。请注意，实际输出可能会因数据和处理方式而有所不同。这些处理后的数据可以作为输入供机器学习模型训练用，从而提高模型的性能和稳定性。

（2）数据转换和规范化。

数据转换和规范化是数据预处理的重要步骤，用于将原始数据转化为适合机器学习模型训练的格式，同时对数据进行标准化处理，以提高模型的性能和稳定性。

数据转换和规范化的具体方法根据数据的类型和问题的需求而有所不同。使用工具如 TensorFlow Transform 可以帮助自动化这些步骤，以保证数据的一致性和准确性。例如实例 3-7 是一个使用 TensorFlow Transform 实现数据转换和规范化的例子，假设我们有一个包含数值特征的 CSV 文件（data.csv），内容如下：

```
feature1,feature2,label
10,0.5,1
20,,0
30,0.2,1
,,0
```

实例 3-7　实现数据转换和规范化处理
源码路径：codes/3/zhuangui.py

实例文件 zhuangui.py 的具体实现代码如下所示。

```python
# 定义 CSV 文件读取和解析函数
def parse_csv(csv_row):
    columns = tf.io.decode_csv(csv_row, record_defaults=[[0], [0.0], [0]])
    return {
        'feature1': columns[0],
        'feature2': columns[1],
        'label': columns[2]
    }

# 定义特征元数据
feature_spec = {
    'feature1': tf.io.FixedLenFeature([], tf.int64),
    'feature2': tf.io.FixedLenFeature([], tf.float32),
    'label': tf.io.FixedLenFeature([], tf.int64),
}

# 定义数据预处理函数,进行特征缩放和归一化处理
def preprocessing_fn(inputs):
    processed_features = {
        'feature1_scaled': tft.scale_to_z_score(inputs['feature1']),
        'feature2_normalized':tft.scale_to_0_1(tft.impute(inputs['feature2'],tft.constants.INT_MIN)),
        'label': inputs['label']
    }
    return processed_features

# 读取 CSV 文件并应用预处理
def preprocess_data(csv_file):
    raw_data = (
        pipeline
        | 'ReadCSV' >> beam.io.ReadFromText(csv_file)
        | 'ParseCSV' >> beam.Map(parse_csv)
    )

    with tft_beam.Context(temp_dir=tempfile.mkdtemp()):
        transformed_data, transformed_metadata = (
```

```
            (raw_data, feature_spec)
            | tft_beam.AnalyzeAndTransformDataset(preprocessing_fn)
        )
    
    return transformed_data, transformed_metadata

# 定义数据管道
with beam.Pipeline() as pipeline:
    transformed_data, transformed_metadata = preprocess_data('data.csv')

# 显示处理后的数据和元数据
for example in transformed_data:
    print(example)
print('Transformed Metadata:', transformed_metadata.schema)
```

对上述代码的具体说明如下。

（1）定义 CSV 文件读取和解析函数：parse_csv 函数用于解析 CSV 行，将其解码为特征字典。这里假设 CSV 文件的每一行包含 3 个字段：feature1、feature2 和 label。

（2）定义特征元数据：feature_spec 指定了特征的名称和数据类型。

（3）定义数据预处理函数：preprocessing_fn 函数对输入特征进行预处理。在这个例子中，feature1 特征被缩放到标准正态分布（Z-Score 标准化），feature2 特征被缩放到 0~1 的范围内（归一化），并且对缺失值进行处理（用 tft.constants.INT_MIN 填充）。

（4）读取 CSV 文件并应用预处理：preprocess_data 函数使用 Beam 数据管道读取 CSV 文件数据，并将数据应用到预处理函数中。使用 TensorFlow Transform 的 AnalyzeAnd Transform Dataset 函数进行数据转换和规范化处理，同时生成转换后的元数据。

（5）定义数据管道：使用 Beam 库创建数据管道，并调用 preprocess_data 函数进行数据预处理。

（6）显示处理后的数据和元数据：遍历处理后的数据，打印每个样本的处理结果，以及转换后的元数据模式。

执行代码后会输出：

```
{'feature1_scaled': 0.0, 'feature2_normalized': 0.5, 'label': 1}
{'feature1_scaled': 0.7071067690849304, 'feature2_normalized': 0.0, 'label': 0}
{'feature1_scaled': 1.4142135381698608, 'feature2_normalized': 0.1, 'label': 1}
{'feature1_scaled': -0.7071067690849304, 'feature2_normalized': 0.0, 'label': 0}
Transformed Metadata: Schema(feature {
  name: "feature1_scaled"
  type: FLOAT
  presence {
    min_fraction: 1.0
    min_count: 1
  }
}
feature {
  name: "feature2_normalized"
  type: FLOAT
  presence {
    min_fraction: 1.0
    min_count: 1
  }
}
feature {
  name: "label"
  type: INT
  presence {
    min_fraction: 1.0
```

```
      min_count: 1
    }
  }
  generated_feature {
    name: "feature1_scaled"
    type: FLOAT
    presence {
      min_fraction: 1.0
      min_count: 1
    }
  }
  generated_feature {
    name: "feature2_normalized"
    type: FLOAT
    presence {
      min_fraction: 1.0
      min_count: 1
    }
  }
  generated_feature {
    name: "label"
    type: INT
    presence {
      min_fraction: 1.0
      min_count: 1
    }
  }
)
```

上述输出显示了经过处理的数据样本和转换后的元数据。注意，处理后的特征值和元数据模式可能会因实际数据的不同而有所变化。

3.1.2 图像预处理

在多模态大模型中，图像预处理是将原始图像数据转换为适合模型输入的关键步骤。主要的图像预处理技术包括图像归一化、去噪、图像裁剪和缩放等，这些预处理技术有助于提高模型的准确性和鲁棒性，使其在融合多模态数据进行分析时表现更佳。

1. 归一化

在多模态大模型的应用中，图像预处理中的归一化是指调整图像像素值的范围，以标准化图像数据。这一步骤非常关键，因为它可以使图像数据在训练过程中更加稳定，提高模型的收敛速度和性能。

（1）归一化的目的。

- 稳定训练过程：通过将像素值缩放到一个标准范围（如0～1或-1～1），可以防止数值过大或过小导致的数值不稳定问题。
- 提高模型性能：归一化可以使不同图像的数据分布更为一致，从而提高模型的泛化能力。
- 加速收敛：标准化的输入数据可以加速梯度下降算法的收敛速度。

（2）归一化的常用方法。

- 像素值缩放：使用 Min-Max 归一化方法将像素值缩放到[0，1]范围。

$$x' = \frac{x - x_{\min}}{x_{\max} - x_{\min}}$$

如果原始像素值范围是0～255，经过 Min-Max 归一化后，所有像素值将缩放到0～1。

- 均值归一化：将像素值调整为零均值和单位方差。

$$x' = \frac{x - \mu}{\sigma}$$

其中，μ是像素值的均值，σ是像素值的标准差。

例如，对于均值为 128、标准差为 64 的图像数据，归一化后像素值的范围为[-2，2]。

- 全局归一化：对整个数据集进行归一化处理，而不是单独对每张图像进行归一化。例如，对整个训练集计算均值和标准差，然后对每张图像进行标准化。

实例 3-8 演示了对指定数据集图像进行归一化处理的过程。

> **实例 3-8** 对指定数据集图像进行归一化处理
> 源码路径：codes/3/gui01.py 和 gui02.py

本实例的具体实现流程如下。

（1）准备 4 张极小型的数据及图像文件 1.jpg、2.jpg、3.jpg、4.jpg。

（2）编写实例文件 gui01.py，对整个图像计算每个通道的均值和标准差。具体实现代码如下所示。

```python
import numpy as np
import cv2
from glob import glob

# 获取所有图像文件的路径
image_paths = glob('path_to_images/*.jpg')

# 初始化均值和标准差
means = np.zeros(3)
stds = np.zeros(3)

# 计算全局均值和标准差
for image_path in image_paths:
    image = cv2.imread(image_path)  # 读取图像
    image = cv2.cvtColor(image, cv2.COLOR_BGR2RGB)  # 转换为 RGB 格式
    means += image.mean(axis=(0, 1))
    stds += image.std(axis=(0, 1))

# 计算均值和标准差
means /= len(image_paths)
stds /= len(image_paths)

print(f'均值: {means}, 标准差: {stds}')
```

（3）编写实例文件 gui02.py，对每张图像进行均值归一化和 Min-Max 归一化处理。具体实现代码如下所示。

```python
def normalize_image(image, means, stds):
    # 均值归一化
    image = (image - means) / stds
    return image

def min_max_normalize(image):
    # Min-Max 归一化
    image = (image - image.min()) / (image.max() - image.min())
    return image

# 对所有图像进行归一化处理
normalized_images = []
for image_path in image_paths:
```

```
image = cv2.imread(image_path)
image = cv2.cvtColor(image, cv2.COLOR_BGR2RGB)
image = normalize_image(image, means, stds)
image = min_max_normalize(image)
normalized_images.append(image)
```

2. 去噪

在多模态大模型应用中,图像去噪是一项重要的预处理步骤,旨在减少图像中的噪声并提高数据质量。噪声可能来自图像采集设备、传输过程或存储过程,影响图像的清晰度和准确性。

(1)均值滤波器。

均值滤波器是一种简单而常用的噪声减少技术。它通过在图像中的每个像素周围取一个固定大小的窗口,计算窗口中所有像素的平均值,并将该平均值赋给中心像素。这种方法对于高斯噪声和均匀噪声的去除效果较好,但可能会导致图像模糊。实例 3-9 演示了使用均值滤波器实现图像噪声减少的过程。

实例 3-9 使用均值滤波器实现图像噪声减少
源码路径:codes\3\junjian.py

实例文件 junjian.py(实例 3-9)的具体实现代码如下。

```python
import cv2
import numpy as np

# 读取图像
image = cv2.imread('888.jpg', cv2.IMREAD_COLOR)

# 将图像转换为灰度图像
gray_image = cv2.cvtColor(image, cv2.COLOR_BGR2GRAY)

# 定义均值滤波器的窗口大小
kernel_size = 5

# 使用均值滤波器进行滤波
filtered_image = cv2.blur(gray_image, (kernel_size, kernel_size))

# 显示原始图像和滤波后的图像
cv2.imshow('Original Image', gray_image)
cv2.imshow('Filtered Image', filtered_image)5x5
cv2.waitKey(0)
cv2.destroyAllWindows()
```

在上述代码中,首先使用 OpenCV 库读取一张彩色图像,并将其转换为灰度图像。然后,我们定义了均值滤波器的窗口大小(在这里是 5×5)。最后,使用 cv2.blur()函数应用均值滤波器进行滤波,并将滤波后的图像显示出来。通过调整 kernel_size 的大小,你可以控制滤波器的窗口大小,从而影响滤波的效果。较大的窗口大小可以更有效地平滑图像,但也会导致细节的丢失。执行效果如图 3-1 所示。

(2)中值滤波器。

中值滤波器是一种非线性滤波器,对于椒盐噪声等脉冲噪声的去除效果较好。它的原理是在窗口中取所有像素的中值,并将中值赋给中心像素。中值滤波器能够有效去除离群值,但可能会导致图像细节的损失。实例 3-10 是一个使用中值滤波器实现图像噪声减少的例子。

图 3-1 执行效果

实例 3-10 使用中值滤波器实现图像噪声减少

源码路径：codes\3\zhong.py

实例文件 3-10 的具体实现代码如下。

```
import cv2
import numpy as np

# 读取图像
image = cv2.imread('image.jpg', cv2.IMREAD_COLOR)

# 将图像转换为灰度图像
gray_image = cv2.cvtColor(image, cv2.COLOR_BGR2GRAY)

# 定义中值滤波器的窗口大小
kernel_size = 5

# 使用中值滤波器进行滤波
filtered_image = cv2.medianBlur(gray_image, kernel_size)

# 显示原始图像和滤波后的图像
cv2.imshow('Original Image', gray_image)
cv2.imshow('Filtered Image', filtered_image)
cv2.waitKey(0)
cv2.destroyAllWindows()
```

在上述代码中，首先使用 OpenCV 库读取一张彩色图像，并将其转换为灰度图像。然后，我们定义了中值滤波器的窗口大小（在这里是 5×5）。最后，我们使用 cv2.medianBlur()函数应用中值滤波器进行滤波，并将滤波后的图像显示出来。通过调整 kernel_size 的大小，可以控制滤波器的窗口大小，从而影响滤波的效果。较大的窗口大小可以更有效地去除椒盐噪声等脉冲噪声，但也会导致细节的丢失。

（3）高斯滤波器。

高斯滤波器是一种线性滤波器，它基于高斯函数对像素进行加权平均。它能够在滤波过程中保留边缘信息，并对高斯噪声有较好的去除效果。调整高斯滤波器的滤波窗口大小和标准差可以在去除噪声的同时尽可能保留图像细节，以达到平衡的效果。实例 3-11 是一个使用高斯滤波器减少图像噪声的例子。

> **实例 3-11** 使用高斯滤波器减少图像噪声
> 源码路径：codes\3\gaolv.py

实例文件 3-11 的具体实现代码如下。

```python
import cv2
import numpy as np

# 读取图像
image = cv2.imread('image.jpg', cv2.IMREAD_COLOR)

# 将图像转换为灰度图像
gray_image = cv2.cvtColor(image, cv2.COLOR_BGR2GRAY)

# 定义高斯滤波器的窗口大小和标准差
kernel_size = 5
sigma = 1.5

# 使用高斯滤波器进行滤波
filtered_image = cv2.GaussianBlur(gray_image, (kernel_size, kernel_size), sigma)

# 显示原始图像和滤波后的图像
cv2.imshow('Original Image', gray_image)
cv2.imshow('Filtered Image', filtered_image)
cv2.waitKey(0)
cv2.destroyAllWindows()
```

在上述代码中，首先使用 OpenCV 库读取一张彩色图像，并将其转换为灰度图像。然后，定义高斯滤波器的窗口大小（在这里是 5×5）和标准差（sigma，控制滤波器的平滑程度）。最后，我们使用 cv2.GaussianBlur()函数应用高斯滤波器进行滤波，并将滤波后的图像显示出来。通过调整 kernel_size 的大小和 sigma 的值，可以控制滤波器的窗口大小和平滑程度，从而影响滤波的效果。较大的窗口大小和较大的标准差可以更有效地平滑图像，但也会导致细节的丢失。

3.1.3 音频预处理

在多模态大模型中，音频预处理是准备音频数据以输入模型进行训练或推理的重要步骤之一。音频数据通常以原始波形的形式存在，需要进行一系列的预处理操作，以提取有用的特征并减少噪声，从而提高模型的性能和鲁棒性。实现音频预处理的基本步骤如下。

1. 采样率调整（Resampling）

在多模态大模型的音频预处理中，采样率调整是一个重要的步骤，用于将音频数据的采样率调整为模型所需的采样率。采样率表示每秒对信号的采样次数，通常以赫兹（Hz）为单位。不同的应用场景和模型可能需要不同的采样率，因此在将音频数据输入模型之前，通常需要将其调整为模型所需的采样率。如果音频数据的采样率过高或过低，可以通过重采样操作将其调整为模型所需的采样率，常见的采样率包括 8 kHz、16 kHz 和 44.1 kHz 等。

采样率调整的过程通常涉及如下两个主要步骤。

（1）降采样（Downsampling）。

如果原始音频数据的采样率高于模型所需的采样率，就需要进行降采样操作。降采样是通过去除原始音频数据中的部分采样点来减少采样率，从而降低数据量和计算复杂度的。通常，降采样会导致信号的频率范围减小，因此需要注意避免频率混叠（aliasing）问题。

（2）上采样（Upsampling）。

如果原始音频数据的采样率低于模型所需的采样率，就需要进行上采样操作。上采样是通过插值等方法在原始音频数据中插入新的采样点来增加采样率,从而提高信号的分辨率和频率范围的。常见的插值方法包括线性插值、样条插值和最近邻插值等。

请看实例 3-12,假设正在开发一个多模态大模型,用于识别音频中的语音指令,并根据指令内容执行相应的操作。在这个任务中需要处理两种数据:音频数据和文本指令数据。可以通过调整音频的采样率来适应模型的需求。

实例 3-12　调整指定的音频的采样率
源码路径:codes\3\caiyin.py

实例文件 caiyin.py 的具体实现代码如下所示。

```python
import librosa
import soundfile as sf

# 加载原始音频文件
audio_path = 'original_audio.wav'
y, sr = librosa.load(audio_path, sr=None)

# 定义目标采样率
target_sr = 16000  # 目标采样率为 16 kHz

# 使用 librosa 进行采样率调整
y_resampled = librosa.resample(y, sr, target_sr)

# 保存调整后的音频文件
resampled_audio_path = 'resampled_audio.wav'
sf.write(resampled_audio_path, y_resampled, target_sr)

print("原始音频采样率: ", sr)
print("调整后音频采样率: ", target_sr)
```

在上述代码中,首先使用 librosa 库加载原始音频文件,并获取其原始采样率(sr)。然后,定义了目标采样率(target_sr),这里设定为 16 kHz。接下来,使用 librosa 的 librosa.resample() 函数对音频进行采样率调整,将其调整为目标采样率。最后,使用 soundfile 库将调整后的音频数据保存为新的音频文件。

在对音频进行采样率调整时,需要注意以下几点。

- 保持音频质量:尽量避免引入失真或伪影,以保持音频数据的质量和完整性。
- 滤波处理:在降采样操作中,通常需要使用低通滤波器来去除高频分量,以防止频率混叠现象的发生。
- 适当的采样率选择:根据模型的需求和任务的特点,选择合适的采样率,以平衡数据质量和计算效率。

总的来说,采样率调整是多模态大模型中音频预处理的重要步骤之一,通过合理调整音频数据的采样率,可以为模型提供适合的输入,从而提高模型的性能和鲁棒性。

2. 音频分帧(Audio Framing)

将长时间的音频信号分割成短的帧,通常采用 20~50ms 的时间窗口。这有助于提取短时特征并提高模型对时序数据的理解能力。请看实例 3-13,假设正在开发一个多模态大模型,用于识别音频中的语音命令,并根据命令内容执行相应的操作。在这个任务中,需要处理两种数据:音频数据和文本命令数据。可以通过音频分帧操作将音频数据分割成小的时间窗口,以便提取时间序列特征。

实例 3-13 对指定的音频进行分帧操作
源码路径：codes\3\fen.py

实例文件 fen.py 的具体实现代码如下所示。

```python
import librosa
import numpy as np

# 加载原始音频文件
audio_path = '123.m4a'
y, sr = librosa.load(audio_path, sr=None)

# 定义帧的参数
frame_length = 0.025   # 帧长为25ms
frame_stride = 0.010   # 帧移为10ms

# 计算帧长和帧移的样本数
frame_length_samples = int(frame_length * sr)
frame_stride_samples = int(frame_stride * sr)

# 计算分帧的总帧数
num_frames =1+int(np.floor((len(y)-frame_length_samples) / frame_stride_samples))

# 初始化帧数组
frames = np.zeros((num_frames, frame_length_samples))

# 矢量化分帧操作
for i in range(num_frames):
    start = i * frame_stride_samples
    frames[i] = y[start:start + frame_length_samples]

# 显示分帧后的结果
print("原始音频长度: ", len(y))
print("分帧后的帧数: ", len(frames))
print("每帧的样本数: ", len(frames[0]))
```

在上述代码中，首先使用 librosa 库加载原始音频文件，并获取其采样率（sr）。然后定义帧的参数，包括帧长（frame_length）和帧移（frame_stride），以 s 为单位。接着，将帧长和帧移转换为采样点数，以便进行分帧操作。然后，使用 for 循环对音频进行分帧操作，并将每个帧存储在 frames 列表中。最后，将 frames 转换为 numpy 数组，并显示分帧后的结果，包括原始音频的长度、分帧后的帧数以及每帧的样本数。

3. 加窗（Windowing）

通过对每一帧的音频信号进行加窗操作，可以减少频谱泄漏并提高频谱分辨率。常见的窗函数包括汉明窗、汉宁窗和布莱克曼窗等。在多模态大模型中，加窗（Windowing）操作通常用于音频信号处理，以便在进行傅里叶变换等操作时减小边界效应。请看实例 3-14，先是对指定的音频信号进行加窗操作，然后提取特征。

实例 3-14 对指定的音频进行加窗操作
源码路径：codes\3\jia.py

实例文件 jia.py 的具体实现代码如下所示。

```python
import librosa
import numpy as np
import matplotlib.pyplot as plt
```

```python
# 加载音频文件
audio_path = '123.m4a'
y, sr = librosa.load(audio_path, sr=None)

# 定义帧的参数
frame_length = 0.025    # 帧长为25ms
frame_stride = 0.010    # 帧移为10ms

# 计算帧长和帧移的样本数
frame_length_samples = int(frame_length * sr)
frame_stride_samples = int(frame_stride * sr)

# 计算分帧的总帧数
num_frames=1+int(np.floor((len(y)- frame_length_samples) / frame_stride_samples))

# 初始化帧数组
frames = np.zeros((num_frames, frame_length_samples))

# 分帧操作
for i in range(num_frames):
    start = i * frame_stride_samples
    frames[i] = y[start:start + frame_length_samples]

# 定义窗函数类型
window_type = 'hann'    # 可以是 'hann', 'hamming', 'blackman' 等

# 生成窗函数
window = np.hanning(frame_length_samples) if window_type == 'hann' else \
         np.hamming(frame_length_samples) if window_type == 'hamming' else \
         np.blackman(frame_length_samples)

# 对每个帧应用窗函数
windowed_frames = frames * window

# 计算每个帧的 STFT（短时傅里叶变换）
stft=np.array([np.fft.rfft(windowed_frame)for windowed_frame in windowed_frames])

# 显示原始音频、加窗后的帧和 STFT 的部分结果
plt.figure(figsize=(15, 10))

# 原始音频信号
plt.subplot(3, 1, 1)
plt.plot(y)
plt.title('Original Audio Signal')

# 加窗后的一个帧
plt.subplot(3, 1, 2)
plt.plot(frames[0], label='Original Frame')
plt.plot(windowed_frames[0], label='Windowed Frame')
plt.title('Windowing Operation on a Frame')
plt.legend()

# STFT 结果的幅度谱
plt.subplot(3, 1, 3)
plt.imshow(np.abs(stft.T), aspect='auto', origin='lower')
plt.title('STFT Magnitude Spectrogram')

plt.tight_layout()
plt.show()

# 显示分帧后的结果
print("原始音频长度: ", len(y))
```

```
print("分帧后的帧数: ", len(frames))
print("每帧的样本数: ", len(frames[0]))
print("加窗后的一个帧: ", windowed_frames[0])
print("STFT 的形状: ", stft.shape)
```

上述代码展示了在多模态大模型应用中实现音频加窗操作的用法，包括分帧、应用窗函数和计算 STFT。加窗操作可以减小边界效应，提高特征提取的效果。假设音频文件 123.wma 的采样率为 22050 Hz，总长度为 10s，则执行代码后会输出：

```
原始音频长度: 220500
分帧后的帧数: 1980
每帧的样本数: 551
加窗后的一个帧: [ 0.           0.00003495  0.00013979 ... -0.00014837 -0.00003549  0.         ]
STFT 的形状: (1980, 276)
```

4. 傅里叶变换（Fourier Transform）

傅里叶变换将每一帧的时域信号转换为频域信号，以便提取频域特征。常见的傅里叶变换包括快速傅里叶变换（FFT）和离散傅里叶变换（DFT）等。实例 3-15 展示了在多模态大模型应用中对音频实现傅里叶变换操作的过程。该例子包含加载音频文件、执行傅里叶变换、提取频率特征，并将其与图像特征结合输入多模态模型等步骤。

实例 3-15 对指定的音频进行傅里叶变换操作
源码路径：codes\3\bian.py

实例文件 bian.py 的具体实现代码如下所示。

```python
import numpy as np
import librosa
import matplotlib.pyplot as plt
from tensorflow.keras.layers import Dense, Input, Concatenate, Flatten
from tensorflow.keras.models import Model

# 加载音频文件
audio_path = 'example.wav'  # 请使用实际的音频文件路径
y, sr = librosa.load(audio_path, sr=None)

# 计算傅里叶变换
fft_result = np.fft.fft(y)
fft_magnitude = np.abs(fft_result)

# 提取前一半频率成分（因为 FFT 结果是对称的）
half_length = len(fft_magnitude) // 2
fft_magnitude = fft_magnitude[:half_length]

# 生成频率轴
frequencies = np.fft.fftfreq(len(y), d=1/sr)[:half_length]

# 绘图
plt.figure(figsize=(12, 6))
plt.subplot(2, 1, 1)
plt.plot(y)
plt.title('Original Audio Signal')

plt.subplot(2, 1, 2)
plt.plot(frequencies, fft_magnitude)
plt.title('Magnitude Spectrum')
plt.xlabel('Frequency (Hz)')
plt.ylabel('Magnitude')
plt.tight_layout()
```

```python
plt.show()

# 构建多模态模型
# 图像输入（假设图像预处理已完成）
image_input = Input(shape=(224, 224, 3))
x = Flatten()(image_input)

# 音频频率特征输入
audio_input = Input(shape=(half_length,))

# 合并图像和音频特征
combined = Concatenate()([x, audio_input])
combined = Dense(128, activation='relu')(combined)
output = Dense(1, activation='sigmoid')(combined)  # 二分类问题（如情感分析）

# 构建模型
model = Model(inputs=[image_input, audio_input], outputs=output)
model.compile(optimizer='adam', loss='binary_crossentropy', metrics=['accuracy'])

# 假设我们有图像和音频频率特征的数据集
image_data = np.random.rand(10, 224, 224, 3)  # 示例图像数据
audio_data = np.random.rand(10, half_length)  # 示例音频频率特征数据
labels = np.random.randint(2, size=10)  # 示例标签

# 训练模型
model.fit([image_data, audio_data], labels, epochs=10)
```

对上述代码的具体说明如下。

（1）加载音频文件：使用 librosa.load 加载音频文件。

（2）计算傅里叶变换：使用 np.fft.fft 计算音频信号的傅里叶变换，并提取幅度谱。

（3）提取前一半频率成分：因为 FFT 结果是对称的，所以我们提取前一半的频率成分即可。

（4）绘制原始音频信号和幅度谱：使用 matplotlib 绘制原始音频信号和傅里叶变换的幅度谱。

（5）构建多模态模型：定义一个简单的多模态模型，接收图像和音频频率特征作为输入。

（6）训练模型：使用示例数据训练模型。

5．声谱图（Spectrogram）

将频域信号转换为二维图像，即声谱图。声谱图可以直观地展示音频信号在时间和频率上的变化，并提供给模型进行处理。实例 3-16 的功能是对指定的音频实现声谱图操作，并将其与图像特征结合，输入多模态模型。

实例 3-16 对指定的音频进行声谱图操作
源码路径：codes\3\sheng.py

实例文件 sheng.py 的具体实现代码如下所示。

```python
import numpy as np
import librosa
import matplotlib.pyplot as plt
import tensorflow as tf
from tensorflow.keras.layers import Dense,Input,Concatenate,Flatten,Conv2D,MaxPooling2D
from tensorflow.keras.models import Model

# 加载音频文件
audio_path = 'example.wav'  # 请使用实际的音频文件路径
y, sr = librosa.load(audio_path, sr=None)

# 计算声谱图
S = librosa.feature.melspectrogram(y=y, sr=sr, n_mels=128)
```

```python
S_dB = librosa.power_to_db(S, ref=np.max)

# 绘制声谱图
plt.figure(figsize=(10, 4))
librosa.display.specshow(S_dB, sr=sr, x_axis='time', y_axis='mel')
plt.colorbar(format='%+2.0f dB')
plt.title('Mel-frequency spectrogram')
plt.tight_layout()
plt.show()

# 声谱图处理成模型输入形状
spectrogram = np.expand_dims(S_dB, axis=-1)  # 增加一个通道维度
spectrogram = np.expand_dims(spectrogram, axis=0)  # 增加一个批处理维度

# 构建多模态模型
# 图像输入(假设图像预处理已完成)
image_input = Input(shape=(224, 224, 3))
x = Conv2D(32, (3, 3), activation='relu')(image_input)
x = MaxPooling2D((2, 2))(x)
x = Flatten()(x)

# 声谱图输入
audio_input = Input(shape=(128, spectrogram.shape[2], 1))
y = Conv2D(32, (3, 3), activation='relu')(audio_input)
y = MaxPooling2D((2, 2))(y)
y = Flatten()(y)

# 合并图像和音频特征
combined = Concatenate()([x, y])
combined = Dense(128, activation='relu')(combined)
output = Dense(1, activation='sigmoid')(combined)  # 二分类问题(如情感分析)

# 构建模型
model = Model(inputs=[image_input, audio_input], outputs=output)
model.compile(optimizer='adam', loss='binary_crossentropy', metrics=['accuracy'])

# 假设我们有图像和声谱图的数据集
image_data = np.random.rand(10, 224, 224, 3)  # 示例图像数据
audio_data = np.random.rand(10, 128, spectrogram.shape[2], 1)  # 示例声谱图数据
labels = np.random.randint(2, size=10)  # 示例标签

# 训练模型
model.fit([image_data, audio_data], labels, epochs=10)
```

对上述代码的具体说明如下。

（1）加载音频文件：使用 librosa.load 加载音频文件。

（2）计算声谱图：使用 librosa.feature.melspectrogram 计算梅尔频谱图，并转换为对数刻度。

（3）绘制声谱图：使用 librosa.display.specshow 绘制声谱图。

（4）预处理声谱图：将声谱图增加一个通道维度和批处理维度，使其适应模型的输入形状。

（5）构建多模态模型。

- 图像输入部分：定义一个简单的卷积神经网络，用于处理图像数据。
- 声谱图输入部分：定义另一个卷积神经网络，用于处理声谱图数据。
- 合并部分：将图像特征和声谱图特征合并，并通过全连接层进行分类。
- 训练模型：使用示例数据训练模型。

注意：上面介绍的是常见的音频预处理步骤，在实际应用中，预处理步骤的选择和参数设置会根据具体任务和数据特点而有所不同。通过合理的音频预处理操作，可以有效地提取音频数据的有用信息，并为模型提供高质量的输入。

3.1.4 视频预处理

在多模态大模型应用中，视频预处理技术是指在处理视频数据以供深度学习模型使用时所采取的一系列技术手段，旨在提取和准备视频数据的特征，以便于模型进行有效的学习和推理。视频预处理在多模态模型中尤其重要，因为视频数据通常是高维、复杂的且包含丰富的信息。常见的视频预处理技术如下。

- 视频分割（Video Segmentation）：将视频分割成较短的片段或帧，以便于处理。这可以是基于时间的分割，比如每隔几秒截取一帧；也可以是基于事件或动作的分割，比如根据场景的变化或者动作的发生。
- 帧提取（Frame Extraction）：从视频中提取关键帧，以降低数据量并保留重要信息。可以使用固定间隔或者根据图像质量、运动等因素选择关键帧。
- 视频压缩和编解码（Video Compression and Encoding/Decoding）：对视频进行压缩以减小存储空间和传输带宽，并在需要时解码还原。常用的视频编码、解码器有 H.264、H.265 等。
- 尺寸调整和裁剪（Resizing and Cropping）：调整视频帧的尺寸，使其适应模型的输入要求。通常模型对输入尺寸有特定的要求，因此需要将视频帧统一到相同的尺寸。
- 光流计算（Optical Flow Computation）：计算视频中相邻帧之间的光流，以捕捉对象的运动信息。光流可以描述相邻帧之间像素的位移关系，有助于用户理解视频中的动态变化。
- 帧间差分（Frame Differencing）：计算相邻帧之间的差异，以便捕获视频中的运动和变化。可以通过计算像素差异或者使用更高级的运动估计技术来实现。
- 颜色空间转换（Color Space Conversion）：将视频帧从 RGB 颜色空间转换到其他颜色空间，如灰度图、HSV（色调、饱和度、亮度）等，以便于特定任务的处理或减小数据的复杂度。
- 数据增强（Data Augmentation）：对视频进行随机变换，如旋转、平移、缩放、翻转等，以增加数据多样性，提升模型的泛化能力。
- 预处理和归一化（Preprocessing and Normalization）：对视频帧进行预处理操作，如均值归一化、标准化等，以减小数据分布的差异，有助于模型更快地收敛。
- 关键帧提取（Key Frame Extraction）：从视频中提取最能代表视频内容的关键帧，可以利用图像内容、运动信息等来选择。
- 语义分割（Semantic Segmentation）：对视频中的每一帧进行像素级别的分类，将不同的像素分为不同的语义类别，以提取视频中的语义信息。
- 音频处理（Audio Processing）：对视频的音频部分进行处理，如音频分离、语音识别等，以获得视频的音频信息，这对于一些多模态任务尤为重要。

上述视频预处理技术可以根据具体任务和模型的需求进行组合和调整，以提取最相关、最有用的信息供多模态大模型使用。

请看下面的例子。实例 3-17 的功能是从视频 123.mp4 中提取每个子视频片段的第一帧作为代表性帧，并保存为图像文件。

实例 3-17 提取指定视频中的图像帧
源码路径：codes\3\zhen.py

实例文件 zhen.py 的具体实现代码如下所示。

```
import cv2
import os
```

```python
# 视频文件路径
video_path = 'example_video.mp4'
# 输出文件夹路径
output_folder = 'processed_frames'
# 每个子视频片段的长度(帧数)
segment_length = 100

def split_video(video_path, output_folder, segment_length):
    # 创建输出文件夹
    if not os.path.exists(output_folder):
        os.makedirs(output_folder)

    # 打开视频文件
    cap = cv2.VideoCapture(video_path)

    frame_count = 0
    segment_count = 0
    success = True
    while success:
        # 读取视频帧
        success, frame = cap.read()

        # 每读取一次 segment_length 帧就保存一帧
        if frame_count % segment_length == 0 and success:
            # 保存帧
            frame_save_path=os.path.join(output_folder,f"segment_{segment_count}_frame_0.jpg")
            cv2.imwrite(frame_save_path, frame)

            segment_count += 1

        frame_count += 1

    cap.release()

# 执行视频分割
split_video(video_path, output_folder, segment_length)

print("视频预处理完成!")
```

上述代码使用 OpenCV 对视频进行预处理，包括视频分割和帧提取。它从给定的视频中提取每个子视频片段的第一帧，并将这些第一帧保存为图像文件。视频分割的间隔由 segment_length 参数控制。这个代码适用于需要对视频进行预处理，并提取代表性帧的场景。

3.2 数据增强

在多模态大模型中，数据增强是一种常用的技术，用于扩充训练数据集以改善模型的泛化能力和性能。数据增强通过对原始数据进行多种变换和处理，生成具有一定差异性的新样本，以增加数据多样性。对于文本模态，可以进行词语替换、添加噪声、随机截断等操作；对于图像模态，数据增强可以包括随机旋转、翻转、缩放、平移、改变亮度和对比度等操作；对于音频模态，可以进行随机增减噪声、变速、变调等处理，等等。数据增强有助于提升模型的鲁棒性和泛化能力，减少过拟合，并能够更好地适应不同的输入数据变化。

3.2.1 文本数据增强

文本数据增强是指对文本数据进行变换和处理，以生成更多的文本样本，增加数据的多样性和丰富性，从而提升模型的泛化能力和性能。文本数据增强的技术可以应用于各种多模态任务，如图文匹配、文本与图像生成等。

在多模态大模型中，常用的文本数据增强技术如下。

- 同义词替换（Synonym Replacement）：将文本中的某些单词替换为其同义词，可以使用词库或者预训练的词向量来进行替换，以生成具有相似语义但略有不同的文本。
- 随机插入（Random Insertion）：在文本中随机插入一些新词，可以从词库中选择词语，或者利用同义词替换生成的新词来插入文本。
- 随机删除（Random Deletion）：随机删除文本中的一些单词，以模拟部分信息缺失的情况，促使模型更加关注重要的词语或上下文。
- 随机交换（Random Swap）：随机交换文本中两个单词的位置，以增加文本的变化性，同时保持原文的语法结构。
- 反转（Reversal）：随机反转文本中的单词顺序，用于增加模型对输入文本顺序的鲁棒性。
- 文本生成模型（Text Generation）：使用文本生成模型（如语言模型、生成对抗网络等）生成新的文本样本，可以根据原始文本生成语法正确、语义相关的新文本。
- 噪声添加（Noise Injection）：向文本中添加噪声，如打字错误、错别字、大小写转换等，以增加模型对噪声的鲁棒性。
- 文本剪裁（Text Cropping）：随机删除文本的一部分内容，以模拟部分信息缺失的情况，引导模型更好地理解文本的上下文。

上述技术可以单独应用或组合使用，根据任务的需求和文本数据的特点来选择合适的文本数据增强策略。文本数据增强有助于提高模型的泛化能力，减少过拟合，并且能够更好地适应不同风格和变体的文本输入。

在自然语言处理领域，数据增强可以应用于文本数据，以扩充训练数据集并提升模型的泛化能力。实例 3-18 是一个使用自然语言工具包（Natural Language Toolkit，NLTK）进行自然语言数据增强的例子，其中包括了文本的同义词替换和随机删除操作。

实例 3–18 使用 NLDK 进行自然语言数据增强
源码路径：codes\3\wenzeng.py

实例文件 wenzeng.py 的具体实现流程如下。

（1）获取英文停用词：从 NLTK 库获取英文停用词集合，这些停用词在自然语言处理任务中通常被过滤掉。对应的实现代码如下所示。

```
stop_words = set(stopwords.words('english'))
```

（2）编写同义词替换函数 synonym_replacement(words, n)，这个函数接收一个文本分词后的单词列表 words 和一个参数 n，表示希望替换的同义词的数量。此函数的具体说明如下。

- 首先，创建一个 new_words 列表，用于存储处理后的单词序列。random_word_list 是一个随机排序的不包含停用词的单词列表。
- 遍历 random_word_list，对每个单词获取其同义词并进行替换。如果同义词列表中有至少一个同义词，则随机选择一个同义词进行替换。这样会对句子中的随机单词进行同义词替换，最多进行 n 次。
- 最终，将新的单词列表连接为句子并返回。

函数 synonym_replacement() 的具体实现代码如下所示。

```
# 同义词替换函数
```

```
def synonym_replacement(words, n):
    new_words = words.copy()
    random_word_list = list(set([word for word in words if word not in stop_words]))
    random.shuffle(random_word_list)
    num_replaced = 0
    for random_word in random_word_list:
        synonyms = get_synonyms(random_word)
        if len(synonyms) >= 1:
            synonym = random.choice(synonyms)
            new_words = [synonym if word == random_word else word for word in new_words]
            num_replaced += 1
        if num_replaced >= n:
            break
    sentence = ' '.join(new_words)
    return sentence
```

（3）编写获取同义词列表函数 get_synonyms(word)，这个函数接收一个单词 word，使用 WordNet 数据库获取该单词的同义词。对于单词的每个同义词集合（synsets），提取每个同义词的词形变体（lemmas）并添加到 synonyms 列表中。具体实现代码如下所示。

```
# 获取同义词列表
def get_synonyms(word):
    synonyms = []
    for syn in wordnet.synsets(word):
        for lemma in syn.lemmas():
            synonyms.append(lemma.name())
    return synonyms
```

（4）编写随机删除函数 random_deletion(words，p)，这个函数接收一个单词列表 words 和一个概率参数 p，表示每个单词被删除的概率。此函数首先遍历单词列表，以概率 p 决定是否删除每个单词，将保留的单词添加到 new_words 列表中。具体实现代码如下所示。

```
# 随机删除函数
def random_deletion(words, p):
    if len(words) == 1:
        return words
    new_words = []
    for word in words:
        if random.uniform(0, 1) > p:
            new_words.append(word)
    return new_words
```

（5）处理原始文本，将原始文本分词，得到一个单词列表 words。具体实现代码如下所示。

```
sentence = "Natural language processing is a subfield of artificial intelligence."
words = word_tokenize(sentence)
```

（6）实现数据增强操作。首先，使用同义词替换函数对原始文本进行处理，替换两个单词的同义词，然后输出处理后的句子。接下来，使用随机删除函数对原始文本进行处理，以 0.2 的概率删除单词，再输出处理后的句子。具体实现代码如下所示。

```
augmented_sentence = synonym_replacement(words, n=2)
print("同义词替换后:", augmented_sentence)
augmented_sentence = random_deletion(words, p=0.2)
print("随机删除后:", ' '.join(augmented_sentence))
```

本实例展示了如何使用 NLTK 库实现自然语言数据增强，包括文本的同义词替换和随机删除操作。这样的数据增强技术有助于增加训练数据的多样性，提高模型的泛化能力。执行代码后输出：

```
同义词替换后: born language litigate is a subfield of artificial intelligence .
随机删除后: Natural language processing a subfield of artificial intelligence .
```

3.2.2　图像数据增强

在多模态大模型中，图像数据增强技术用于扩充图像数据集以提高模型的泛化能力和性能。图像数据增强可以通过对原始图像进行多种变换和处理来生成新的图像样本，增加数据的多样性。下面列出了一些常见的图像数据增强技术，它们可以被应用于多模态大模型中。

- 随机旋转（Random Rotation）：随机旋转图像一定角度，以增加图像的不同视角，使模型对旋转具有鲁棒性。
- 随机翻转（Random Flip）：随机水平或垂直翻转图像，以增加图像在水平或垂直方向上的变化。
- 随机裁剪（Random Crop）：随机裁剪图像的一部分区域，以模拟不同尺度和位置的目标出现，同时调整图像尺寸。
- 缩放和平移（Scaling and Translation）：对图像进行随机缩放和平移操作，以模拟不同距离和视角下的图像变化。
- 色彩扭曲（Color Distortion）：对图像进行色彩变换，如亮度、对比度、饱和度的调整等，以增加图像的多样性。
- 添加噪声（Add Noise）：向图像中添加随机噪声，如高斯噪声、椒盐噪声等，以提高模型对噪声的鲁棒性。
- 随机旋转、缩放和裁剪（Random Rotation, Scaling, and Cropping）：组合应用多种变换，增加图像的多样性。
- Mixup：将两张图像按一定比例混合在一起，生成新的图像，可以提高模型的泛化能力。
- Cutout：在图像中随机遮挡一部分区域，以减少模型对局部特征的依赖。

上述图像数据增强技术可以单独应用或组合使用，根据任务的需求和数据集的特点来选择合适的增强策略。数据增强有助于提高模型的泛化能力、减少过拟合，并且能够更好地适应不同风格和变体的图像输入。

在 PyTorch 程序中，可以使用 transforms 模块中的各种数据增强方法来对数据集进行数据增强操作。例如，可以使用类 RandomCrop 对图像进行随机裁剪，以提取不同的局部区域并增加数据的多样性；可以使用类 RandomHorizontalFlip 和类 RandomRotation 对图像进行随机翻转和旋转，以增加数据的多样性；可以使用类 transforms 中的方法对图像进行亮度、对比度和饱和度的调整，例如 AdjustBrightness、AdjustContrast 和 AdjustSaturation 等方法。实例 3-19 是一个使用 PyTorch 调整数据集的亮度、对比度和饱和度的例子。

> **实例 3-19**　调整数据集的亮度、对比度和饱和度
> 源码路径：codes\3\liang1.py

实例文件 liang1.py 的具体实现代码如下所示。

```
import torch
import torchvision.transforms as transforms
from torchvision.datasets import CIFAR10
import matplotlib.pyplot as plt

# 定义转换操作列表,包括调整亮度、对比度和饱和度
```

```
transform = transforms.Compose([
    transforms.ColorJitter(brightness=0.2, contrast=0.2, saturation=0.2),
    transforms.ToTensor(),
])

# 创建CIFAR-10数据集实例并应用转换操作
dataset = CIFAR10(root='data/', train=True, download=True, transform=transform)

# 获取第一个样本
sample = dataset[0]

# 将张量转换为图像并显示
image = transforms.ToPILImage()(sample[0])
plt.imshow(image)
plt.axis('off')
plt.show()
```

上述代码定义了一个名为 transform 的转换操作列表,其中包括了 ColorJitter 操作。通过调整亮度、对比度和饱和度的参数,可以改变图像的外观。然后,创建 CIFAR-10 数据集实例时应用了这个转换操作。最后,将样本的图像张量转换为 PIL 图像,并显示出来。运行上述代码后会看到第一个样本图像的亮度、对比度和饱和度发生了变化,增加了数据的多样性。执行效果如图 3-2 所示。

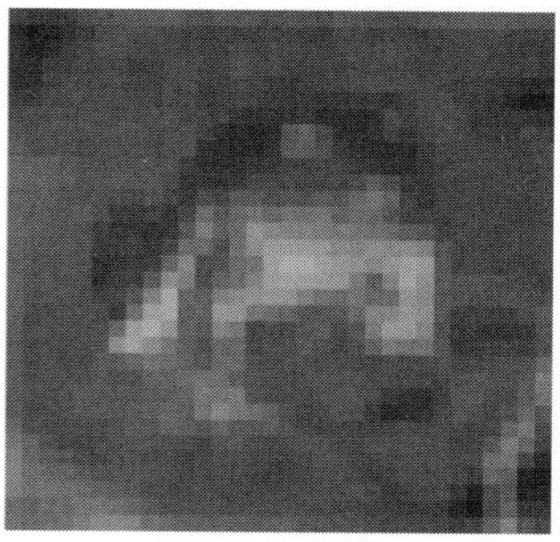

3.2.3 音频数据增强

图 3-2 执行效果

在多模态大模型中,音频数据增强是指对音频数据进行变换和处理,以增加数据的多样性和丰富性,提高模型的泛化能力和性能的一种重要技术。音频数据增强可以帮助模型更好地学习数据的特征,使模型对于噪声、变化和不确定性具有更好的鲁棒性。在实际应用中,常见的音频数据增强技术如下。

- 音频剪切(Audio Cropping):随机剪切音频片段,以模拟不同长度的语音输入,引导模型对不同长度的语音进行处理。
- 音频增益(Audio Gain):调整音频的增益或音量水平,以模拟不同的录音设备或录音环境。
- 白噪声添加(Add White Noise):向音频中添加白噪声,以增加数据的多样性和模型对噪声的鲁棒性。
- 音频速度变化(Audio Speed Variation):随机改变音频的播放速度,以模拟不同的语速和语调。
- 音频平移(Audio Shift):随机平移音频信号的时间轴,以模拟不同的录制时间和声音起始时间。
- 音调变化(Pitch Shift):随机改变音频的音调,以模拟不同的声音高低。
- 时间拉伸(Time Stretch):改变音频的时间尺度,以延长或缩短音频的持续时间。
- 频谱扭曲(Spectral Distortion):对音频进行频谱变换,如频谱平移、拉伸等,以增加数据的多样性。
- 混合(Mixing):将两个或多个音频信号混合在一起,生成新的音频样本。
- 声音掩蔽(Sound Masking):在音频中添加环境声音或背景噪声,以提高模型对环境干扰的适应能力。

可以单独使用或组合使用音频数据增强技术,根据任务的需求和数据集的特点选择合适的增强策略。数据增强有助于提高模型的泛化能力、减少过拟合，并且能够更好地适应不同环境和录音条件下的音频输入。实例 3-20 演示了结合使用库 Librosa 和库 PyDub 实现音频数据增强的用法，本实例将实现以下几种数据增强技术。

- 音频剪切：随机剪切音频片段。
- 白噪声添加：向音频中添加白噪声。
- 音频速度变化：随机改变音频的播放速度。

实例 3-20　对指定的音频实现数据增强
源码路径：codes\3\yinzeng.py

实例文件 yinzeng.py 的具体实现代码如下所示。

```python
import librosa
import numpy as np
from pydub import AudioSegment
from pydub.generators import WhiteNoise

# 音频文件路径
audio_file = "example_audio.wav"

def audio_cropping(audio_file, crop_duration):
    # 加载音频文件
    y, sr = librosa.load(audio_file, sr=None)

    # 计算音频时长
    audio_duration = librosa.get_duration(y, sr)

    # 随机选择剪切的起始时间
    start_time = np.random.uniform(0, audio_duration - crop_duration)

    # 剪切音频
    cropped_audio = y[int(start_time * sr): int((start_time + crop_duration) * sr)]

    return cropped_audio, sr

def add_white_noise(audio, noise_level=0.005):
    # 生成白噪声
    noise = WhiteNoise().to_audio_segment(duration=len(audio))
    noise = noise - noise.dBFS + audio.dBFS - 20   # 调整白噪声的音量

    # 将白噪声添加到音频中
    augmented_audio = audio.overlay(noise)

    return augmented_audio

def change_audio_speed(audio, speed_factor):
    # 改变音频的播放速度
    augmented_audio = audio.speedup(playback_speed=speed_factor)

    return augmented_audio

# 随机剪切音频片段
cropped_audio, sr = audio_cropping(audio_file, crop_duration=5)

# 加载原始音频文件
original_audio = AudioSegment.from_file(audio_file)

# 添加白噪声
augmented_audio_with_noise = add_white_noise(original_audio)
```

```
# 随机改变音频的播放速度
speed_factor = np.random.uniform(0.7, 1.3)  # 随机选择速度变化因子
augmented_audio_with_speed = change_audio_speed(original_audio, speed_factor)

# 保存增强后的音频文件
cropped_audio_file = "cropped_audio.wav"
augmented_audio_with_noise_file = "augmented_audio_with_noise.wav"
augmented_audio_with_speed_file = "augmented_audio_with_speed.wav"

# 保存剪切后的音频
librosa.output.write_wav(cropped_audio_file, cropped_audio, sr)

# 保存添加白噪声后的音频
augmented_audio_with_noise.export(augmented_audio_with_noise_file, format="wav")

# 保存改变速度后的音频
augmented_audio_with_speed.export(augmented_audio_with_speed_file, format="wav")

print("音频数据增强完成！")
```

上述代码演示了在多模态大模型应用中实现音频数据增强的过程，联合使用库 Librosa 和库 PyDub 实现了 3 种音频数据增强技术：随机剪切音频片段、添加白噪声以及随机改变音频的播放速度。这些技术可以帮助扩充训练数据集，增加音频数据的多样性，提高模型的泛化能力和鲁棒性。

3.2.4　视频数据增强

在多模态大模型应用中，视频数据增强是对视频数据进行变换和处理，以增加数据的多样性和丰富性，从而提高模型的泛化能力和性能的一种重要技术。视频数据增强可以帮助模型更好地学习到视频的特征，使模型对于不同场景、视角和噪声的适应能力更强。在现实应用中，常用的视频数据增强技术如下。

- 随机裁剪（Random Cropping）：随机裁剪视频的一部分区域，以模拟不同尺度和位置的目标出现，同时调整视频尺寸。
- 随机翻转（Random Flipping）：随机水平或垂直翻转视频，增加数据的多样性。
- 随机剪切（Random Cutting）：随机选择视频片段并进行剪切，以模拟不同长度和内容的视频输入。
- 随机旋转（Random Rotation）：随机旋转视频一定角度，增加视频的不同视角。
- 添加噪声（Add Noise）：向视频中添加随机噪声，如高斯噪声、椒盐噪声等，以提高模型对噪声的鲁棒性。
- 调整亮度和对比度（Adjust Brightness and Contrast）：随机调整视频的亮度和对比度，以模拟不同光照条件下的视频输入。
- 帧采样（Frame Sampling）：随机选择视频帧并进行采样，以减少帧率或增加帧率，改变视频的播放速度。
- 视频混合（Video Mixing）：将两个或多个视频混合在一起，生成新的视频，以增加数据的多样性。
- 时域变换（Temporal Transformation）：对视频进行时域变换，如时间扭曲、时间反转等，增加数据的多样性。
- 空间变换（Spatial Transformation）：对视频进行空间变换，如图像旋转、缩放等，增加数据的多样性。

上面列出的这些视频数据增强技术可以单独应用或组合使用，根据任务的需求和数据集的特点选择合适的增强策略。实例 3-21 演示了在多模态大模型应用中实现视频数据增强的过程。本实例将使用一些高级技术实现数据增强功能，如随机裁剪、添加噪声、调整亮度和对比度、随机旋转和随机翻转。

实例 3-21　对指定的视频实现数据增强
源码路径：codes\3\yinzeng.py

实例文件 yinzeng.py 的具体实现代码如下所示。

```python
import cv2
import numpy as np
import random
from moviepy.editor import VideoFileClip, vfx

def random_crop(frame, crop_size):
    h, w, _ = frame.shape
    top = np.random.randint(0, h - crop_size)
    left = np.random.randint(0, w - crop_size)
    frame = frame[top:top + crop_size, left:left + crop_size]
    return frame

def add_noise(frame):
    noise = np.random.normal(0, 25, frame.shape).astype(np.uint8)
    frame = cv2.add(frame, noise)
    return frame

def adjust_brightness_contrast(frame, brightness=0, contrast=0):
    frame = cv2.convertScaleAbs(frame, alpha=1 + contrast/100, beta=brightness)
    return frame

def random_rotation(frame):
    angle = random.randint(-30, 30)
    h, w = frame.shape[:2]
    center = (w // 2, h // 2)
    matrix = cv2.getRotationMatrix2D(center, angle, 1.0)
    frame = cv2.warpAffine(frame, matrix, (w, h))
    return frame

def random_flip(frame):
    flip_code = random.choice([-1, 0, 1])
    frame = cv2.flip(frame, flip_code)
    return frame

def augment_frame(frame, crop_size):
    frame = random_crop(frame, crop_size)
    frame = add_noise(frame)
    frame = adjust_brightness_contrast(frame,brightness=random.randint(-50, 50),contrast=random.randint(-50, 50))
    frame = random_rotation(frame)
    frame = random_flip(frame)
    return frame

def process_video(video_path, output_path, crop_size=224):
    cap = cv2.VideoCapture(video_path)
    fourcc = cv2.VideoWriter_fourcc(*'XVID')
    out = cv2.VideoWriter(output_path, fourcc, 20.0, (crop_size, crop_size))

    while(cap.isOpened()):
        ret, frame = cap.read()
```

```
        if ret:
            augmented_frame = augment_frame(frame, crop_size)
            out.write(augmented_frame)
        else:
            break

    cap.release()
    out.release()

# 使用 moviepy 添加一些额外的视频增强效果
def additional_video_effects(input_video_path, output_video_path):
    clip = VideoFileClip(input_video_path)
    augmented_clip = clip.fx(vfx.colorx, 1.2)    # 调整颜色
    augmented_clip = augmented_clip.fx(vfx.lum_contrast,lum=10,contrast=50,contrast_thr
=127)    # 调整亮度和对比度
    augmented_clip.write_videofile(output_video_path)

# 路径
video_path = 'example_video.mp4'
output_path = 'augmented_video.avi'
final_output_path = 'final_augmented_video.mp4'

# 处理视频
process_video(video_path, output_path)

# 添加额外效果
additional_video_effects(output_path, final_output_path)

print("视频数据增强完成！")
```

上述代码的实现流程如下。

（1）定义增强函数：定义多个用于视频数据增强的函数，包括随机裁剪、添加噪声、调整亮度和对比度、随机旋转和随机翻转。这些函数将分别对视频帧进行不同的增强处理。

（2）实现随机裁剪：随机选择视频帧的一个区域进行裁剪，以增加数据的多样性。

（3）添加噪声：向视频帧中添加随机噪声，使模型对噪声具有鲁棒性。

（4）调整亮度和对比度：随机调整视频帧的亮度和对比度，模拟不同光照条件下的视频输入。

（5）随机旋转：随机旋转视频帧一定角度，增加视频的不同视角。

（6）随机翻转：随机水平或垂直翻转视频帧，增加数据的多样性。

（7）增强单帧图像：定义一个 augment_frame 函数，结合上述所有增强技术，对每一帧进行增强处理。

（8）处理视频文件：创建一个 process_video 函数，用于处理整个视频文件。该函数读取视频文件，逐帧应用增强技术，并将增强后的帧写入新的视频文件中。

（9）使用 moviepy 添加额外效果：创建一个 additional_video_effects 函数，使用 moviepy 库对处理后的视频添加额外的增强效果，如颜色调整和对比度调整。

（10）执行视频处理：指定输入视频文件路径和输出文件路径，调用 process_video 函数处理视频，然后调用 additional_video_effects 函数添加额外效果。

（11）保存和输出增强后的视频：保存增强后的视频文件，并输出处理完成的提示信息。

3.3 特征提取

在多模态大模型中，特征提取是指从不同模态的数据（如图像、文本、音频、视频等）中提

取出有代表性的特征或描述符。这些特征能够捕捉输入数据的核心信息和模式，并将其转换为模型可处理的向量表示。

3.3.1 特征在大模型中的关键作用

特征在大模型中起着至关重要的作用，它们直接影响了模型的性能、泛化能力和对数据的理解。特征在大模型中的主要作用如下。

- 信息表示和提取：特征是原始数据的抽象表示，能够捕捉数据中的关键信息和模式。好的特征能够帮助模型更有效地区分不同类别、理解数据的含义和上下文。
- 降低维度和计算复杂度：大模型通常需要大量的计算资源，但原始数据可能具有高维度。特征提取可以帮助将数据映射到更低维度的空间，从而减少计算复杂度并提高模型的效率。
- 泛化能力：好的特征能够捕捉数据的一般性质，使模型能够更好地泛化到未见过的数据。通过在特征中保留重要的、有意义的信息，模型可以更准确地处理新的样本。
- 对抗性防御：在安全性方面，一些特征提取方法可以帮助模型更好地识别和抵御对抗性攻击，从而提高模型的鲁棒性。
- 领域适应和迁移学习：在不同领域之间，数据分布可能有所不同。好的特征可以帮助模型更好地适应新领域的数据，从而实现迁移学习。
- 解释性：一些特征提取方法可以提高模型的解释性，使人们更容易理解模型的决策过程和推理基础。
- 处理缺失数据：特征提取可以通过合理的方法处理缺失数据，从而避免模型因缺失数据而降低性能。
- 序列建模：在序列数据中，特征提取有助于将序列数据转化为模型能够处理的表示形式，如在自然语言处理中将句子转化为嵌入向量。

总之，特征在大模型中的关键作用在于将原始数据转化为更具有信息含量和表达能力的形式，从而使模型能够更好地理解数据、学习模式并进行预测、分类、生成等任务。选择适当的特征提取方法是大模型开发中的一个关键决策，能够直接影响模型的性能和实际应用效果。

3.3.2 文本特征提取

在文本数据的处理中，特征提取是将原始文本转换为数值表示，以便于机器学习算法的使用。文本数据具有高维度和稀疏性的特点，因此需要对文本进行适当的处理，以获取有用的特征表示。

1. 嵌入

在序列建模中，嵌入（Embedding）是将离散的符号（如单词、字符、类别等）映射到连续向量空间的过程。嵌入是将高维离散特征转换为低维连续特征的一种方式，这种转换有助于提取序列数据中的语义和上下文信息，从而改善序列模型的性能。

嵌入层是深度学习中的一种常见层类型，通常用于自然语言处理和推荐系统等任务，其中输入数据通常是符号序列。通过嵌入，每个符号（例如单词）被映射为一个稠密向量，这个向量可以捕捉到符号的语义和语境信息。

下面列出了嵌入在序列建模中的一些重要应用场景。

- 自然语言处理：在文本处理任务中，嵌入可以将单词或字符映射到连续的向量表示，使模型能够捕获词语之间的语义关系和上下文信息。Word2Vec、GloVe 和 BERT 等模型都使用了嵌入技术。

- 推荐系统：在推荐系统中，嵌入可以用于表示用户和物品（如商品、电影等），从而构建用户–物品交互矩阵的表示。这种表示可以用于预测用户对未知物品的兴趣。
- 时间序列预测：对于时间序列数据，嵌入可以用于将时间步和历史数据映射为连续向量，以捕获序列中的趋势和模式。
- 序列标注：在序列标注任务中，嵌入可以用于将输入的序列元素（如字母、音素等）映射为向量，供序列标注模型使用。
- 图像描述生成：在图像描述生成任务中，嵌入可以将图像中的对象或场景映射为向量，作为生成描述的输入。

当使用 PyTorch 进行文本数据的特征提取时，可以使用嵌入层来将单词映射为连续向量表示。实例 3-22 包含一个完整的示例代码，演示了在 PyTorch 中使用嵌入层进行文本数据的特征提取的过程。

实例 3-22　使用嵌入层提取文本数据的特征

源码路径：codes\3\qian.py

实例文件 qian.py 的具体实现代码如下所示。

```python
# 生成一些示例文本数据
texts = ["this is a positive sentence",
         "this is a negative sentence",
         "a positive sentence here",
         "a negative sentence there"]

labels = [1, 0, 1, 0]

# 构建词汇表
word_counter = Counter()
for text in texts:
    tokens = text.split()
    word_counter.update(tokens)

vocab = sorted(word_counter, key=word_counter.get, reverse=True)
word_to_index = {word: idx for idx, word in enumerate(vocab)}

# 文本数据预处理和转换为索引
def preprocess_text(text, word_to_index):
    tokens = text.split()
    token_indices = [word_to_index[token] for token in tokens]
    return token_indices

texts_indices = [preprocess_text(text, word_to_index) for text in texts]

# 划分训练集和验证集
train_data, val_data, train_labels, val_labels = train_test_split(texts_indices, labels,
test_size=0.2, random_state=42)

# 自定义数据集和数据加载器
class CustomDataset(Dataset):
    def __init__(self, data, labels):
        self.data = data
        self.labels = labels

    def __len__(self):
        return len(self.data)

    def __getitem__(self, idx):
```

```python
        return torch.tensor(self.data[idx]), torch.tensor(self.labels[idx])

# 获取最长文本序列的长度
max_seq_length = max([len(text) for text in train_data])

# 填充数据,使每个文本序列长度相同
train_data_padded = [text + [0] * (max_seq_length - len(text)) for text in train_data]
val_data_padded = [text + [0] * (max_seq_length - len(text)) for text in val_data]

train_dataset = CustomDataset(train_data_padded, train_labels)
val_dataset = CustomDataset(val_data_padded, val_labels)

train_loader = DataLoader(train_dataset, batch_size=2, shuffle=True)

# 定义模型
class TextClassifier(nn.Module):
    def __init__(self, vocab_size, embedding_dim, output_dim):
        super(TextClassifier, self).__init__()
        self.embedding = nn.Embedding(vocab_size, embedding_dim)
        self.fc = nn.Linear(embedding_dim, output_dim)

    def forward(self, x):
        embedded = self.embedding(x)
        pooled = torch.mean(embedded, dim=1)
        return self.fc(pooled)


# 设置参数和优化器
vocab_size = len(vocab)
embedding_dim = 10
output_dim = 1
learning_rate = 0.01
num_epochs = 10

model = TextClassifier(vocab_size, embedding_dim, output_dim)
criterion = nn.BCEWithLogitsLoss()
optimizer = optim.Adam(model.parameters(), lr=learning_rate)

# 训练模型
for epoch in range(num_epochs):
    for batch_data, batch_labels in train_loader:
        optimizer.zero_grad()
        predictions = model(batch_data)

        # 将标签调整为向量形式,与模型输出维度相匹配
        batch_labels = batch_labels.unsqueeze(1).float()

        loss = criterion(predictions, batch_labels)
        loss.backward()
        optimizer.step()
    print(f'Epoch [{epoch + 1}/{num_epochs}], Loss: {loss.item():.4f}')

# 在验证集上评估模型性能
with torch.no_grad():
    val_data_tensor = pad_sequence([torch.tensor(text) for text in val_data_padded],batch_first=True)
    val_predictions = model(val_data_tensor)
    val_predictions = torch.round(torch.sigmoid(val_predictions))
    accuracy=(val_predictions==torch.tensor(val_labels).unsqueeze(1)).sum().item() / len(val_labels)
    print(f'Validation accuracy: {accuracy:.2f}')
```

总的来说,这段代码演示了如何使用 PyTorch 进行文本分类任务,其中包括数据预处理、模

型定义、训练和评估过程。请注意，这个实例是一个简化版的文本分类流程，实际应用中可能需要更多的步骤和技术来处理更复杂的文本数据和任务。

2. 词袋模型

词袋模型是一种常用的文本特征提取方法，用于将文本数据转换为数值表示。它的基本思想是将文本看作由单词构成的"袋子"（即无序集合），然后统计每个单词在文本中出现的频次或使用其他权重方式来表示单词的重要性。这样，每个文本都可以用一个向量表示，其中向量的每个维度对应于一个单词，并记录了该单词在文本中的出现次数或权重。

在 TensorFlow 中使用词袋模型进行文本特征提取时需要一些预处理步骤，例如，实例 3-23 是一个 TensorFlow 使用词袋模型进行文本特征提取的例子。

实例 3-23　使用词袋模型提取文本特征
源码路径：codes\3\ci.py

实例文件 ci.py 的具体实现代码如下所示。

```python
# 生成示例文本数据和标签
texts = ["this is a positive sentence",
         "this is a negative sentence",
         "a positive sentence here",
         "a negative sentence there"]

labels = [1, 0, 1, 0]

# 划分训练集和验证集
train_texts, val_texts, train_labels, val_labels = train_test_split(texts, labels, test_size=0.2, random_state=42)

# 创建分词器并进行分词
tokenizer = Tokenizer()
tokenizer.fit_on_texts(train_texts)
train_sequences = tokenizer.texts_to_sequences(train_texts)
val_sequences = tokenizer.texts_to_sequences(val_texts)

# 填充文本序列,使其长度相同
max_seq_length = max(len(seq) for seq in train_sequences)
train_data= pad_sequences(train_sequences, maxlen=max_seq_length, padding='post')
val_data = pad_sequences(val_sequences, maxlen=max_seq_length, padding='post')

# 构建词袋特征表示
train_features = tokenizer.sequences_to_matrix(train_sequences, mode='count')
val_features = tokenizer.sequences_to_matrix(val_sequences, mode='count')

# 创建朴素贝叶斯分类器
classifier = MultinomialNB()
classifier.fit(train_features, train_labels)

# 预测并评估模型性能
predictions = classifier.predict(val_features)
accuracy = accuracy_score(val_labels, predictions)
print(f'Validation accuracy: {accuracy:.2f}')
```

在这个例子中，首先使用 Tokenizer 对文本进行分词和索引化，然后使用 pad_sequences 对文本序列进行填充。接着，我们使用 sequences_to_matrix 方法将文本序列转换为词袋特征表示，模式设置为 count，表示计算单词出现的频次。然后，使用 MultinomialNB 创建朴素贝叶斯分类器，对词袋特征进行训练和预测，并使用 accuracy_score 计算模型在验证集上的准确率。

3. TF-IDF 特征

词频–逆文档频率（Term Frequency-Inverse Document Frequency，TF-IDF）是一种用于文本特征提取的常用方法，它结合了词频和逆文档频率，用于衡量单词在文本中的重要性。TF-IDF 考虑了一个单词在文本中的频率（TF），以及它在整个文集中的稀有程度（IDF）。

在 PyTorch 中，TF-IDF 特征提取需要借助 Scikit-learn 来计算 TF-IDF 值，然后将结果转换为 PyTorch 张量进行模型训练。例如，实例 3-24 是一个 PyTorch 使用 TF-IDF 特征进行文本特征提取的例子。

实例 3-24　使用 TF-IDF 特征提取文本特征
源码路径：codes\3\ti.py

实例文件 ti.py 的具体实现代码如下所示。

```python
# 生成示例文本数据和标签
texts = ["this is a positive sentence",
         "this is a negative sentence",
         "a positive sentence here",
         "a negative sentence there"]

labels = [1, 0, 1, 0]

# 划分训练集和验证集
train_texts, val_texts, train_labels, val_labels = train_test_split(texts, labels, test_size=0.2, random_state=42)

# 创建 TF-IDF 特征表示
vectorizer = TfidfVectorizer()
train_features = vectorizer.fit_transform(train_texts).toarray()
val_features = vectorizer.transform(val_texts).toarray()

# 转换为 PyTorch 张量
train_features_tensor = torch.tensor(train_features, dtype=torch.float32)
train_labels_tensor = torch.tensor(train_labels, dtype=torch.float32)
val_features_tensor = torch.tensor(val_features, dtype=torch.float32)
val_labels_tensor = torch.tensor(val_labels, dtype=torch.float32)

# 创建朴素贝叶斯分类器
classifier = MultinomialNB()
classifier.fit(train_features, train_labels)

# 预测并评估模型性能
predictions = classifier.predict(val_features)
accuracy = accuracy_score(val_labels, predictions)
print(f'Validation accuracy: {accuracy:.2f}')
```

在这个例子中，首先使用 TfidfVectorizer 创建 TF-IDF 特征表示，然后将结果转换为 NumPy 数组，并将其转换为 PyTorch 张量。接着，我们创建了一个朴素贝叶斯分类器，对 TF-IDF 特征进行训练和预测，最后使用 accuracy_score 计算模型在验证集上的准确率。

3.3.3　图像特征提取

在图像数据中，特征提取通常涉及将原始像素数据转换为更高级别的表征，以便在后续任务中使用，如图像分类、目标检测、图像生成等。

1. 预训练的图像特征提取模型

预训练的图像特征提取模型是在大规模图像数据集上训练的深度卷积神经网络模型，这些模型可以有效地提取图像中的高级特征。通过在大量图像上训练，这些模型能够学习到通用的视觉特征，这些特征可以在多种图像相关任务中重复使用，例如图像分类、目标检测、图像生成等。下面是一些常见的预训练的图像特征提取模型。

- VGG16 和 VGG19：VGGNet 是一个经典的卷积神经网络，它通过堆叠多个卷积和池化层来提取图像特征。VGG16 和 VGG19 分别具有 16 和 19 个卷积层，在 ImageNet 数据集上进行了训练。
- ResNet：ResNet（Residual Network）引入了残差连接，允许网络更深地进行训练，避免了梯度消失问题。ResNet 以不同深度的变体存在，如 ResNet-50、ResNet-101 等。
- Inception：Inception 模型系列采用了多尺度卷积和不同卷积核尺寸的并行操作，从而提高了网络的感受野和特征提取能力。InceptionV3 和 InceptionResNetV2 是其中的代表。
- MobileNet：MobileNet 是一系列轻量级的模型，适用于移动设备和嵌入式系统。它通过深度可分离卷积来降低参数数量和计算成本，从而实现更高效的特征提取。
- EfficientNet：EfficientNet 是一系列综合了多种网络设计技术的模型，旨在实现更好的性能和计算效率。它通过网络深度、宽度和分辨率的平衡来优化模型性能。

例如，实例 3-25 是一个使用预训练的卷积神经网络模型 VGG16 模型进行图像特征提取的示例，其中使用 TensorFlow 和 Keras 进行演示。

实例 3-25　使用 VGG16 模型进行图像特征提取
源码路径：codes\3\ttu.py

实例文件 ttu.py 的具体实现代码如下所示。

```python
# 加载预训练的 VGG16 模型（不包括顶部分类器）
base_model = VGG16(weights='imagenet', include_top=False)
# 要处理的图像路径
image_path = 'path_to_your_image.jpg'
# 加载图像并预处理
img = image.load_img(image_path, target_size=(224, 224))
x = image.img_to_array(img)
x = np.expand_dims(x, axis=0)
x = preprocess_input(x)

# 使用预训练的 VGG16 模型提取特征
features = base_model.predict(x)

print(features.shape)    # 输出特征张量的形状
```

在这个例子中，使用预训练的 VGG16 模型来进行图像特征提取。我们首先加载了 VGG16 模型，并在加载模型时通过设置 include_top=False 来去除模型的顶部分类器部分。然后，我们加载了待处理的图像，使用预处理函数 preprocess_input 对图像进行预处理，使其与在 ImageNet 数据集上训练的模型期望的输入匹配。最后，使用 VGG16 模型对预处理后的图像进行预测，得到了提取的特征张量。执行代码后输出：

```
(1, 7, 7, 512)
```

这说明执行成功了，并且输出了（1，7，7，512）的形状，这是预训练的 VGG16 模型在输入图像上提取的特征张量的形状。这意味着该模型的输出是一个（1，7，7，512）的四维张量，表示一个图像经过卷积神经网络后在某一层得到的特征图。其中第一个维度是批次大小，这里是 1，表示当前只处理了一个图像。后面的三个维度（7，7，512）分别代表特征图的高度、宽度和通道数。这里，模型在某一层上提取了一个7×7大小的特征图，每个位置上有512个通道的特征。这个特征张量可以被用作后续任务的输入，比如分类、目标检测、图像生成等，从而利用 VGG16 模型在大规模图像数据上学到的特征来提升任务的性能。

2. 基本图像特征：边缘检测、颜色直方图等

基本图像特征是从图像的原始像素值中提取的简单特征，这些特征可以帮助我们理解图像的基本视觉属性。例如下面是两个基本图像特征的示例：边缘检测和颜色直方图。

（1）边缘检测。

边缘是图像中像素值变化剧烈的地方，通常表示物体的边界或纹理变化。边缘检测可以帮助我们识别图像中的物体轮廓和形状。

（2）颜色直方图。

颜色直方图表示图像中不同颜色值的频率分布。它可以帮助我们了解图像中的颜色分布情况，有助于图像分类、检索和分割。

实例 3-26 使用 PyTorch 实现了图像的特征提取功能，包括边缘检测和颜色直方图。

实例 3-26	实现图像的边缘检测和颜色直方图功能
	源码路径：codes\3\jianzhi.py

实例文件 jianzhi.py 的具体实现代码如下所示。

```python
import torch
import cv2
import numpy as np
from torchvision.transforms import ToTensor
from matplotlib import pyplot as plt

# 加载图像
image_path = 'your_image_path.jpg'
image = cv2.imread(image_path)
image_rgb = cv2.cvtColor(image, cv2.COLOR_BGR2RGB)

# 数据预处理
transform = ToTensor()
image_tensor = transform(image_rgb)

# 边缘检测
edges = cv2.Canny(image, threshold1=100, threshold2=200)
edges_tensor = torch.tensor(edges).unsqueeze(0).unsqueeze(0).float()

# 颜色直方图
hist_r = cv2.calcHist([image_rgb], [0], None, [256], [0, 256]).squeeze()
hist_g = cv2.calcHist([image_rgb], [1], None, [256], [0, 256]).squeeze()
hist_b = cv2.calcHist([image_rgb], [2], None, [256], [0, 256]).squeeze()

# 可视化结果
plt.figure(figsize=(12, 4))

plt.subplot(131), plt.imshow(image_rgb)
plt.title('Original Image'), plt.xticks([]), plt.yticks([])

plt.subplot(132), plt.imshow(edges, cmap='gray')
plt.title('Edge Image'), plt.xticks([]), plt.yticks([])

plt.subplot(133)
plt.plot(hist_r, color='r', label='Red')
plt.plot(hist_g, color='g', label='Green')
plt.plot(hist_b, color='b', label='Blue')
```

```
plt.title('Color Histogram')
plt.legend()

plt.tight_layout()
plt.show()
```

上述代码从加载图像开始,通过 OpenCV 库进行图像处理,使用 PyTorch 的张量进行数据处理,执行边缘检测,计算颜色直方图,最后通过 Matplotlib 库进行可视化。这个例子展示了将图像处理与特征提取结合起来,并进行可视化展示的过程。执行效果如图 3-3 所示。

  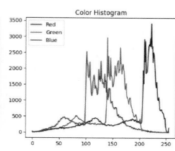

图 3-3　执行效果

3.3.4　音频特征提取

在多模态大模型中,音频特征提取是指从原始音频数据中提取出能够代表其主要信息和模式的特征,这些特征可以用于各种任务,如语音识别、情感分析、声音分类等。音频特征提取的主要目的是将复杂的音频信号转换为模型能够处理的结构化数据,从而提升模型的性能和鲁棒性。下面列出了常见的音频特征提取方法和技术。

1. **时域特征(Time-Domain Features)**
- 短时能量(Short-Time Energy):反映音频信号的能量变化,用于检测语音段落和静音段落。
- 零交叉率(Zero-Crossing Rate,ZCR):音频信号通过零点的次数,反映信号的频率成分和噪声特性。
2. **频域特征(Frequency-Domain Features)**
- 傅里叶变换(Fourier Transform):将时域信号转换为频域信号,用于分析音频信号的频率成分。
- 梅尔频率倒谱系数(Mel-Frequency Cepstral Coefficients,MFCCs):通过模拟人耳听觉特性提取的特征,广泛用于语音识别和音频分类。
3. **时频域特征(Time-Frequency Features)**
- 短时傅里叶变换(Short-Time Fourier Transform,STFT):在不同时间窗口内计算音频信号的频率成分,生成频谱图。
- 梅尔频谱(Mel-Spectrogram):将频谱图映射到梅尔刻度上,更符合人耳的听觉特性。
4. **其他高级特征**
- 色谱图(Chromagram):表示音频信号在不同音调上的能量分布,用于音乐处理。
- 谱质心(Spectral Centroid):频谱的中心位置,反映音频信号的亮度。
- 谱带宽(Spectral Bandwidth):频谱的宽度,反映音频信号的复杂性。
- 谱滚降点(Spectral Roll-off):频谱能量累积到一定比例时的频率位置,用于检测高频内容。

第 3 章 多模态数据处理

音频特征提取在多模态大模型中扮演着关键角色，它将复杂的音频信号转化为模型可以处理的结构化数据。这些特征不仅能提高模型的性能，还能增强模型在不同场景下的鲁棒性。请看实例 3-27，功能是使用库 Librosa 提取音频特征。

实例 3-27　使用库 Librosa 提取音频特征
源码路径：codes\3\yinti.py

实例文件 yinti.py 的具体实现代码如下所示。

```python
import librosa
import numpy as np
import matplotlib.pyplot as plt

# 加载音频文件
audio_path = 'example_audio.wav'
y, sr = librosa.load(audio_path)

# 计算短时傅里叶变换（STFT）
stft = np.abs(librosa.stft(y))

# 计算梅尔频谱
mel_spectrogram = librosa.feature.melspectrogram(y, sr=sr)

# 计算梅尔频率倒谱系数（MFCCs）
mfccs = librosa.feature.mfcc(y, sr=sr, n_mfcc=13)

# 计算色谱图
chromagram = librosa.feature.chroma_stft(y, sr=sr)

# 计算谱质心
spectral_centroid = librosa.feature.spectral_centroid(y, sr=sr)

# 计算谱带宽
spectral_bandwidth = librosa.feature.spectral_bandwidth(y, sr=sr)

# 计算谱滚降点
spectral_rolloff = librosa.feature.spectral_rolloff(y, sr=sr)

# 绘制梅尔频谱图
plt.figure(figsize=(10, 4))
librosa.display.specshow(librosa.power_to_db(mel_spectrogram, ref=np.max), sr=sr, y_axis='mel', x_axis='time')
plt.colorbar(format='%+2.0f dB')
plt.title('Mel Spectrogram')
plt.tight_layout()
plt.show()
```

上述代码演示了使用库 Librosa 对音频数据进行特征提取的操作。其从音频文件中提取了多种特征，包括短时傅里叶变换、梅尔频谱、梅尔频率倒谱系数、色谱图、谱质心、谱带宽和谱滚降点，并通过绘制梅尔频谱图可视化部分特征。这些特征能够捕捉音频信号的核心信息，为多模态大模型提供结构化的输入数据，从而增强模型的性能和鲁棒性。

3.3.5　视频特征提取

在多模态大模型中，视频特征提取是指从视频数据中提取有代表性的特征，用于模型的训练和推理。在实际应用中，常用的视频特征提取方法如下。

1. 时空特征提取

- 卷积神经网络（CNN）：主要用于从视频帧中提取空间特征。通过对每一帧使用预训练的CNN（如 ResNet、VGG），可以获得图像的高层次特征表示。
- 3D 卷积神经网络（3D CNN）：可以同时捕捉视频的空间和时间特征。3D 卷积在时空体积（即帧序列）上进行卷积操作，可以直接从视频中提取时空特征。
- 时空残差网络（C3D, I3D）：这类网络结合了 2D CNN 和 3D CNN 的优势，能够高效地提取视频的时空特征。I3D（Inflated 3D）是一种将 2D CNN 模型扩展到 3D 的方法，能够利用在图像数据上预训练的权重。

2. 序列模型

- 循环神经网络（RNN）和长短期记忆网络（LSTM）：RNN 和 LSTM 主要用于处理时间序列数据。通过将视频帧的特征序列输入 RNN 或 LSTM，可以捕捉视频的时间依赖关系和动态变化。
- Transformer：Transformer 模型通过自注意力机制，能够并行处理帧序列中的长程依赖关系。它在视频分类、动作识别等任务中表现出色。

3. 预训练模型

预训练的深度学习模型使用在大规模视频数据集（如 Kinetics、UCF-101）上预训练的模型，可以有效提取视频特征。这些模型可以用作特征提取器，将视频帧输入到模型并提取中间层的特征表示。

4. 光流

光流法用于捕捉视频中的运动信息，常用于动作识别等任务。光流法通过计算相邻帧之间的像素运动，可以获得视频中的动态特征。

5. 视频特征聚合

- 平均池化和最大池化：对于提取的每帧特征，可以使用平均池化或最大池化对特征进行聚合，生成整个视频的全局特征表示。
- 注意力机制：使用注意力机制来加权不同帧的特征，使模型能够聚焦于视频中重要的帧。

实例 3-28 演示了使用深度学习模型从指定视频中提取特征的过程。本实例将分别使用视频分段、提取帧等操作对每个帧进行特征提取，并将这些特征整合成视频级别的特征表示。

实例 3-28　使用深度学习模型从指定视频中提取特征

源码路径：codes\3\shiti.py

实例文件 shiti.py 的具体实现代码如下所示。

```python
import cv2
import numpy as np
from keras.applications import ResNet50
from keras.applications.resnet50 import preprocess_input

# 加载预训练的 ResNet-50 模型
model = ResNet50(weights='imagenet', include_top=False, pooling='avg')

def extract_frames(video_path, frame_count=16, target_size=(224, 224)):
    cap = cv2.VideoCapture(video_path)
    frames = []
    total_frames = int(cap.get(cv2.CAP_PROP_FRAME_COUNT))
    for i in range(frame_count):
        frame_no = int(total_frames / frame_count) * i
        cap.set(cv2.CAP_PROP_POS_FRAMES, frame_no)
        ret, frame = cap.read()
```

```
            if ret:
                frame = cv2.resize(frame, target_size)
                frames.append(frame)
    cap.release()
    return np.array(frames)

def extract_video_features(video_path, segment_count=4, frames_per_segment=16):
    cap = cv2.VideoCapture(video_path)
    total_frames = int(cap.get(cv2.CAP_PROP_FRAME_COUNT))
    segment_length = total_frames  //计算每个视频片段的长度
    
    video_features = []
    
    for i in range(segment_count):
        frames = []
        start_frame = i * segment_length
        for j in range(frames_per_segment):
            frame_no = start_frame + int(segment_length / frames_per_segment) * j
            cap.set(cv2.CAP_PROP_POS_FRAMES, frame_no)
            ret, frame = cap.read()
            if ret:
                frame = cv2.resize(frame, (224, 224))
                frames.append(frame)
        frames = np.array(frames)
        frames = preprocess_input(frames)   # 预处理帧
        segment_features = model.predict(frames)  # 使用 ResNet50 提取特征
        segment_features = np.mean(segment_features, axis=0)  # 对每个段内的帧特征求平均
        video_features.append(segment_features)
    
    cap.release()
    return np.array(video_features)

# 提取视频特征
video_path = 'example_video.mp4'
features = extract_video_features(video_path)
print("视频特征形状:", features.shape)
```

上述代码首先将视频分为几个段,然后从每个段中提取一定数量的帧(在这里是 16 帧),对这些帧进行特征提取,并对每个段内的帧特征取平均,最终得到整个视频的特征表示。执行代码后会输出:

```
1/1 [==============================] - 1s 1s/step
1/1 [==============================] - 0s 260ms/step
1/1 [==============================] - 0s 270ms/step
1/1 [==============================] - 0s 260ms/step
视频特征形状: (4, 2048)
```

上面的输出结果表明视频特征的形状为(4,2048),有 4 个段,每个段的特征维度为 2048。

第 4 章 多模态表示学习

多模态表示学习是一种机器学习方法，旨在通过融合来自不同模态的数据（如文本、图像、音频等）来提高模型的理解和表现能力。这种方法利用各模态的互补信息，构建统一的表示，以更全面地捕捉和理解复杂的语义信息，从而在任务（如图像描述生成、跨模态检索和多模态分类）中取得更好的效果。本章将详细讲解多模态表示学习的知识和用法。

4.1 多模态表示学习介绍

多模态表示学习是一种先进的机器学习方法，旨在综合处理和理解来自多种模态的数据，例如文本、图像、音频和视频等。这些不同模态的数据通常具有互补的信息，通过融合这些信息，可以增强模型的表现能力。

4.1.1 多模态表示学习简介

多模态表示学习的核心思想是通过学习一个共同的表示空间，使不同模态的数据能够在这个空间中进行对齐和融合，从而充分利用各模态的优势。这样可以提高模型在各种任务中的性能，如图像描述生成、跨模态检索和多模态分类等。

在现实应用中，多模态表示学习的主要应用领域如下。
- 图像描述生成：根据图像内容生成自然语言描述。
- 跨模态检索：根据文字描述检索相应的图片或视频。
- 多模态情感分析：结合语音、面部表情和文本进行情感分析。

尽管多模态表示学习在人工智能大模型领域中得到了飞速发展，但是目前却面临着巨大的挑战，具体如下。
- 模态间差异：不同模态的数据有着不同的特性和结构，如图像是像素数据、文本是序列数据，如何在一个统一的表示空间中有效地融合这些异构数据是一个挑战。
- 数据对齐：需要确保不同模态的数据在时间和空间上的对齐，如在视频中对齐图像帧和对应的音频或字幕。
- 信息冗余和噪声：不同模态的数据可能包含冗余信息或噪声，如何有效地提取有用的信息并过滤掉无关的信息也是一个难题。

4.1.2 多模态表示学习的主要方法

通过多模态表示学习，可以充分利用各模态的信息，提高模型的理解和预测能力，解决单一模态模型无法处理的复杂问题。在实际应用中，多模态表示学习的主要方法如下。

1. **特征提取**
- 卷积神经网络：用于从图像中提取高层次特征。
- 循环神经网络和 Transformer：用于从文本或序列数据中提取特征。

2. **表示融合**
- 拼接（Concatenation）：将来自不同模态的特征向量直接拼接在一起，形成一个长向量。
- 加权平均：对不同模态的特征进行加权平均，权重可以通过训练学习得到。
- 注意力机制：通过计算不同模态特征之间的相关性，动态调整特征的权重。
- 多模态 Transformer：使用多头自注意力机制融合不同模态的特征。
- 张量融合（Tensor Fusion）：利用张量运算进行高阶的特征交互，增强表示能力。

3. **联合学习**
- 多任务学习（Multi-Task Learning）：同时训练多个相关任务，通过共享表示来提升模型性能。
- 对抗训练（Adversarial Training）：使用生成对抗网络或对抗性损失来提升模型的泛化能力。
- 协同训练（Co-Training）：利用不同模态的互补信息，通过迭代的方式交替训练模型。

4. **表示对齐**
- 互信息最大化（Maximizing Mutual Information）：通过最大化不同模态表示之间的互信息来确保对齐。
- 对比学习（Contrastive Learning）：使用对比损失函数使得相同样本的不同模态表示更接近，不同样本的表示更远离。

5. **模态间关系建模**
- 图神经网络（Graph Neural Network，GNN）：构建图结构来表示模态间的关系，通过图卷积进行特征传播和融合。
- 跨模态注意力（Cross-Modal Attention）：在不同模态间计算注意力权重，捕捉模态间的关系和依赖。

4.2 多模态表示学习方法

本章前面介绍过，多模态表示学习方法包括特征提取、表示融合、联合学习、表示对齐和模态间关系建模。通过使用卷积神经网络和 Transformer 模型等技术提取各模态特征，然后通过简单拼接、注意力机制、多模态 Transformer 等方法融合特征，再结合多任务学习、对比学习等联合训练技术，确保不同模态特征在统一表示空间中的对齐和有效融合，从而提升模型在图像描述生成、跨模态检索和情感分析等任务上的表现。本节将详细讲解这些多模态表示学习方法的使用。

4.2.1 表示融合

在多模态表示学习中，表示融合是指将来自不同模态的特征整合到一个统一的表示空间的关键步骤。这一步骤通过有效地结合多种模态的信息，提升模型在各种任务上的表现。在实际应用中，常见的表示融合方法如下。

1. **简单融合方法**

（1）拼接。
- 方法：将来自不同模态的特征向量直接拼接成一个长向量。
- 优点：实现简单，直观地整合各模态特征。
- 缺点：可能导致高维度特征空间，计算开销大。

（2）加权平均。
- 方法：对不同模态的特征进行加权平均，权重可以通过训练过程学习得到。
- 优点：计算简单，可以调节各模态的贡献。

- 缺点：无法捕捉模态间的复杂关系。

2. 复杂融合方法

（1）线性变换和投影（Linear Transformation and Projection）。
- 方法：将不同模态的特征通过线性变换映射到同一维度，再进行融合。
- 优点：统一了特征维度，便于后续处理。
- 缺点：线性变换可能无法捕捉复杂的模态间关系。

（2）多模态 Transformer。
- 方法：使用自注意力机制对不同模态特征进行加权组合，捕捉各模态之间的相互依赖关系。
- 优点：能够灵活地建模模态间的复杂关系，具有强大的表达能力。
- 缺点：计算复杂度高，需要大量数据和计算资源。

（3）注意力机制。
- 方法：通过计算不同模态特征之间的相关性，动态调整每个模态特征的权重。
- 优点：能够有效地捕捉模态间的依赖关系，适应性强。
- 缺点：需要额外的计算资源，模型复杂度增加。

（4）张量融合。
- 方法：使用张量运算（如张量积）进行高阶的特征交互，生成更丰富的表示。
- 优点：可以捕捉到多模态特征之间更复杂的关系。
- 缺点：计算复杂度高，容易导致维度爆炸。

3. 模态间对齐

（1）互信息最大化。
- 方法：通过最大化不同模态表示之间的互信息来确保对齐。
- 优点：增强模态间的关联性，提升融合效果。
- 缺点：计算互信息较为复杂，优化过程可能较慢。

（2）对比学习。
- 方法：使用对比损失函数，使相同样本的不同模态表示更接近，不同样本的表示更远离。
- 优点：有效提高模态间的表示对齐，增强泛化能力。
- 缺点：需要构造正负样本对，训练过程较为复杂。

4. 联合表示学习

（1）多任务学习。
- 方法：同时训练多个相关任务，通过共享表示来提升模型性能。
- 优点：共享表示空间，促进不同任务之间的知识共享。
- 缺点：需要设计合理的多任务框架，任务之间可能存在冲突。

（2）对抗训练。
- 方法：使用生成对抗网络或对抗性损失来提升模型的泛化能力。
- 优点：增强模型的鲁棒性，提升对未见数据的泛化能力。
- 缺点：训练过程复杂，容易出现不稳定性。

通过上面列出的表示融合方法，多模态表示学习能够充分利用不同模态的信息，提升模型在复杂任务上的表现，解决单一模态模型无法处理的挑战。实例 4-1 实现了一个多模态情感识别模型，利用文本（BERT）、音频（ResNet-18）和视频（ResNet-18）数据进行训练和推理。将文本、音频和视频特征向量沿着特定维度（dim=1）进行拼接（torch.cat），形成一个融合的特征向量。这个融合的特征向量通过另一个线性层（self.fc_combined）进行处理，以减少特征的维度并准备进行分类。

实例 4-1 实现一个多模态情感识别模型
源码路径：codes/4/qing.py

实例文件 qing.py 的具体实现代码如下所示。

```python
import torch
import torch.nn as nn
import torch.nn.functional as F
from transformers import BertModel, BertTokenizer
import torchvision.models as models

# 模型定义
class MultimodalEmotionRecognition(nn.Module):
    def __init__(self):
        super(MultimodalEmotionRecognition, self).__init__()

        # 文本特征提取
        self.bert = BertModel.from_pretrained('bert-base-uncased')
        self.tokenizer = BertTokenizer.from_pretrained('bert-base-uncased')

        # 音频特征提取
        self.audio_model = models.resnet18(pretrained=True)
        self.audio_model.conv1 = nn.Conv2d(3, 64, kernel_size=7, stride=2, padding=3, bias=False)
        self.audio_model.fc = nn.Linear(self.audio_model.fc.in_features, 512)

        # 视频特征提取
        self.video_model = models.resnet18(pretrained=True)
        self.video_model.fc = nn.Linear(self.video_model.fc.in_features, 512)

        # 特征融合
        self.fc_text = nn.Linear(768, 512)
        self.fc_audio = nn.Linear(512, 512)
        self.fc_video = nn.Linear(512, 512)

        # 分类器输入维度
        self.fc_combined = nn.Linear(1536, 512)

        # 分类器
        self.classifier = nn.Linear(512, 7)   # 假设情感分类有 7 个类别

    def forward(self, text, audio, video):
        # 文本特征提取
        inputs = self.tokenizer(text, return_tensors='pt', padding=True, truncation=True)
        text_features = self.bert(**inputs).last_hidden_state[:, 0, :]   # 取[CLS] 标记的输出
        text_features = self.fc_text(text_features)

        # 音频特征提取
        audio_features = self.audio_model(audio)
        audio_features = self.fc_audio(audio_features)

        # 视频特征提取
        video_features = self.video_model(video)
        video_features = self.fc_video(video_features)

        # 表示融合
        combined_features = torch.cat([text_features, audio_features, video_features], dim=1)
        combined_features = self.fc_combined(combined_features)

        # 分类
        output = self.classifier(combined_features)
```

```python
        return output

# 数据处理示例
text_inputs = ["I'm so happy to see you!", "This is a sad movie.", "I love this song!",
"The weather is gloomy today."]
audio = torch.randn(4, 3, 224, 224)   # 示例音频数据（4个样本,每个样本3通道,224×224）
video = torch.randn(4, 3, 224, 224)   # 示例视频数据（4个样本,每个样本3通道,224×224）
labels = torch.tensor([0, 1, 2, 3])   # 示例标签

# 模型实例化
model = MultimodalEmotionRecognition()

# 训练设置
criterion = nn.CrossEntropyLoss()
optimizer = torch.optim.Adam(model.parameters(), lr=1e-4)
epochs = 10

# 训练过程
model.train()
for epoch in range(epochs):
    optimizer.zero_grad()
    output = model(text_inputs, audio, video)
    loss = criterion(output, labels)
    loss.backward()
    optimizer.step()
    print(f'Epoch {epoch + 1}, Loss: {loss.item()}')

# 测试模式下的输出
model.eval()
output = model(text_inputs, audio, video)
print("Raw output:", output)

# softmax 转换为概率
probs = F.softmax(output, dim=1)
print("Probabilities:", probs)
```

上述代码的实现流程如下。

（1）定义一个名为 MultimodalEmotionRecognition 的类，包括文本、音频和视频特征提取器，以及特征融合层和分类器。

（2）模型的初始化过程中实现如下功能。

- 文本特征提取器：使用预训练的 BERT 模型和 Tokenizer，用于提取文本特征。
- 音频特征提取器：使用预训练的 ResNet-18 模型，但修改了第一层卷积以适应音频数据，并添加了一个线性层进行特征转换。
- 视频特征提取器：使用预训练的 ResNet-18 模型，并添加了一个线性层进行特征转换。
- 特征融合层：包括3个线性层，用于将文本、音频和视频特征融合为一个统一的特征表示。
- 分类器：最后是一个用于情感分类的线性层。

（3）在模型的前向传播过程中实现如下功能。

- 文本特征提取：对文本数据进行 BERT 编码，提取文本特征，并通过线性层进行转换。
- 音频特征提取：对音频数据经过 ResNet-18 提取音频特征，并通过线性层进行转换。
- 视频特征提取：对视频数据经过 ResNet-18 提取视频特征，并通过线性层进行转换。
- 表示融合：将3种特征在特征维度上拼接起来，通过一个线性层进行融合得到最终的特征表示。
- 分类：使用分类器进行情感分类。

（4）在数据处理中提供了文本数据、音频数据和视频数据的示例，并定义了示例标签。

（5）在训练过程中使用交叉熵损失函数进行训练，使用 Adam 优化器更新模型参数。然后进行多个 epoch 的训练，打印每个 epoch 的损失。

（6）在测试模式下将模型设置为评估模式，然后对示例数据进行推理，并打印输出原始输出和通过 softmax 转换的概率输出。执行代码后会输出：

```
Epoch 1, Loss: 1.9965693950653076
Epoch 2, Loss: 1.1554456949234009
Epoch 3, Loss: 0.6599863171577454
Epoch 4, Loss: 0.33530205488204956
Epoch 5, Loss: 0.170254647731781
Epoch 6, Loss: 0.08373984694480896
Epoch 7, Loss: 0.0419047512114048
Epoch 8, Loss: 0.022740745916962624
Epoch 9, Loss: 0.011885522864758968
Epoch 10, Loss: 0.006920659448951483
Raw output: tensor([[ 1.9783,  0.4665,  0.7782,  0.4492, -1.2149, -1.1478, -1.2899],
        [ 0.4907,  2.0613,  0.6375,  0.6668, -1.2085, -1.1130, -1.2313],
        [ 0.3111,  0.3041,  2.3677,  0.3611, -1.1806, -1.0315, -1.0692],
        [ 0.0910,  0.2284,  0.3165,  2.6712, -0.8218, -0.8218, -0.8132]],
       grad_fn=<AddmmBackward0>)
Probabilities: tensor([[0.5372, 0.1185, 0.1618, 0.1164, 0.0220, 0.0236, 0.0205],
        [0.1146, 0.5514, 0.1328, 0.1367, 0.0210, 0.0231, 0.0205],
        [0.0862, 0.0856, 0.6740, 0.0906, 0.0194, 0.0225, 0.0217],
        [0.0562, 0.0644, 0.0704, 0.7412, 0.0225, 0.0225, 0.0227]],
       grad_fn=<SoftmaxBackward0>)
```

4.2.2 联合学习

在多模态表示学习中，联合学习是指通过同时处理多种数据类型（如文本、图像、音频等）来学习它们之间的相关性和共同的特征表示。联合学习旨在从不同模态的数据中提取信息，并在一个统一的模型中进行融合，以更好地理解和利用多模态数据之间的关联。在多模态表示学习中实现联合学习的一般流程如下。

（1）特征提取：针对每种数据类型，首先需要设计合适的特征提取器。对于文本数据，可以使用词嵌入或预训练的模型（如 BERT）进行特征提取；对于图像和视频数据，可以使用卷积神经网络进行特征提取；对于音频数据，可以使用声谱图或类似方法进行特征提取。

（2）模态融合：联合学习的关键是将不同模态的特征有效融合。这可以通过简单的拼接、加权求和、注意力机制等方式实现。融合后的特征应该能够保留每种数据类型的重要信息，并提供一个更加综合的表示。

（3）联合学习模型：设计一个能够同时处理多种数据类型的模型架构。这个模型可以包括多个分支，每个分支处理一种数据类型的特征，然后将它们融合在一起；也可以采用单一的模型结构，在不同的层级或分支中处理不同的数据类型。

（4）损失函数设计：定义一个综合考虑所有数据类型的损失函数。这个损失函数可以由各个数据类型的损失函数组合而成，也可以设计特定的联合损失函数，以鼓励模型学习到数据之间的相关性。

（5）训练和优化：使用联合损失函数进行模型训练。通过反向传播算法和优化器对模型参数进行更新，使模型能够更好地学习到不同数据类型之间的关联和共同表示。

（6）评估和推断：在训练完成后，对模型进行评估并在新数据上进行推断。评估指标可以包括准确率、多样性、一致性等，以及针对具体任务的特定指标。

联合学习的优势在于能够更全面地利用多模态数据之间的相关性，提高了对数据的理解和表征能力，有助于更好地解决多模态任务，如多模态情感分析、视觉问答等。接下来介绍一个实现联合学习的例子（见实例 4-2）：假设我们有一个多模态模型，同时处理文本和图像，执行情感分析和图像分类任务。我们将实现联合学习，以最大化情感分析任务和图像分类任务的性能。

实例 4-2　实现联合学习

源码路径：codes/4/lian.py

实例文件 lian.py 的具体实现代码如下所示。

```python
import torch
import torch.nn as nn
import torch.optim as optim

# 定义多模态模型
class MultimodalModel(nn.Module):
    def __init__(self, text_input_size, image_input_size, text_num_classes,
 image_num_classes):
        super(MultimodalModel, self).__init__()

        # 文本特征提取层
        self.text_fc = nn.Linear(text_input_size, 512)

        # 图像特征提取层
        self.image_fc = nn.Linear(image_input_size, 512)

        # 全连接层
        self.fc_combined = nn.Linear(1024, 256)

        # 情感分析分类器
        self.text_classifier = nn.Linear(256, text_num_classes)

        # 图像分类分类器
        self.image_classifier = nn.Linear(256, image_num_classes)

    def forward(self, text_input, image_input):
        # 文本特征提取
        text_features = torch.relu(self.text_fc(text_input))

        # 图像特征提取
        image_features = torch.relu(self.image_fc(image_input))

        # 合并特征
        combined_features = torch.cat((text_features, image_features), dim=1)

        # 全连接层
        combined_features = torch.relu(self.fc_combined(combined_features))

        # 情感分析分类
        text_output = self.text_classifier(combined_features)

        # 图像分类
        image_output = self.image_classifier(combined_features)

        return text_output, image_output

# 示例数据
text_input_size = 300   # 文本输入特征大小
image_input_size = 512  # 图像输入特征大小
text_num_classes = 5    # 情感分析类别数
image_num_classes = 3   # 图像分类类别数

text_input = torch.randn(4, text_input_size)    # 4 个样本的文本特征
image_input = torch.randn(4, image_input_size)  # 4 个样本的图像特征
```

```python
text_labels = torch.randint(0, text_num_classes, (4,))   # 文本标签
image_labels = torch.randint(0, image_num_classes, (4,))  # 图像标签

# 实例化模型
model = MultimodalModel(text_input_size, image_input_size, text_num_classes,
image_num_classes)

# 定义损失函数和优化器
criterion_text = nn.CrossEntropyLoss()
criterion_image = nn.CrossEntropyLoss()
optimizer = optim.Adam(model.parameters(), lr=0.001)

# 联合学习训练过程
def joint_learning_train(model, text_input, image_input, text_labels, image_labels, optimizer):
    model.train()
    optimizer.zero_grad()

    text_output, image_output = model(text_input, image_input)

    # 计算文本任务损失并反向传播
    loss_text = criterion_text(text_output, text_labels)
    loss_text.backward(retain_graph=True)

    # 计算图像任务损失并反向传播
    loss_image = criterion_image(image_output, image_labels)
    loss_image.backward()

    optimizer.step()

    return loss_text.item(), loss_image.item()

# 训练模型
epochs = 5
for epoch in range(epochs):
    loss_text, loss_image = joint_learning_train(model, text_input, image_input,
text_labels, image_labels, optimizer)
    print(f'Epoch {epoch + 1}: Text Loss: {loss_text}, Image Loss: {loss_image}')
```

上述代码的实现流程如下。

（1）定义一个名为 MultimodalModel 的多模态模型类，该模型包括文本特征提取层、图像特征提取层、全连接层和两个分类器。

（2）定义训练函数 joint_learning_train，用于联合学习的训练过程。在这个函数中执行了以下步骤。

- 将模型设置为训练模式，然后将优化器的梯度清零。
- 将文本和图像输入传递给模型，得到文本输出和图像输出。
- 计算文本任务和图像任务的损失函数。
- 分别对文本任务和图像任务的损失进行反向传播。
- 使用优化器更新模型参数。

（3）训练模型。在每个 epoch 中，我们调用 joint_learning_train 函数来执行联合学习的训练过程，并打印出当前 epoch 的文本任务损失和图像任务损失。

这样，整个实例实现了一个简单的多模态大模型应用中的联合学习过程，同时处理文本和图像任务。执行代码后会输出：

```
Epoch 1: Text Loss: 1.5934516191482544, Image Loss: 1.0635684728622437
Epoch 2: Text Loss: 0.9369609951972961, Image Loss: 0.6992965936660767
Epoch 3: Text Loss: 0.4882076680660248, Image Loss: 0.41088801622390747
Epoch 4: Text Loss: 0.21009638905525208, Image Loss: 0.1823546588420868
```

Epoch 5: Text Loss: 0.07330022752285004, Image Loss: 0.059179119765758514

4.2.3 表示对齐

在多模态表示学习中,表示对齐是指将不同模态(例如文本、图像、音频等)的特征表示映射到一个共享的表示空间中,使不同模态之间的表示能够在这个共享空间中对齐或接近。表示对齐的目标是使不同模态的表示能够在语义上保持一致,从而可以更好地进行跨模态任务的处理,比如多模态检索、多模态分类等。

在实际应用中,常用的表示对齐的方法如下。

- 共享表示学习(Shared Representation Learning):通过共享的神经网络层或共享的损失函数,将不同模态的输入映射到共享的表示空间中。这种方法通常采用多模态神经网络结构,如多输入、多输出的神经网络,以确保学到的表示能够在不同模态之间对齐。
- 跨模态嵌入学习(Cross-Modal Embedding Learning):通过学习将不同模态的输入映射到低维连续向量空间中,使不同模态的向量表示在这个空间中距离较近。常用的方法包括使用自编码器、Siamese 网络等。
- 对抗性训练(Adversarial Training):利用对抗性训练框架,通过生成对抗性损失来最大化不同模态的表示在共享空间中的对齐程度。例如,生成对抗网络可以用来生成逼真的跨模态数据,以促进表示对齐。
- 正则化方法(Regularization):在表示学习过程中引入正则化项,以约束不同模态的表示在共享空间中的一致性。例如,最大化不同模态的表示的互信息,或者最小化它们的距离损失。
- 联合训练(Joint Training):在联合训练中,同时考虑多个模态的数据,共同训练模型以使它们的表示在共享空间中对齐。这种方法能够通过端到端的方式直接学习到多模态之间的对应关系。

上述方法可以单独或者组合使用,以实现不同任务下的多模态表示对齐。表示对齐的好处在于可以提取出语义更加丰富的跨模态特征,从而在多模态任务中取得更好的性能。实例 4-3 演示了使用对抗性训练实现多模态表示对齐的过程。本实例将使用文本和图像作为两种模态,使用对抗生成网络来实现它们之间的表示对齐。

实例 4-3 使用对抗生成网络实现多模态之间的表示对齐
源码路径:codes/4/dui.py

实例文件 dui.py 的具体实现代码如下所示。

```
import torch
import torch.nn as nn
import torch.optim as optim
import torchvision.models as models
import torchvision.transforms as transforms
from torch.utils.data import DataLoader, Dataset
from transformers import BertModel, BertTokenizer
import numpy as np

# 定义文本数据集
class TextDataset(Dataset):
    def __init__(self, texts, labels, tokenizer, max_length=128):
        self.texts = texts
        self.labels = labels
        self.tokenizer = tokenizer
        self.max_length = max_length
```

```python
        def __len__(self):
            return len(self.texts)

        def __getitem__(self, idx):
            text = self.texts[idx]
            label = self.labels[idx]

            # 文本处理
            encoding = self.tokenizer(text, return_tensors='pt', padding='max_length',
            truncation=True,
            max_length=self.max_length)
            input_ids = encoding['input_ids'].squeeze(0)
            attention_mask = encoding['attention_mask'].squeeze(0)

            return input_ids, attention_mask, label

# 定义图像数据集
class ImageDataset(Dataset):
    def __init__(self, images, labels, transform=None):
        self.images = images
        self.labels = labels
        self.transform = transform

    def __len__(self):
        return len(self.images)

    def __getitem__(self, idx):
        image = self.images[idx]
        label = self.labels[idx]
        # 将示例张量图像转换为 PIL Image 或 ndarray
        image = transforms.ToPILImage()(image)
        if self.transform:
            image = self.transform(image)
        return image, label

# 定义生成器和鉴别器
class Generator(nn.Module):
    def __init__(self, latent_dim, output_dim):
        super(Generator, self).__init__()
        self.fc = nn.Linear(latent_dim, output_dim)

    def forward(self, x):
        return self.fc(x)

class Discriminator(nn.Module):
    def __init__(self, input_dim):
        super(Discriminator, self).__init__()
        self.fc = nn.Linear(input_dim, 1)

    def forward(self, x):
        return torch.sigmoid(self.fc(x))

# 参数设置
latent_dim = 100
text_input_dim = 768    # BERT 隐藏层大小
image_input_dim = 512   # 图像特征维度
num_classes = 10
batch_size = 64
num_epochs = 20
lr = 0.0002

# 加载预训练的 BERT 模型和 tokenizer
```

```python
tokenizer = BertTokenizer.from_pretrained('bert-base-uncased')
bert_model = BertModel.from_pretrained('bert-base-uncased')

# 加载图像数据和文本数据(示例)
texts = ["I'm so happy to see you!", "This is a sad movie.", "I love this song!",
"The weather is gloomy today."]
images = torch.randn(100, 3, 224, 224)    # 示例图像数据
labels = torch.randint(0, num_classes, (100,))

# 定义数据集和数据加载器
text_dataset = TextDataset(texts, labels, tokenizer)
image_dataset = ImageDataset(images, labels, transform=transforms.ToTensor())
text_loader = DataLoader(text_dataset, batch_size=batch_size, shuffle=True)
image_loader = DataLoader(image_dataset, batch_size=batch_size, shuffle=True)

# 初始化生成器和鉴别器
generator = Generator(latent_dim, text_input_dim)
discriminator = Discriminator(text_input_dim)

# 定义优化器和损失函数
optimizer_G = optim.Adam(generator.parameters(), lr=lr)
optimizer_D = optim.Adam(discriminator.parameters(), lr=lr)
criterion = nn.BCELoss()

# 训练过程
for epoch in range(num_epochs):
    for (text_inputs, attention_masks, labels), (images, _) in zip(text_loader,
    image_loader):
        # 训练鉴别器
        optimizer_D.zero_grad()

        # 生成器产生文本特征
        z = torch.randn(len(text_inputs), latent_dim)
        fake_text_features = generator(z)

        # 真实文本特征通过鉴别器
        real_text_features = bert_model(input_ids=text_inputs, attention_mask=
        attention_masks)[0][:, 0, :]

        # 计算损失并更新鉴别器
        real_pred = discriminator(real_text_features)
        fake_pred = discriminator(fake_text_features.detach())
        d_loss = criterion(real_pred, torch.ones_like(real_pred)) + criterion
        (fake_pred, torch.zeros_like(fake_pred))
        d_loss.backward()
        optimizer_D.step()

        # 训练生成器
        optimizer_G.zero_grad()

        # 生成器生成文本特征
        fake_text_features = generator(z)

        # 计算生成器损失并更新生成器
        fake_pred = discriminator(fake_text_features)
        g_loss = criterion(fake_pred, torch.ones_like(fake_pred))
        g_loss.backward()
        optimizer_G.step()

    # 打印损失
    print(f"Epoch [{epoch+1}/{num_epochs}], d_loss: {d_loss.item():.4f}, g_loss:
    {g_loss.item():.4f}")
```

本实例通过交替训练生成器和鉴别器，不断优化两者的性能，直到达到指定的 epoch 数目为止。上述代码的实现流程如下。

（1）定义一个生成器（Generator）和一个鉴别器（Discriminator），构成一个简单的生成对抗网络模型。

（2）定义生成器和鉴别器的网络结构。其中生成器负责生成样本，鉴别器负责判断样本的真假。在代码中，生成器和鉴别器都是简单的全连接神经网络。

（3）实现设置损失函数和优化器：使用 Binary Cross Entropy 损失函数来衡量鉴别器对真假样本的判断能力，同时定义生成器的损失。使用 Adam 优化器来更新生成器和鉴别器的参数。

（4）开始进行 GAN 的训练，在每个 epoch 中，遍历数据集，对生成器和鉴别器进行交替训练。对鉴别器的具体说明如下。

- 使用真实样本计算鉴别器的损失（真实样本的损失）。
- 使用生成器生成的样本计算鉴别器的损失（生成样本的损失）。
- 将这两个损失相加作为鉴别器的总损失。
- 反向传播并更新鉴别器的参数。

对生成器的具体说明如下。

- 使用生成器生成样本。
- 使用鉴别器判断生成样本的真假，并计算生成器的损失（希望生成样本被判断为真实样本）。
- 反向传播并更新生成器的参数。

（5）在每个 epoch 结束时打印输出每个 epoch 的鉴别器和生成器的损失值，用于监控训练过程中模型的性能和收敛情况。执行代码后会输出：

```
Epoch [1/20], d_loss: 1.4986, g_loss: 0.6088
Epoch [2/20], d_loss: 1.3680, g_loss: 0.7148
Epoch [3/20], d_loss: 1.2888, g_loss: 0.8233
Epoch [4/20], d_loss: 1.4514, g_loss: 0.6012
Epoch [5/20], d_loss: 1.2995, g_loss: 0.7505
Epoch [6/20], d_loss: 1.3592, g_loss: 0.6738
Epoch [7/20], d_loss: 1.3191, g_loss: 0.6798
Epoch [8/20], d_loss: 1.2948, g_loss: 0.6969
Epoch [9/20], d_loss: 1.3280, g_loss: 0.6833
Epoch [10/20], d_loss: 1.1946, g_loss: 0.7614
Epoch [11/20], d_loss: 1.1697, g_loss: 0.7801
Epoch [12/20], d_loss: 1.1853, g_loss: 0.7539
Epoch [13/20], d_loss: 1.1804, g_loss: 0.7478
Epoch [14/20], d_loss: 1.2256, g_loss: 0.6968
Epoch [15/20], d_loss: 1.1276, g_loss: 0.7873
Epoch [16/20], d_loss: 1.1915, g_loss: 0.7019
Epoch [17/20], d_loss: 1.2723, g_loss: 0.6247
Epoch [18/20], d_loss: 1.1606, g_loss: 0.7013
Epoch [19/20], d_loss: 1.1214, g_loss: 0.7273
Epoch [20/20], d_loss: 1.2670, g_loss: 0.5817
```

上面的输出表示已经成功实现了训练过程，并且每个 epoch 都输出了鉴别器（d_loss）和生成器（g_loss）的损失值。这个过程是 GAN 的训练过程，其中鉴别器的损失（d_loss）用于衡量鉴别器对真实样本和生成样本的判断能力，生成器的损失（g_loss）用于衡量生成器生成样本的质量。

模态特征对齐是表示对齐的一种具体应用，专注于多模态数据中的特征对齐。前面介绍的表示对齐，目的是在将不同来源或形式的数据表示对齐到一个共同的表示空间，使它们可以有效结合和比较。模态特征对齐则进一步解决了如何在多模态学习中将来自不同模态（如图像、文本、音频等）的特征对齐，从而实现不同模态数据有效融合和利用的问题。因此可以将模态特征对齐看作表示对齐在多模态环境中的一个具体实现。

模态特征对齐是多模态学习中的关键技术，旨在将来自不同模态的特征对齐到一个共享的表示空间中，以便在该空间中进行有效的融合和比较。这一过程对于提高多模态模型的性能至关重要。下面介绍实现模态特征对齐的一些关键技术和方法。

1. 线性层

线性层（Linear Layer）是一种简单而常用的对齐技术，通过将不同模态的特征投影到相同维度的空间来实现对齐。该方法通过将每种模态的特征映射到一个共享的表示空间中，确保所有模态的特征可以在相同的维度上进行比较。

（1）优点。
- 简单易实现：线性层的实现和计算非常直接，易于集成到现有模型中。
- 计算效率高：线性层计算复杂度低，可以快速进行特征对齐。

（2）缺点。
- 能力有限：线性层可能无法捕捉复杂的模态间关系和非线性特征的对齐需求。
- 表达能力不足：对于高维或复杂数据，线性层可能无法提供足够的表示能力。

2. Qformer

Qformer（Query-based Transformer）是一种基于 Transformer 的技术，利用查询机制来对齐不同模态的特征。Qformer 通过引入查询向量（query vector）来对齐和融合不同模态的特征，通常使用自注意力机制来捕捉模态间的复杂关系。

（1）优点。
- 表达能力强大：通过自注意力机制，Qformer 能够捕捉模态间复杂的关系和依赖。
- 灵活性高：可以处理非线性关系和高维特征，适用于复杂的多模态任务。

（2）缺点。
- 计算开销大：Transformer 结构复杂，计算和存储开销较大。
- 数据需求大：Qformer 的训练需要大量数据和计算资源，可能不适合资源有限的环境。

3. 其他模态对齐技术

除了上面介绍的线性层和 Qformer 外，还有一些其他技术可以用于模态特征对齐。
- 对比学习（Contrastive Learning）：利用对比损失函数，推动相似模态的数据靠近，不相似的模态数据远离，从而实现有效的模态对齐。
- 跨模态嵌入（Cross-Modal Embeddings）：使用共同的嵌入空间来表示不同模态的数据。这种技术通过训练模型，将不同模态的数据映射到一个共享的嵌入空间中，以便它们在同一空间内进行对齐和融合。
- 多模态对齐网络（Multimodal Alignment Networks）：专门设计的网络架构，用于对齐和融合不同模态的特征。这些网络可能结合了多种机制，如注意力机制、对比损失和嵌入学习，以实现高效的模态对齐。

总之，模态特征对齐是多模态学习中的一个核心问题，通过将不同模态的特征对齐到一个共享的表示空间中，实现信息的有效融合。上面介绍的线性层、Qformer 和其他技术各有优缺点，适用于不同的应用场景和需求。选择合适的对齐技术取决于任务的复杂性、计算资源和数据规模等因素。

4.2.4 模态间关系建模

在多模态表示学习方法中，模态间关系建模是指在处理多模态数据时考虑不同模态之间的相关性和交互关系，以更好地学习数据的表示。实现模态间关系建模的常见方法如下。
- 共享表示模型：将不同模态的数据映射到一个共享的表示空间中，以便于模态之间的交互和信息传递。通过共享参数或权重来实现不同模态之间的表示一致性，从而使不同模态的信息可以互相影响和补充。
- 多模态融合：将不同模态的信息进行融合，生成一个统一的表示形式。融合方法包括简单的拼接、加权求和、逐元素相乘等，也可以使用更复杂的神经网络结构进行学习。
- 注意力机制：使用注意力机制来动态地调整不同模态信息的重要性。可以根据任务需求和输入数据的特点，学习不同模态之间的注意力权重，以便于更有效地整合不同模态的信息。
- 联合学习：同时在多个模态上进行学习，通过联合训练来学习模态之间的关系。联合学习可以通过共享部分网络层或者交替训练的方式来实现。
- 图模型：使用图模型来建模多模态数据中的关系，将不同模态的数据表示为图的节点，模态间的关系表示为图的边。可以利用图神经网络等方法来学习模态之间的复杂关系。
- 生成式方法：使用生成式模型来生成多模态数据，如生成对抗网络等。生成式方法可以通过生成器和鉴别器之间的博弈来学习模态之间的相关性。

上述方法既可以单独应用，也可以结合使用，以更好地建模和利用多模态数据中的模态间关系。例如，实例 4-4 是一个使用生成对抗网络生成多模态数据的例子，这个例子实现了一个基于生成对抗网络的图像生成模型。首先定义了生成器和判别器网络，然后设置了损失函数和优化器。生成器通过学习将随机噪声转换为与真实图像相似的图像，而判别器则试图区分生成的假图像和真实图像。在训练过程中，生成器和判别器交替训练以提高性能。

实例 4-4　基于生成对抗网络的图像生成模型
源码路径：codes/4/guanxi.py

实例文件 guanxi.py 的具体实现代码如下所示。

```python
import torch
import torch.nn as nn
import torch.optim as optim
from torchvision.utils import save_image
import numpy as np

# 定义生成器网络
class Generator(nn.Module):
    def __init__(self, latent_dim, img_shape):
        super(Generator, self).__init__()
        self.img_shape = img_shape

        self.model = nn.Sequential(
            nn.Linear(latent_dim, 128),
            nn.LeakyReLU(0.2, inplace=True),
            nn.Linear(128, 256),
            nn.BatchNorm1d(256),
            nn.LeakyReLU(0.2, inplace=True),
            nn.Linear(256, int(np.prod(img_shape))),
            nn.Tanh()
        )

    def forward(self, z):
        img = self.model(z)
        img = img.view(img.size(0), *self.img_shape)
        return img

# 定义判别器网络
class Discriminator(nn.Module):
    def __init__(self, img_shape):
        super(Discriminator, self).__init__()
```

```python
        self.model = nn.Sequential(
            nn.Linear(int(np.prod(img_shape)), 512),
            nn.LeakyReLU(0.2, inplace=True),
            nn.Linear(512, 256),
            nn.LeakyReLU(0.2, inplace=True),
            nn.Linear(256, 1),
            nn.Sigmoid()
        )

    def forward(self, img):
        img_flat = img.view(img.size(0), -1)
        validity = self.model(img_flat)
        return validity

# 参数设置
latent_dim = 100
img_shape = (3, 64, 64)   # 图像形状: 3 通道,64×64

# 初始化生成器和判别器
generator = Generator(latent_dim,img_shape)
discriminator = Discriminator(img_shape)

# 定义损失函数和优化器
adversarial_loss = nn.BCELoss()
optimizer_G = optim.Adam(generator.parameters(), lr=0.0002,betas=(0.5,0.999))
optimizer_D = optim.Adam(discriminator.parameters(), lr=0.0002,betas=(0.5,0.999))

# 数据生成函数
def generate_data(batch_size):
    z = torch.randn(batch_size, latent_dim)
    fake_images = generator(z)
    return fake_images

# 训练
num_epochs = 20
batch_size = 64

for epoch in range(num_epochs):
    for i in range(100):    # 每个 epoch 训练 100 个 batch
        # 训练判别器
        optimizer_D.zero_grad()

        real_imgs = torch.rand(batch_size, *img_shape)   # 使用随机噪声作为真实图像的替代
        real_validity = discriminator(real_imgs)
        d_loss_real = adversarial_loss(real_validity, torch.ones(batch_size, 1))

        fake_imgs = generate_data(batch_size)
        fake_validity = discriminator(fake_imgs.detach())
        d_loss_fake = adversarial_loss(fake_validity, torch.zeros(batch_size, 1))

        d_loss = (d_loss_real + d_loss_fake) / 2
        d_loss.backward()
        optimizer_D.step()

        # 训练生成器
        optimizer_G.zero_grad()
        z = torch.randn(batch_size, latent_dim)
        gen_imgs = generator(z)
        validity = discriminator(gen_imgs)
        g_loss = adversarial_loss(validity, torch.ones(batch_size, 1))
        g_loss.backward()
        optimizer_G.step()
```

```
    print("[Epoch %d/%d] [D loss: %f] [G loss: %f]" % (epoch, num_epochs, d_loss.
item(), g_loss.item()))

    # 保存生成的图像
    if epoch % 5 == 0:
        save_image(gen_imgs.data[:25], "images/%d.png" % epoch, nrow=5, normalize= True)
```

上述代码的实现流程如下。

（1）分别定义生成器（Generator）和判别器（Discriminator）的网络结构，生成器负责将随机噪声转换为图像，而判别器则负责区分真实图像和生成的假图像。

（2）设置损失函数为二元交叉熵损失（BCELoss），优化器为 Adam。

（3）定义函数 generate_data，用于生成假图像，以供训练使用。

（4）开始训练模型，在每个 epoch 中训练判别器和生成器。判别器首先对一批真实图像进行训练，然后对生成器生成的假图像进行训练，以区分真假。生成器则通过生成图像来尽量欺骗判别器。

（5）打印输出每个 epoch 的判别器和生成器的损失，并且每 5 个 epoch 保存生成的图像。执行代码后会输出：

```
[Epoch 0/20] [D loss: 0.012912] [G loss: 5.349892]
[Epoch 1/20] [D loss: 0.003672] [G loss: 6.851644]
[Epoch 2/20] [D loss: 0.002546] [G loss: 7.304008]
[Epoch 3/20] [D loss: 0.001948] [G loss: 7.917387]
[Epoch 4/20] [D loss: 0.001359] [G loss: 8.194927]
[Epoch 5/20] [D loss: 0.000876] [G loss: 8.345054]
[Epoch 6/20] [D loss: 0.000951] [G loss: 8.820522]
[Epoch 7/20] [D loss: 0.000733] [G loss: 8.965182]
[Epoch 8/20] [D loss: 0.000686] [G loss: 8.888860]
[Epoch 9/20] [D loss: 0.001133] [G loss: 9.221610]
[Epoch 10/20] [D loss: 0.000814] [G loss: 8.844706]
[Epoch 11/20] [D loss: 0.000511] [G loss: 9.142951]
[Epoch 12/20] [D loss: 0.000454] [G loss: 8.656384]
[Epoch 13/20] [D loss: 0.000411] [G loss: 8.793859]
[Epoch 14/20] [D loss: 0.000322] [G loss: 9.148103]
[Epoch 15/20] [D loss: 0.227680] [G loss: 1.171384]
[Epoch 16/20] [D loss: 0.279097] [G loss: 0.962872]
[Epoch 17/20] [D loss: 0.275200] [G loss: 0.988554]
[Epoch 18/20] [D loss: 0.183633] [G loss: 1.257497]
[Epoch 19/20] [D loss: 0.247085] [G loss: 1.088501]
```

4.3 基于 S3D MIL-NCE 的多模态文本到视频检索

本项目旨在开发一个高效的文本到视频检索系统，基于多模态模型，结合自然语言处理和计算机视觉技术，使用 S3D 模型和多实例学习（MIL-NCE）方法。通过将文本查询转换为词向量，该系统从视频帧中提取视觉特征，能够在视频数据库中找到与输入文本最相关的视频片段。该系统广泛应用于视频搜索和内容推荐等领域，具体实现如实例 4-5 所示。

实例 4–5 使用 S3D MIL-NCE 进行多模态文本到视频检索

源码路径：codes/4/mil-s3d-video.ipynb

4.3.1 项目介绍

本项目实现了一个文本到视频的检索系统，旨在根据用户提供的文本查询，从视频库中检索出与之相关的视频。该系统利用了多模态技术，结合了自然语言处理和计算机视觉领域的方法。

1. 功能特点

- **多模态检索**：实现了文本到视频的检索功能，用户可以输入文本进行查询，系统返回与之相关的视频。
- **使用预训练模型**：使用了预训练的大型模型，将文本和视频转换为语义表示向量。
- **交互式界面**：通过交互式输入表单，用户可以轻松地输入文本查询和示例视频链接。
- **结果展示**：将检索结果以清晰的 HTML 格式呈现，显示了输入查询、排名靠前的视频以及它们的相似度分数。

2. 技术细节

- **多模态表示学习**：使用预训练的模型将文本和视频转换为嵌入向量，以捕获它们的语义信息。
- **相似度计算**：通过计算文本查询与视频嵌入向量之间的相似度分数来确定检索结果的相关性。
- **交互式界面**：使用表单输入和 HTML 展示技术，使用户能够直观地查看检索结果。

3. 应用领域

- **视频检索系统**：在视频库管理和搜索领域有着广泛的应用，例如视频编辑、内容管理系统等。
- **多模态应用**：为了实现更智能的搜索和推荐功能，在多模态应用中有着重要的作用，例如智能视频推荐系统、多模态搜索引擎等。

4.3.2 准备工作

（1）导入多个关键库，包括用于操作系统交互的 os、用于深度学习的 TensorFlow 及其 Hub 模块、用于数值计算的 NumPy、用于图像和视频处理的 OpenCV，以及用于显示的 IPython 显示模块。这些库共同支持项目中的视频帧提取、预处理、特征提取和文本到视频检索的实现。

```
import os
import tensorflow.compat.v2 as tf
import tensorflow_hub as hub
import numpy as np
import cv2
from IPython import display
import math
```

（2）定义函数 generate_embeddings，该函数的功能是使用从 TensorFlow Hub 加载的 S3D 模型，分别从视频帧和输入文本中生成对应的嵌入向量。视频帧需要先进行规范化处理，函数返回视频和文本的嵌入向量，用于后续的文本到视频的检索任务。

```
def generate_embeddings(model, input_frames, input_words):
    """从视频帧和输入单词中生成模型嵌入。"""
    # 输入帧必须在[0, 1]范围内规范化,形状为 Batch × T×H×W×3
    vision_output = model.signatures['video'](tf.constant(tf.cast(input_frames, dtype=tf.float32)))
    text_output = model.signatures['text'](tf.constant(input_words))
    return vision_output['video_embedding'], text_output['text_embedding']
```

4.3.3 视频加载和可视化

下面这段代码定义了几个函数，用于加载和可视化视频数据，包括裁剪视频帧为正方形、加载视频并返回帧序列、以表格形式显示多个视频，以及显示查询和结果视频及其分数。

```
def crop_center_square(frame):
    """将视频帧裁剪为正方形。"""
    y, x = frame.shape[0:2]
    min_dim = min(y, x)
    start_x = (x // 2) - (min_dim // 2)
    start_y = (y // 2) - (min_dim // 2)
```

```python
        return frame[start_y:start_y+min_dim, start_x:start_x+min_dim]

def load_video(video_url, max_frames=32, resize=(224, 224)):
    """加载视频并返回帧序列。"""
    # 获取视频文件路径
    path = tf.keras.utils.get_file(os.path.basename(video_url)[-128:], video_url)
    cap = cv2.VideoCapture(path)
    frames = []
    try:
        while True:
            ret, frame = cap.read()
            if not ret:
                break
            # 裁剪视频帧为正方形
            frame = crop_center_square(frame)
            # 调整帧大小
            frame = cv2.resize(frame, resize)
            # 调整通道顺序
            frame = frame[:, :, [2, 1, 0]]
            frames.append(frame)

            if len(frames) == max_frames:
                break
    finally:
        cap.release()
    frames = np.array(frames)
    # 如果帧数不足,重复帧直至达到最大帧数
    if len(frames) < max_frames:
        n_repeat = int(math.ceil(max_frames / float(len(frames))))
        frames = frames.repeat(n_repeat, axis=0)
    frames = frames[:max_frames]
    return frames / 255.0

def display_video(urls):
    """以表格形式显示多个视频。"""
    html = '<table>'
    html += '<tr><th>Video 1</th><th>Video 2</th><th>Video 3</th></tr><tr>'
    for url in urls:
        html += '<td>'
        html += '<img src="{}" height="224">'.format(url)
        html += '</td>'
    html += '</tr></table>'
    return display.HTML(html)

def display_query_and_results_video(query, urls, scores):
    """显示文本查询和排名靠前的结果视频及其分数。"""
    sorted_ix = np.argsort(-scores)
    html = ''
    html += '<h2>输入查询: <i>{}</i> </h2><div>'.format(query)
    html += '结果: <div>'
    html += '<table>'
    html += '<tr><th>排名第一,得分:{:.2f}</th>'.format(scores[sorted_ix[0]])
    html += '<th>排名第二,得分:{:.2f}</th>'.format(scores[sorted_ix[1]])
    html += '<th>排名第三,得分:{:.2f}</th></tr><tr>'.format(scores[sorted_ix[2]])
    for i, idx in enumerate(sorted_ix):
        url = urls[sorted_ix[i]]
        html += '<td>'
        html += '<img src="{}" height="224">'.format(url)
        html += '</td>'
    html += '</tr></table>'
    return html
```

在上述代码中定义了如下函数。

（1）crop_center_square(frame)。
- 描述：该函数以一个视频帧为输入，通过取最小的维度并居中裁剪，将其裁剪为正方形。
- 目的：确保帧在进一步处理中的维度一致性。

（2）load_video(video_url，max_frames=32，resize=(224，224))。
- 描述：该函数从给定的 URL 加载视频，将其裁剪为正方形，然后调整大小并返回一系列帧。
- 目的：通过调整大小和标准化尺寸，为模型输入准备视频数据。

（3）display_video(urls)。
- 描述：该函数以表格格式并排显示多个视频。
- 目的：允许对多个视频进行视觉比较。

（4）display_query_and_results_video(query，urls，scores)。
- 描述：该函数显示输入查询以及排名靠前的结果视频及其相应的分数。
- 目的：提供查询和检索结果的视觉表示，以便进行评估。

上述函数用于加载和预处理视频数据，并以可视化方式展示视频及其检索结果，从而支持文本到视频检索任务的评估和优化。

4.3.4　加载视频并定义文本查询

下面这段代码允许用户加载示例视频并为每个视频定义文本查询,通过交互式表单输入视频 URL 和文本查询。加载的视频会经过预处理，存储在 all_videos 中，文本查询存储在'all_queries_video'中，同时显示加载的所有示例视频以供查看。

```
# @title 加载示例视频并定义文本查询 { display-mode: "form" }

# 注意,您可以在下面粘贴任何 gif/小视频链接,然后检查结果
# 这些 gif 是我从谷歌网站上获取的（如果有问题,请告诉我）

video_1_url = 'https://i.pinimg.com/originals/96/d8/98/96d89818c091b500c3a41ffb05d858ff.gif' # @param {type:"string"}
video_2_url = 'https://mir-s3-cdn-cf.behance.net/project_modules/max_1200/43d68193783231.62642905a82e2.gif' # @param {type:"string"}
video_3_url = 'https://i.gifer.com/8tVa.gif' # @param {type:"string"}

# 加载视频
video_1 = load_video(video_1_url)
video_2 = load_video(video_2_url)
video_3 = load_video(video_3_url)
all_videos = [video_1, video_2, video_3]

# 定义文本查询
query_1_video = 'Cube' # @param {type:"string"}
query_2_video = 'Dance' # @param {type:"string"}
query_3_video = 'Cycling' # @param {type:"string"}
all_queries_video = [query_1_video, query_2_video, query_3_video]

# 所有示例视频的 URL
all_videos_urls = [video_1_url, video_2_url, video_3_url]

# 显示所有示例视频
display_video(all_videos_urls)
```

对上述代码的具体说明如下。

- video_1_url，video_2_url，video_3_url：允许输入 3 个示例视频的 URL。
- video_1，video_2，video_3：分别加载了 3 个示例视频，使用了 load_video 函数。
- query_1_video，query_2_video，query_3_video：允许为每个视频输入文本查询。

- 📼 all_videos：包含了加载的所有视频。
- 📁 all_queries_video：包含了每个视频的文本查询。
- 🔗 all_videos_urls：包含了所有示例视频的 URL。

执行后会显示如下所示的下载过程，并显示加载的示例视频。上述代码使用了 display_video 函数来显示所有示例视频，因此执行代码后会在输出中显示加载的 3 个示例视频，如图 4-1 所示。

图 4-1　加载的 3 个示例视频

```
Downloading data from https://i.pinimg.com/originals/96/d8/98/96d89818c091b500c3a41ffb05
d858ff.gif
1302879/1302879 [==============================] - 0s 0us/step
Downloading data from https://mir-s3-cdn-cf.behance.net/project_modules/max_1200/
43d68193783231.62642905a82e2.gif
96255/96255 [==============================] - 0s 0us/step
Downloading data from https://i.gifer.com/8tVa.gif
1009275/1009275 [==============================] - 1s 1us/step
```

4.3.5　预处理视频和查询

下面这段代码的功能是将加载的视频和文本查询准备好，并使用提供的 TensorFlow Hub 模型生成它们的嵌入向量。然后，通过计算文本与视频嵌入向量之间的点积，得到它们的相似度分数，用于衡量文本与视频之间的相关性。

```
# 准备视频输入
videos_np = np.stack(all_videos, axis=0)

# 准备文本输入
words_np = np.array(all_queries_video)

# 生成视频和文本嵌入向量
video_embd, text_embd = generate_embeddings(hub_model, videos_np, words_np)

# 使用点积计算文本和视频嵌入向量之间的相似度分数
all_scores = np.dot(text_embd, tf.transpose(video_embd))
```

上述代码的实现流程如下。

（1）将所有加载的视频沿着指定轴堆叠成一个单独的 NumPy 数组。

（2）将文本查询列表转换为一个 NumPy 数组。

（3）使用提供的 TensorFlow Hub 模型，通过 generate_embeddings() 函数为视频和文本生成嵌入向量。

（4）使用点积计算文本和视频嵌入向量之间的相似度分数。每行对应一个文本查询，每列对应一个视频，得分越高表示相似度越高。

4.3.6　展示结果

下面这段代码用于遍历每个文本查询，生成相应的 HTML 内容，并显示每个查询的检索结果，

4.3 基于 S3D MIL-NCE 的多模态文本到视频检索

包括输入查询、排名靠前的视频以及它们的分数,最后以 HTML 格式呈现,确保以清晰、有序的方式展示文本到视频检索的结果。

```
html = ''
for i, words in enumerate(words_np):
    html += display_query_and_results_video(words, all_videos_urls, all_scores[i, :])
    html += '<br>'
display.HTML(html)
```

执行代码后会显示每个文本的检索结果,包括以下几项。
- 输入的文本查询。
- 排名靠前的视频以及它们的预览图像。
- 每个视频与查询之间的相似度分数。

检索结果如图 4-2 所示,这些结果以 HTML 格式呈现在输出中,让用户能够清晰地查看每个查询的检索结果。

图 4-2 检索结果

第5章 多模态嵌入表示

多模态嵌入表示是一种机器学习技术，通过将来自不同模态（如文本、图像、音频等）的数据映射到同一个高维向量空间中，实现多模态数据的融合与统一表示。这种方法不仅能够捕捉不同模态之间的语义关系，还可以在各种模态数据之间进行高效的信息检索和相似性计算。本章将详细讲解多模态嵌入表示的知识。

5.1 多模态嵌入基础

多模态嵌入表示在自然语言处理、计算机视觉和多媒体分析等领域具有广泛的应用，促进了跨模态任务（如图像描述生成、视频内容理解和跨模态检索）的研究与发展。

5.1.1 多模态嵌入介绍

多模态嵌入（Multimodal Embedding）是一种将来自不同模态的数据（如文本、图像、音频等）映射到一个共同的向量空间中的技术，这种技术的主要目的是通过统一的表示方式融合不同模态的信息，从而实现更丰富和全面的理解和处理。

1. 核心概念
- 模态：指不同类型的数据源，例如文本、图像、音频、视频等。
- 嵌入：指将高维或非结构化的数据转换为低维的、结构化的向量表示。

2. 关键技术
- 联合嵌入空间（Joint Embedding Space）：不同模态的数据通过学习共享一个嵌入空间，使相同语义的信息在该空间中具有相近的表示。
- 对比学习：一种常用的方法，通过最大化相关数据对的相似性和最小化非相关数据对的相似性来学习嵌入。
- 神经网络模型：通常采用深度神经网络，如卷积神经网络用于图像处理，循环神经网络或 Transformer 用于文本处理。

3. 优点
- 信息融合：能够综合利用多种数据源的信息，提高理解和处理信息的准确性。
- 多样性处理：适用于多种任务和应用场景，具有广泛的适应性。

总之，多模态嵌入技术在近年来的研究和应用中展示了其强大的能力和潜力，尤其是在人工智能领域，为实现更智能和人性化的系统提供了重要支持。

5.1.2 多模态嵌入的应用

多模态嵌入技术通过将不同模态的数据融合在一起，提供了更加全面和准确的信息处理能力，在多个领域中展现出了强大的优势和潜力。

- 图像描述生成（Image Captioning）：利用多模态嵌入，系统可以将图像内容转换为自然语言描述。这对于辅助盲人、自动生成图像标签以及内容管理等非常有用。
- 跨模态检索（Cross-modal Retrieval）：用户可以用一种模态的数据（如文本描述）来检索另一种模态的数据（如图像或视频）。例如，根据一段文本描述找到相关的图像，或根据图片找到相关的文本资料。
- 视频内容理解与生成（Video Understanding and Generation）：通过结合视频帧和音频信息，系统可以对视频内容进行更深层次的理解，如生成视频摘要、内容标注、动作识别等。
- 多模态情感分析（Multimodal Sentiment Analysis）：融合文本、语音、面部表情等多模态数据，进行更精准的情感分析。这在市场调查、客户服务、社交媒体分析等方面有广泛应用。
- 语音与图像识别（Speech and Image Recognition）：结合语音和图像数据，提高识别的准确性。例如，在智能助理中，通过语音命令和摄像头图像共同识别用户的意图和周围环境。
- 医学影像分析（Medical Image Analysis）：在医学影像分析中，结合患者的文本病历、图像数据（如X光片、MRI等）和其他诊断信息，提高疾病诊断的准确性和全面性。
- 虚拟现实和增强现实（VR/AR）：在 VR/AR 应用中，多模态嵌入可以结合用户的视线、动作、语音指令等多种模态数据，为用户提供更加沉浸和互动的体验。
- 自动驾驶（Autonomous Driving）：自动驾驶系统需要融合来自不同传感器的数据，如摄像头、雷达、激光雷达（LiDAR）等，通过多模态嵌入技术，增强对环境的感知和理解，从而提高驾驶的安全性和决策能力。
- 安全监控（Security Monitoring）：结合视频监控、音频监控、文本报告等多种数据源，进行更高效的异常行为检测和安全事件分析。

5.2 图像嵌入

图像嵌入（Image Embedding）是将图像数据转换为低维向量表示的技术，这些向量捕捉了图像的语义和内容信息。图像嵌入的目的是将高维的图像数据映射到一个低维的向量空间，以便高效执行图像检索、分类、聚类等任务。

5.2.1 图像嵌入介绍

在多模态大模型应用中，图像嵌入是将图像数据转换为低维向量表示的关键步骤，这些向量能够捕捉图像的语义和内容信息，并在多模态融合中与其他模态（如文本、音频等）进行对齐和联合表示。多模态大模型如 CLIP、DALL-E、ALIGN 等，广泛应用于图像和文本的联合理解与生成任务。以下是关于多模态大模型中图像嵌入的详细介绍。

1. **核心概念**
   - 嵌入向量（Embedding Vector）：图像被表示为一个低维向量，通常通过深度神经网络（如卷积神经网络）提取图像的特征。
   - 特征提取网络：用于从图像中提取高层次的语义特征，常见的网络结构包括 ResNet、EfficientNet 等。
2. **在多模态大模型中的作用**
   - 特征提取：使用预训练的卷积神经网络从图像中提取特征，并将这些特征映射到一个向量空间。
   - 模态对齐：通过训练，使图像和其他模态（如文本）的嵌入向量在同一个向量空间中表示，从而实现跨模态的对齐。例如，图像和对应描述文本的向量在这个空间中距离很近。

- 联合学习：通过联合训练数据，使模型能够同时学习图像和文本的表示，并理解两者之间的关系。

3. **多模态大模型中的图像嵌入技术**

（1）CLIP。
- 方法：通过对比学习训练模型，使图像和文本的嵌入在相同的向量空间中对齐。具体来说，CLIP 使用图像-文本对，通过最大化正确配对的相似性和最小化错误配对的相似性来学习嵌入。
- 应用：图像分类、图像搜索、图像生成等任务。

（2）DALL-E。
- 方法：结合图像生成和文本描述，DALL-E 从文本描述生成图像。通过训练，DALL-E 学习到如何从文本描述中生成对应的图像嵌入，再通过生成模型输出图像。
- 应用：文本到图像生成、图像创作等。

（3）ALIGN（A Large-scale ImaGe and Noisy-text embedding）。
- 方法：使用大规模的图像-文本对进行对比学习，ALIGN 优化图像和文本嵌入，使得相匹配的图像和文本在嵌入空间中接近。
- 应用：图像分类、图像检索、跨模态检索等。

4. **训练方法**
- 对比学习：通过成对的正样本（正确的图像-文本对）和负样本（错误的图像-文本对）训练模型，使正确图像-文本对的相似性最大化，错误图像-文本对的相似性最小化。
- 大规模数据训练：利用海量的图像-文本对进行训练，以捕捉更多的语义关系和特征。

5. **应用场景**
- 图像分类和检索：通过嵌入向量实现高效的图像分类和检索，能够根据文本描述找到对应的图像，或根据图像找到相关的文本。
- 跨模态生成：如文本生成图像或图像生成文本，通过多模态嵌入实现复杂的生成任务。
- 增强现实和虚拟现实：结合多模态嵌入技术，实现更自然和智能的交互体验。

总之，多模态大模型中的图像嵌入技术通过统一的向量空间表示图像和其他模态的数据，实现了跨模态的数据对齐和融合，为多种应用提供了强大的技术支持。

5.2.2 图像特征提取

特征提取是指从图像中提取高层次的语义特征，这通常通过预训练的深度神经网络（如卷积神经网络）来实现。在多模态大模型中，使用图像嵌入技术进行图像特征提取是关键步骤之一。这种方法将图像数据转换为低维向量表示，使图像的语义和内容信息能够在向量空间中进行表示和处理。

使用图像嵌入实现图像特征提取的基本流程如下。

（1）准备预训练模型。通常使用在大型数据集（如 ImageNet）上预训练的模型，如 ResNet、Inception、VGG 等。这些模型能够捕捉丰富的视觉特征。从预训练模型的中间层或最后一层提取特征，这些特征向量能够有效地表示图像的内容。

（2）生成特征向量。在卷积神经网络中通过一系列卷积层、池化层和全连接层，提取图像的高层特征。通过全局平均池化或全局最大池化，将特征图转化为固定长度的特征向量。

（3）嵌入空间的构建。通过全连接层或其他线性变换，将特征向量映射到指定的嵌入空间。对特征向量进行归一化处理，以便在向量空间中进行对比学习或相似性计算。

实例 5-1 实现了一个基于深度学习的图像检索系统。本实例结合了自然语言处理和计算机视觉（CV）技术，能够根据文本描述检索相似的图像。

> **实例 5-1** 根据文本描述检索相似的图像
> 源码路径：codes/5/tuteti.py

实例文件 tuteti.py 的具体实现代码如下所示。

```python
from transformers import BertModel, BertTokenizer
import torch
from PIL import Image
import torch
import torchvision.transforms as transforms
from torchvision.models import resnet50, ResNet50_Weights    # 假设使用 ResNet-50 模型
import faiss    # 用于相似性搜索
import numpy as np
import os
os.environ["KMP_DUPLICATE_LIB_OK"]="TRUE"

# 初始化 BERT 分词器和模型
tokenizer = BertTokenizer.from_pretrained('bert-base-uncased')
model = BertModel.from_pretrained('bert-base-uncased')

def text_to_feature(text_description):
    # 对文本进行分词
    inputs= tokenizer(text_description, return_tensors='pt', padding=True, truncation=True, max_length=512)

    # 获取 BERT 模型的输出
    with torch.no_grad():
        outputs = model(**inputs)

    # 使用最后一层的隐藏状态作为特征向量
    last_hidden_states = outputs.last_hidden_state
    feature_vector = last_hidden_states.mean(dim=1)    # 取平均值作为特征向量

    # 将特征向量转换为 NumPy 数组
    feature_array = feature_vector.squeeze().numpy()

    return feature_array


# 演示文本到特征的转换
text_description = "A beautiful landscape with mountains and a lake."
feature_vector = text_to_feature(text_description)
print("Feature vector shape:", feature_vector.shape)

# 加载预训练的模型，这里使用 ResNet-50 作为特征提取器
model1 = resnet50(weights=ResNet50_Weights.IMAGENET1K_V1)
model1.eval()

# 定义图像转换，将图像转换为模型输入所需的格式
transform = transforms.Compose([
    transforms.Resize(256),
    transforms.CenterCrop(224),
    transforms.ToTensor(),
    transforms.Normalize(mean=[0.485, 0.456, 0.406], std=[0.229, 0.224, 0.225]),
])

# 假设有一个图像数据库和相应的特征向量数据库
image_database = ['image1.png', 'image2.png','image1.jpg','image2.jpg','image3.png', 'image4.png','image5.png']    # 图像文件列表
# 存储图像特征的列表
```

```python
feature_database = []

for image_path in image_database:
    image = Image.open(image_path).convert('RGB')  # 确保图像是 RGB 格式
    image = transform(image).unsqueeze(0)  # 增加一个批次维度
    print("Image shape:", image.shape)  # 打印图像形状

    with torch.no_grad():
        features = model1(image)  # 直接使用返回的特征图
        # 由于 ResNet-50 默认只返回特征图，我们不需要解包，features 已经是我们需要的特征图
    feature_database.append(features.squeeze(0).cpu().numpy())
    # 移除批次维度，转换为 NumPy 数组并移动到 CPU

# 将特征列表转换为 NumPy 数组
feature_database = np.array(feature_database)

# 确保特征数据库的维度是正确的
print("Feature database shape:", feature_database.shape)

# ...（后续代码不变）

# 使用 faiss 创建索引
# 特征维度应该是第二维，且确保 feature_database 是二维数组
index = faiss.IndexFlatL2(feature_database.shape[1])
index.add(feature_database)

def retrieve_similar_images(text_description):
    # 获取文本的特征向量
    feature_vector = text_to_feature(text_description)

    # 扩展 BERT 特征向量到 1000 维，这里简单地用零填充
    feature_vector_expanded = np.zeros((1000,))
    feature_vector_expanded[:feature_vector.shape[0]] = feature_vector

    # 将查询特征向量转换为二维数组，以满足 faiss 搜索的要求
    desired_feature = feature_vector_expanded[np.newaxis, :]

    # 使用 faiss 进行搜索
    _, indices = index.search(desired_feature, 5)  # 检索 5 个最相似的图像

    # 确保 indices 是一维数组并且转换为 Python 整数
    indices = indices.squeeze().tolist()

    # 使用 indices 中的每个值作为索引来访问 image_database
    return [image_database[int(idx)] for idx in indices]

# 在函数外部创建 faiss 索引并添加特征数据
index = faiss.IndexFlatL2(1000)  # 确保这里的维度与 feature_database 的列数一致
index.add(feature_database)

# 演示检索功能
text_description = "这是一只小狗"
similar_images = retrieve_similar_images(text_description)
print("检索到的相似图像:", similar_images)
```

上述代码实现了一个多模态检索框架，将文本和图像内容通过特征向量的方式关联起来，实现了基于文本的图像检索功能。对上述代码的具体说明如下。

（1）使用 transformers 库中的 BERT 模型来处理文本描述，将其转换为特征向量。BERT 分词器（BertTokenizer）用于将文本分词，并将其转换为模型可处理的输入格式。然后，BERT 模型（BertModel）用于获取文本的隐藏状态表示。

（2）利用 torchvision 的 ResNet-50 预训练模型来提取图像的视觉特征。图像首先通过一系列变换（transforms）来适配模型输入要求，然后通过 ResNet-50 模型提取特征。

（3）将所有图像特征存储在 feature_database 中，并转换为 NumPy 数组。接着，使用 faiss 库创建一个向量索引，用于高效相似性搜索。

（4）定义函数 retrieve_similar_images，它接收一个文本描述，通过 BERT 模型转换为特征向量，并将其扩展到与图像特征相同的维度。然后，使用 faiss 索引来找到与文本特征最相似的图像特征，最终返回最相似的图像路径列表。

（5）为了避免 OpenMP 运行时的初始化错误，在代码中设置了环境变量 KMP_DUPLICATE_LIB_OK=TRUE。

执行代码后将根据文本描述"这是一只小狗"检索相似图像，并打印出检索到的图像路径：

```
Feature vector shape: (768,)
Image shape: torch.Size([1, 3, 224, 224])
Image shape: torch.Size([1, 3, 224, 224])
Image shape: torch.Size([1, 3, 224, 224])
Image shape: torch.Size([1, 3, 224, 224])
Image shape: torch.Size([1, 3, 224, 224])
Image shape: torch.Size([1, 3, 224, 224])
Image shape: torch.Size([1, 3, 224, 224])
Feature database shape: (7, 1000)
检索到的相似图像：['image1.png', 'image4.png', 'image1.jpg', 'image2.png','image2.jpg']
```

5.2.3 模态对齐

在多模态大模型应用中，模态对齐技术是确保不同类型的数据（如图像、文本、音频等）能够在一个统一的特征空间中表示的关键步骤。这种对齐使模型能够理解和关联不同模态之间的语义信息。常用的模态对齐技术如下。

- 共享嵌入空间：在多模态模型中，不同模态的数据被映射到一个共享的特征空间中，使不同模态之间的相似性可以直接比较。
- 特征融合：通过融合来自不同模态的特征来实现对齐，例如，将图像特征和文本特征通过串联（concatenation）或加权和（weighted sum）的方式组合在一起。
- 协变对齐：确保不同模态的特征向量在统计上是协变的，即它们的变化趋势是一致的。这可以通过设计损失函数来实现，比如最大均值差异（Maximum Mean Discrepancy，MMD）。
- 对抗性训练：使用对抗性网络来训练模态之间的对齐。例如，一个判别器尝试区分特征是否来自同一模态，而生成器则试图欺骗判别器，使其认为不同模态的特征来自同一模态。
- 自适应实例归一化：通过对每个模态的特征进行归一化处理，使不同模态的特征具有相似的分布，从而实现对齐。
- 跨模态注意力机制：利用注意力机制来学习不同模态之间的关联，例如，文本中的某些词汇可能与图像中的特定区域相关联。
- Zero-Shot 学习：在没有直接监督的情况下，通过模态对齐技术来实现跨模态的迁移学习。例如，使用文本描述来识别未见过的图像类别。
- 跨模态对比学习：通过最小化正样本对之间的距离和最大化负样本对之间的距离来学习跨模态的一致性表示。
- 多任务学习：在多任务框架下训练模型，使其同时学习到不同模态的特征表示，并通过任务间的交互来实现模态对齐。
- 图神经网络：将不同模态的数据视为图中的节点，通过图神经网络来捕捉节点间的复杂关系，实现模态对齐。
- **Transformer 架构**：利用 Transformer 模型的自注意力机制来处理多模态数据，它可以捕捉不同模态之间的长距离依赖关系。
- **模态特定编码器**：对每种模态使用特定的编码器（如图像编码器、文本编码器）来提取特征，然后通过某种方式将这些特征融合在一起。

实例 5-2 展示了实现一个多模态对齐模型的过程。本实例使用图像和文本数据生成了嵌入向量，并计算三元组损失。

实例 5-2	实现一个多模态对齐模型
	源码路径：codes/5/duiqi.py

实例文件 duiqi.py 的具体实现代码如下所示。

```python
import torch
from torch import nn
from transformers import BertModel, BertTokenizer
from torchvision import models, transforms
from PIL import Image

# 初始化 BERT 分词器和模型
tokenizer = BertTokenizer.from_pretrained('bert-base-uncased')
text_model = BertModel.from_pretrained('bert-base-uncased')

# 初始化图像处理流程
image_transforms = transforms.Compose([
    transforms.Resize((224, 224)),
    transforms.CenterCrop(224),
    transforms.ToTensor(),
    transforms.Normalize((0.485, 0.456, 0.406), (0.229, 0.224, 0.225)),
])

# 加载和预处理图像
image_path = 'image3.jpg'
text_description = "A photo of a dog"

image = Image.open(image_path).convert('RGB')
image = image_transforms(image)


class ModalAligner(nn.Module):
    def __init__(self, embed_size=256):
        super(ModalAligner, self).__init__()
        self.image_encoder = models.resnet18(weights=models.ResNet18_Weights.DEFAULT)
        self.image_encoder = nn.Sequential(*list(self.image_encoder.children())[:-1])
        self.image_fc = nn.Linear(512, embed_size)

        self.text_encoder = BertModel.from_pretrained('bert-base-uncased')
        self.text_fc = nn.Linear(768, embed_size)

    def forward(self, image, input_ids, attention_mask):
        image_features = self.image_encoder(image).view(image.size(0), -1)
        image_embeds = self.image_fc(image_features)

        text_outputs=self.text_encoder(input_ids=input_ids,attention_mask=attention_mask)
        text_features = text_outputs.last_hidden_state[:, 0, :]
        text_embeds = self.text_fc(text_features)
```

```
        return image_embeds, text_embeds

def triplet_loss(anchor, positive, negative, margin=1.0):
    pos_dist = (anchor - positive).pow(2).sum(1)
    neg_dist = (anchor - negative).pow(2).sum(1)
    losses = torch.relu(pos_dist - neg_dist + margin)
    return losses.mean()


# 实例化模型
embed_size = 256
model = ModalAligner(embed_size)

# 准备图像输入
image = image.unsqueeze(0)

# 对文本进行编码
inputs = tokenizer(text_description, return_tensors='pt', padding=True, truncation=True,
 max_length=512)

# 前向传播计算图像和文本的嵌入
with torch.no_grad():
    image_embeds,text_embeds=model(image,inputs['input_ids'],inputs['attention_mask'])

# 假设我们有负样本(实际应用中需要提供负样本)
negative_text_description = "A photo of a cat"
negative_inputs=tokenizer(negative_text_description, return_tensors='pt', padding=True,
 truncation=True, max_length=512)

with torch.no_grad():
    _,negative_text_embeds=model(image,negative_inputs['input_ids'],negative_inputs
      ['attention_mask'])

# 计算三元组损失
loss = triplet_loss(image_embeds, text_embeds, negative_text_embeds)
print(f"三元组损失: {loss.item()}")

# 打印嵌入
print(f"图像嵌入: {image_embeds}")
print(f"文本嵌入: {text_embeds}")
print(f"负样本文本嵌入: {negative_text_embeds}")
```

上述代码的实现流程如下。

(1)初始化 BERT 分词器和模型,以及图像处理流程。
(2)定义一个多模态对齐模型 ModalAligner,其中包括图像编码器和文本编码器。
(3)定义三元组损失函数 triplet_loss,用于衡量正样本和负样本之间的距离。
(4)实例化 ModalAligner 模型,并准备图像和文本输入。
(5)通过模型计算图像和文本的嵌入向量,并处理负样本,得到负样本文本的嵌入向量。
(6)计算三元组损失,并打印输出结果。执行代码后会输出:

```
三元组损失: 0.62823486328125
图像嵌入: tensor([[-0.6760,  0.7779,  0.7018,  1.0671,  1.0474,  0.6971,
          0.2806,  0.2390,
         -0.5168, -1.1336,  0.0964, -0.4932, -0.9681,  0.4497, -0.3325,  0.6357,
         -0.3097, -1.2380, -0.2106, -1.1434, -0.7401, -0.1034,  0.5712,  0.4939,
         -0.1541, -0.2631, -0.4091, -0.2827, -0.5295, -1.3269, -0.9495,  0.2895,
         -0.2263,  0.3431, -0.1014,  0.4806, -0.1248,  0.1142,  0.5384,  0.4486,
         -0.9371,  0.2124,  0.3097,  0.5017,  0.5946,  0.3089, -0.5132, -0.4608,
```

```
           -0.9980, -0.0657,  0.5342,  0.4430, -0.4219,  0.3910,  1.0996,  0.2197,
           -0.3107,  0.2340,  0.0259,  0.5239, -0.2728, -0.7826,  0.0273,  0.8058,
####省略部分结果
            2.3806e-01,  5.9179e-02,  2.2546e-01,  4.6646e-03,  1.0846e-02,
           -2.6939e-02, -7.8928e-01,  1.3646e-01,  2.1889e-01,  1.7625e-01,
           -2.8915e-01, -4.1227e-01, -8.3739e-02, -6.9160e-02, -2.6744e-03,
            6.1476e-01,  1.8131e-01,  2.5329e-01, -2.2697e-01,  7.0750e-03,
           -2.6410e-01]])
```

注意：在实际应用中，模态对齐技术的选择取决于具体的应用场景和可用数据的性质，在使用时需要结合多种技术来达到最佳的对齐效果。通过有效的模态对齐，多模态模型能够更准确地理解数据，提高在诸如图像检索、情感分析、自动标注等任务上的性能。

5.2.4　CLIP 模型

CLIP 是由 OpenAI 提出的一种多模态模型，是一种重要的图像嵌入技术。CLIP 模型使用了大量的图像-文本对进行预训练，学习将图像内容与文本描述映射到同一个特征空间。这样，给定一个图像，CLIP 可以生成一个特征向量；同样地，给定一个文本描述，它也可以生成一个特征向量。这些特征向量可以用于多种任务，包括但不限于图像检索、文本到图像的搜索、零样本分类等。

CLIP 的主要特点如下。

- 多模态学习：CLIP 同时处理图像和文本数据，使两种模态的特征表示能够相互对应。
- 对比学习：CLIP 使用对比学习的方法，通过最小化正样本对（即匹配的图像-文本对）之间的距离，同时最大化负样本对（即不匹配的图像-文本对）之间的距离，来优化特征表示。
- 大规模数据集：CLIP 在大规模的图像-文本对数据集上进行训练，以学习丰富的特征表示。
- 零样本能力：由于 CLIP 学习了通用的视觉概念和语言表示，因此能够在没有看过特定类别的图像的情况下，根据文本描述识别图像。

CLIP 模型为多模态应用提供了强大的图像和文本嵌入技术，能够支持各种复杂的视觉和语言任务。实例 5-3 演示了使用 CLIP 模型实现图文格式的相似度匹配操作的过程。

实例 5-3　使用 CLIP 模型实现图文格式的相似度匹配操作
源码路径：codes/5/cl.py

实例文件 cl.py 的具体实现代码如下所示。

```python
from PIL import Image
import requests
from transformers import CLIPProcessor, CLIPModel

model = CLIPModel.from_pretrained("openai/clip-vit-base-patch16")
processor = CLIPProcessor.from_pretrained("openai/clip-vit-base-patch16")
url = "http://images.cocodataset.org/val2017/000000039769.jpg"
image = Image.open(requests.get(url, stream=True).raw)
inputs = processor(text=["a photo of a cat", "a photo of a dog"], images=image, return_tensors="pt", padding=True)
outputs = model(**inputs)
logits_per_image = outputs.logits_per_image
probs = logits_per_image.softmax(dim=1)
for prob in probs_numpy[0]:
    print(f"Probability: {prob:.4f}")
```

上述代码的实现流程如下。

（1）加载 CLIP 模型和处理器：使用 from_pretrained 方法从预定义的模型仓库中加载 CLIP 模型和处理器。这里加载的是"openai/clip-vit-base-patch16"模型，这是一个基于 Vision Transformer（ViT）

的 CLIP 模型。

（2）获取图像。
- 使用 requests.get(url, stream=True).raw 从给定的 URL 中获取图像数据。这里，URL 是一个示例图像的 URL。
- 使用 PIL 的 Image.open 方法打开从 URL 获取的图像数据。

（3）准备输入数据。
- 使用 CLIP 处理器 processor 将文本和图像数据转换为模型可以接受的输入格式。这里，文本是"a photo of a cat"和"a photo of a dog"两个句子，图像是之前从 URL 获取的图像。
- return_tensors="pt"表示输出应为 PyTorch 张量。
- padding=True 确保文本输入被适当地填充，使其长度与批处理中的最长文本相匹配。

（4）执行模型推理：使用模型 model 和输入的文本-图像数据执行推理。结果 outputs 包含多个属性，其中一个是我们关心的 logits_per_image。

（5）获取图像-文本相似度分数和概率。
- logits_per_image 是模型为每个文本-图像对生成的未归一化的相似度分数。这些分数是线性的，表示图像与每个文本描述的相似度。
- 使用 softmax 函数对 logits_per_image 进行归一化，得到每个文本描述与图像匹配的概率（probs）。这样，每个概率值都在 0 和 1 之间，并且所有概率值的和等于 1。

（6）得到相似度概率值：使用 probs.numpy()将张量转换为 NumPy 数组，并使用循环或列表推导式来打印每个概率值。执行后会输出下面的结果，这表示 CLIP 模型认为图像与文本描述 "a photo of a dog" 的相似度非常高（概率为 0.9992），而与 "a photo of a cat" 的相似度非常低（概率为 0.0008）。

```
Probability: 0.9992
Probability: 0.0008
```

5.3 文本嵌入

文本嵌入是将文本信息（如单词、短语或句子）转换为固定长度的连续实数向量的过程。这些向量通常被称为词向量或文本向量，它们能够捕捉文本中的语义和语法信息。文本嵌入的目标是将文本表示为计算机可以理解和处理的数学形式，从而支持各种自然语言处理任务。在多模态嵌入表示应用中，文本嵌入是一个关键组成部分，它涉及将文本信息转换为数学向量表示，以便计算机能够更有效地处理和理解。

5.3.1 多模态模型中的文本嵌入

在多模态嵌入表示应用中，实现文本嵌入的技术和方法多种多样，每种方法都有其特点和适用场景。常用的实现文本嵌入的技术和方法如下。

- 词袋模型（Bag of Words，BoW）：最简单的文本表示方法之一，不考虑词序，只关注词频。将文本转换为词频向量，每个维度代表一个词的出现次数。
- TF-IDF：考虑了词在文档中出现的频率（TF）以及在语料库中出现的稀有程度（IDF）。能够突出文档中的重要词汇，降低常见词汇的权重。
- Word2Vec：一种预测词的上下文的模型，可以生成词的向量表示，包括连续词袋（CBOW）和 Skip-gram 两种模型。
- GloVe（Global Vectors for Word Representation）：利用全局统计信息来学习词的向量表示，结合了 Word2Vec 的局部上下文信息和矩阵分解技术。

- **FastText**：由 Facebook 开发，与 Word2Vec 类似，但增加了对词的子词（subword）信息的捕捉。通过考虑词的内部结构，提高了模型对未知词的泛化能力。
- **BERT**：一种基于 Transformer 的预训练语言表示模型，能够捕捉词的双向上下文信息。通过预训练和微调，BERT 在多种 NLP 任务上取得了突破性的性能。
- **ELMo（Embeddings from Language Models）**：一种深层双向语言模型，能够生成上下文相关的词嵌入。通过学习词在不同上下文中的不同含义，提高了语义理解的深度。
- **Transformer 和自注意力机制**：Transformer 架构的核心是自注意力机制，它允许模型在处理序列时考虑所有位置的信息。这使得模型能够生成更加丰富和动态的文本表示。
- **句子和文档级别的嵌入技术**：如 Sent2Vec 和 Doc2Vec，这些技术通常基于词嵌入，通过聚合句子或文档中的词向量来生成更高级别的文本表示。
- **池化技术（Pooling Techniques）**：在句子或文档嵌入中，池化技术用于从多个词向量中提取一个代表性的向量。常见的池化方法包括最大池化、平均池化和加权池化。
- **对抗性训练**：通过引入对抗性样本来训练模型，提高模型对噪声和异常值的鲁棒性。
- **多任务学习**：在预训练阶段同时学习多个任务，以提高模型的泛化能力和灵活性。
- **跨语言嵌入（Cross-lingual Embeddings）**：学习不同语言之间的共同表示，以便于跨语言的文本分析和应用。
- **零样本和少样本学习（Zero-shot and Few-shot Learning）**：在没有或只有少量标注数据的情况下，利用预训练的文本嵌入执行任务。

上述文本嵌入技术和方法在不同的应用场景下有各自的优势和局限性。随着深度学习的发展，文本嵌入技术也在不断进步，以更好地适应复杂的语言现象和多样化的应用需求。实例 5-4 的功能是使用 transformers 库来演示使用 BERT 模型生成文本的嵌入表示的过程。

实例 5-4　使用 BERT 模型生成文本的嵌入表示
源码路径：codes/5/wenqian.py

实例文件 wenqian.py 的具体实现代码如下所示。

```python
from transformers import BertTokenizer, BertModel
import torch

# 初始化分词器和模型
tokenizer = BertTokenizer.from_pretrained('bert-base-uncased')
model = BertModel.from_pretrained('bert-base-uncased')

# 要处理的文本
text = "Hello, how are you?"

# 使用分词器将文本转换为分词后的标记序列
encoded_input = tokenizer(text, return_tensors='pt')

# 通过 BERT 模型获取嵌入表示
with torch.no_grad():
    output = model(**encoded_input)

# 获取最后一层的隐藏状态
last_hidden_states = output.last_hidden_state

# 打印输出的嵌入向量
print(last_hidden_states)
```

在上述代码中，首先加载了 BERT 的分词器和模型。接着，将一个简单的问候语文本转换为 BERT 模型可以理解的格式，并通过模型生成嵌入表示。最后，打印输出这些嵌入向量。执行代码后会输出：

```
tensor([[[-0.0824,  0.0667, -0.2880,  ..., -0.3566,  0.1960,  0.5381],
         [ 0.0310, -0.1448,  0.0952,  ..., -0.1560,  1.0151,  0.0947],
         [-0.8935,  0.3240,  0.4184,  ..., -0.5498,  0.2853,  0.1149],
         ...,
         [-0.2812, -0.8531,  0.6912,  ..., -0.5051,  0.4716, -0.6854],
         [-0.4429, -0.7820, -0.8055,  ...,  0.1949,  0.1081,  0.0130],
         [ 0.5570, -0.1080, -0.2412,  ...,  0.2817, -0.3996, -0.1882]]])
```

上面的输出结果是一个包含文本嵌入的张量（tensor），这个张量是多维的，其维度通常是 [batch_size, sequence_length, hidden_size]，其中：

- batch_size 是输入的文本序列的数量，在这个例子中是 1。
- sequence_length 是每个序列中标记的数量，包括实际的词和一些特殊的标记，如序列开始标记[CLS]和序列结束标记[SEP]。
- hidden_size 是模型隐藏层的大小，对于 bert-base-uncased 模型，这个值通常是 768。

注意：本实例使用了 bert-base-uncased 模型，这是一个预训练的 BERT 模型，没有考虑词汇的大小写。在实际应用中，可能会根据需要选择不同的模型和分词器。

5.3.2 基于 CLIP 模型的文本嵌入

实例 5-3 讲解过使用 CLIP 模型实现对图像和文本的联合嵌入操作，展示了使用 CLIP 模型识别图像中内容（例如，判断图像是"猫"还是"狗"的照片）的方法。如果想要演示使用 CLIP 模型实现文本嵌入操作，只需要对相关代码进行些许调整即可。实例 5-5 演示了使用 CLIP 模型实现文本嵌入的过程。

实例 5-5	使用 CLIP 模型实现文本嵌入
	源码路径：codes/5/clwen.py

实例文件 clwen.py 的具体实现代码如下所示。

```python
from transformers import CLIPProcessor, CLIPModel
from PIL import Image
import requests
import torch

# 加载 CLIP 模型和处理器
model = CLIPModel.from_pretrained("openai/clip-vit-base-patch16")
processor = CLIPProcessor.from_pretrained("openai/clip-vit-base-patch16")

# 要嵌入的文本
text = "a photo of a cat"

# 由于 CLIP 模型需要图像输入，我们可以使用一个占位图像
# 这里使用一个空白图像，因为我们只关注文本嵌入
blank_image = Image.new("RGB", (224, 224))

# 将文本和图像转换为模型需要的格式
inputs = processor(text=[text], images=blank_image, return_tensors="pt", padding="max_length", max_length=77)

# 获取文本嵌入
with torch.no_grad():
    outputs = model(**inputs)
    text_embeddings = outputs.text_embeds   # 修正后的属性名称
```

```
# 打印文本嵌入的维度
print("Text embedding shape:", text_embeddings.shape)

# 打印文本嵌入的前几个值
print("Text embedding values:", text_embeddings[0][:10])
```

本实例演示了使用 CLIP 模型生成文本的嵌入表示，并展示了打印出嵌入向量的基本信息的用法。这种嵌入可以用于多种下游任务，如文本相似性度量、分类或多模态匹配任务。上述代码的实现流程如下。

（1）加载 CLIP 模型和处理器。
- CLIPModel.from_pretrained 加载预训练的 CLIP 模型，这里使用的是"openai/clip-vit-base-patch16"，这是一个基于 Vision Transformer 的 CLIP 模型变体。
- CLIPProcessor.from_pretrained 加载与模型配套的处理器，它负责将文本和图像转换为模型所需的格式。

（2）定义要嵌入的文本：定义一个字符串 text，包含我们想要生成嵌入的文本内容。

（3）创建一个空白图像作为占位图像：由于 CLIP 模型需要同时接收文本和图像输入，即使我们只对文本嵌入感兴趣，也需要提供一个图像输入。这里创建了一个 224 像素×224 像素的空白 RGB 图像。

（4）使用处理器准备输入数据：processor 对象的 __call__ 方法被用来处理文本和图像，将它们转换为模型可以理解的格式。text=[text]是一个包含单条文本的列表，images=blank_image 是占位图像，return_tensors="pt"指定输出为 PyTorch 张量，padding="max_length"确保文本序列被填充到最大长度以匹配模型的输入要求。

（5）获取文本嵌入。
- 使用 with torch.no_grad()上下文管理器来禁用梯度计算，这样在推理时可以减少内存使用并提高计算速度。
- 调用模型的 forward 方法（通过**inputs 语法将输入数据解包）来获取输出。
- 从模型输出中获取 text_embeds 属性，这是文本的嵌入表示。

（6）打印输出文本嵌入的维度和值。
- text_embeddings.shape 给出了嵌入向量的维度，这里应该是[1, 512]，表示一个批次大小为 1、维度为 512 的嵌入向量。
- text_embeddings[0][:10]打印了嵌入向量的前 10 个元素，以查看嵌入的具体数值。

执行代码后输出：

```
Text embedding shape: torch.Size([1, 512])
Text embedding values: tensor([ 0.0413, -0.0037,  0.0096,  0.0096, -0.0124,  0.0342,
0.0092, -0.0036, 0.0224, -0.0395])
```

对上面输出结果的具体说明如下。
- Text embedding shape：torch.Size（[1, 512]）表示文本嵌入的维度是[1, 512]。这里 1 表示批次大小（batch size），即我们输入的文本数量是 1 条；512 是文本嵌入的维度，意味着每条文本被转换成一个 512 维的向量。
- Text embedding values: tensor（[0.0413, -0.0037, 0.0096, ... , 0.0224, -0.0395]）显示了生成的文本嵌入向量的前 10 个值。这些值是浮点数，代表了文本在 512 维空间中的坐标。每个值都是通过模型的文本编码器计算得出的，反映了文本在该维度上的特征。

文本嵌入可以用于多种下游任务，如文本相似性度量、文本分类、信息检索等。在多模态应用中，这些文本嵌入可以与图像或音频的嵌入一起使用，以实现跨模态的匹配或联合表示学习。

5.4 音频嵌入

在多模态嵌入表示应用中，音频嵌入通常用于将音频数据转换为低维向量表示，以便执行后续的分析、检索或分类任务。本节将详细讲解音频嵌入的基本知识和用法。

5.4.1 音频特征提取

在多模态嵌入表示应用中，音频特征提取是将原始音频数据转换为计算机可理解的表示形式的关键步骤。这些特征捕捉了音频数据的重要信息，使计算机可以更好地理解和处理音频内容。在多模态嵌入应用中，常用的音频特征的提取方法如下。

1. 时域特征
- 时域波形：是音频数据的原始波形表示，采样率越高，时域波形中包含的信息越丰富。
- 能量：音频信号的能量分布，常用的能量特征包括短时能量（Short-time Energy）和过零率（Zero Crossing Rate）。

2. 频域特征
- 傅里叶变换（FFT）：将时域信号转换为频域表示，得到频谱信息。
- 梅尔频谱图（Mel Spectrogram）：通过将频谱图映射到梅尔频率刻度，提取更符合人类听觉感知的频谱特征。
- 功率谱密度（Power Spectral Density）：频域内各频率成分的能量分布。
- 梅尔频率倒谱系数（MFCC）：基于梅尔频率刻度的频谱包络系数，常用于语音识别和音频分类任务。

3. 时频域特征
- 短时傅里叶变换（STFT）：将音频信号分成短时段，并对每个时段进行傅里叶变换，得到时频谱信息。
- 色度特征：将音频信号转换为音乐色度空间，通常用于与音乐相关的任务。

4. 时序特征
- 声学特征：基于声学事件的时序模式，如语音的音素、音节、语义单元等。
- 时间序列建模：使用循环神经网络或 Transformer 等模型对音频时序数据进行建模，从中提取时序特征。

5. 其他特征
- 语音端点检测：识别语音信号的起始和结束点。
- 声音事件识别（Audio Event Detection）：识别音频中的特定事件或环境声音。

实例 5-6 演示了使用 librosa 库来处理音频文件的过程，本实例提取两种常见的音频特征：梅尔频谱图和 MFCC。

实例 5-6 使用 TFT 处理 CSV 文件中的缺失值
源码路径：codes/5/que.py

实例文件 que.py 的具体实现代码如下所示。

```
import librosa.display
import matplotlib.pyplot as plt
import numpy as np
```

```python
# 定义音频文件路径
audio_file = "888.wav"

# 加载音频文件
y, sr = librosa.load(audio_file)

# 提取梅尔频谱图
mel_spec = librosa.feature.melspectrogram(y=y, sr=sr, n_mels=128, fmax=8000)

# 转换为对数刻度
log_mel_spec = librosa.power_to_db(mel_spec, ref=np.max)

# 提取 MFCC 特征
mfccs = librosa.feature.mfcc(y=y, sr=sr, n_mfcc=20)

# 绘制梅尔频谱图
plt.figure(figsize=(10, 6))
librosa.display.specshow(log_mel_spec, sr=sr, x_axis='time', y_axis='mel')
plt.colorbar(format='%+2.0f dB')
plt.title('Mel Spectrogram')
plt.tight_layout()
plt.show()

# 绘制 MFCC 特征
plt.figure(figsize=(10, 4))
librosa.display.specshow(mfccs, sr=sr, x_axis='time')
plt.colorbar()
plt.title('MFCC')
plt.tight_layout()
plt.show()
```

上述代码的实现流程如下。

（1）定义音频文件路径：变量 audio_file 存储了要处理的音频文件的名称或路径。

（2）加载音频文件：librosa.load 函数加载音频文件，并返回音频信号 y（时间序列）和采样率 sr。

（3）提取梅尔频谱图：函数 librosa.feature.melspectrogram 用于计算音频信号的梅尔频谱图。n_mels=128 指定了梅尔滤波器的数量，fmax=8000 指定了最高频率限制。

（4）转换为对数刻度：函数 librosa.power_to_db 将梅尔频谱图的功率谱转换为对数刻度，这通常可以更好地表示人耳对声音强度的感知。

（5）提取 MFCC 特征：函数 librosa.feature.mfcc 用于计算音频信号的 MFCC 特征。n_mfcc=20 指定了提取的 MFCC 系数的数量。

（6）绘制梅尔频谱图：使用 matplotlib 库绘制梅尔频谱图，librosa.display.specshow 函数用于显示频谱图，plt.colorbar 函数用于添加颜色条以表示数值范围，plt.title 函数用于设置图表标题。

（7）绘制 MFCC 特征：类似于梅尔频谱图，这里绘制 MFCC 特征图。MFCC 通常用于语音识别和音频分类任务。

（8）显示图表：通过 plt.show 函数显示可视化图。

注意：在多模态嵌入应用中，通常会将这些特征提取方法结合到一个流水线中，以获取更全面的音频表示。选择哪些特征提取方法取决于具体的应用场景和任务要求。一般来说，MFCC 和梅尔频谱图是语音相关任务中最常用的特征表示方法，而梅尔频谱图也被广泛用于与音乐相关的任务。

5.4.2 常用音频嵌入模型

在现实应用中，常用的音频嵌入模型包括传统的基于特征提取的模型以及近年来基于深度学习的大模型和预训练模型。

1. 传统模型

- MFCC：基于人类听觉系统对声音的感知，能够提取声音频谱的特征，常用于语音识别和音频分类等任务。
- 梅尔频谱图：将声音波形图转换为梅尔频谱图，基于频谱的特征表示，在音频处理任务中广泛使用。
- Spectrogram+CNN：使用卷积神经网络（CNN）处理声谱图像，提取特征，可以在分类、检测等任务中使用。

2. 深度学习模型

- VGGish：基于 VGG 网络结构，专门用于音频分类和音频特征提取，可以提取高层次的语义特征。
- OpenL3：基于深度卷积神经网络，提取音频的时频特征。提供了预训练模型，支持多种任务，如音频检索、分类和相似性匹配等。
- YAMNet：由 Google 开发的轻量级模型，用于音频事件识别，能够识别数百种常见的音频事件。

3. 预训练模型

- Wave2Vec：由 Facebook 提出的预训练模型，用于学习音频表示，利用自监督任务从未标记的音频数据中学习表示。
- VGGVox：基于 VGG 网络的预训练模型，用于说话人识别和语音验证，在说话人识别任务中表现出色。
- DeepSpeech：由 Mozilla 开发的端到端语音识别模型，基于深度学习，可以实现将语音转换为文本的任务。
- Wav2Vec2：Hugging Face 提出的预训练模型，用于自监督学习任务。能够学习音频的有用表示，适用于多种语音任务。

上面列出的这些模型可以用于各种音频处理任务，如音频分类、说话人识别、语音转文本等。选择模型时，可以根据任务的复杂度、数据量以及计算资源等因素进行选择。

实例 5-7 的功能是从指定的 WAV 文件中转录语音，并利用预训练的 Wav2Vec2.0 模型进行音频转文本操作，最后将转录结果保存到文本文件中。

实例 5-7　将指定音频文件的内容转换为文本
源码路径：codes/5/zhuan/transcribe.py

实例文件 transcribe.py 的具体实现代码如下所示。

```python
import argparse
import soundfile as sf
import torch
from transformers import Wav2Vec2ForCTC, Wav2Vec2Processor
from pydub import AudioSegment
import os
from tqdm import tqdm


def convert_and_split_audio(input_file, output_dir, chunk_length_ms=1*60*1000,target_sample_rate=16000):
    audio = AudioSegment.from_file(input_file)

    # 转换采样率
    audio = audio.set_frame_rate(target_sample_rate)
```

```python
    # 将音频分成多个块
    chunks = [audio[i:i + chunk_length_ms] for i in range(0, len(audio), chunk_length_ms)]

    output_files = []
    for idx, chunk in enumerate(chunks):
        chunk_name = os.path.join(output_dir, f"{os.path.splitext(os.path.basename(input_file))[0]}_chunk{idx}.wav")
        chunk.export(chunk_name, format="wav")
        output_files.append(chunk_name)

    return output_files


def main():
    # 设置参数解析器
    parser = argparse.ArgumentParser(description="使用Wav2Vec2.0 将WAV文件中的语音转录成文本。")
    parser.add_argument(
        "wav_file",
        type=str,
        nargs='?',
        default="default.wav",
        help="要转录的WAV文件路径（默认：'default.wav'）"
    )
    parser.add_argument(
        "--model_name",
        type=str,
        default="facebook/wav2vec2-large-xlsr-53",
        help="要使用的预训练模型名称（默认：'facebook/wav2vec2-large-xlsr-53'）"
    )
    args = parser.parse_args()

    # 转换并分割音频文件
    output_dir = "chunks"
    os.makedirs(output_dir, exist_ok=True)
    chunk_files = convert_and_split_audio(args.wav_file, output_dir)

    # 加载预训练模型和处理器
    try:
        processor = Wav2Vec2Processor.from_pretrained(args.model_name)
        model = Wav2Vec2ForCTC.from_pretrained(args.model_name)
    except Exception as e:
        print(f"加载模型时出错：{e}")
        return

    transcriptions = []

    for chunk_file in tqdm(chunk_files, desc="处理音频块"):
        # 加载音频文件
        try:
            audio_input, sample_rate = sf.read(chunk_file)
        except Exception as e:
            print(f"读取音频文件 {chunk_file} 时出错：{e}")
            continue

        # 确保音频采样率为16kHz，这是模型的预期值
        if sample_rate != 16000:
            print(f"文件 {chunk_file} 的采样率应为16000，但实际为 {sample_rate}")
            continue

        # 预处理音频文件
        input_values = processor(audio_input, sampling_rate=sample_rate, return_tensors="pt").input_values
```

```
        # 进行推理
        try:
            with torch.no_grad():
                logits = model(input_values).logits

            # 获取预测的 ID
            predicted_ids = torch.argmax(logits, dim=-1)

            # 将 ID 解码成文本
            transcription = processor.decode(predicted_ids[0])
            transcriptions.append(transcription)
        except Exception as e:
            print(f"文件 {chunk_file} 推理过程中出错: {e}")

    # 将转录结果保存到文本文件
    with open("transcriptions.txt", "w") as f:
        for idx, transcription in enumerate(transcriptions):
            f.write(f"块 {idx}:\n{transcription}\n\n")

    print("转录结果已保存到 transcriptions.txt")

if __name__ == "__main__":
    main()
```

上述代码的实现流程如下。

（1）通过解析命令行参数获取输入的 WAV 文件路径和预训练模型的名称，若未提供参数，则使用默认的 WAV 文件和模型。

（2）加载音频文件，并将其转换为目标采样率，并分割成每块 1 分钟的音频段。

（3）加载指定的预训练 Wav2Vec2 模型和处理器，对每个音频块进行预处理、推理，生成转录文本。

（4）将所有转录结果保存到一个文本文件中。

5.5 多模态图像搜索引擎

本项目是一个基于 CLIP 大模型的图像搜索引擎（见实例 5-8），针对孟加拉语文本描述进行图像检索。它包括了训练 CLIP 模型、准备数据集、构建图像搜索 Web 应用程序等功能，用户可以通过输入文本描述来搜索与描述相匹配的图像，并查看相似度分数。

实例 5-8　基于 CLIP 模型的文搜图系统
源码路径：codes/5/bangla-CLIP

5.5.1 项目介绍

本项目是一个基于多模态学习的图像搜索引擎，利用 CLIP 模型实现文本描述到图像的检索功能。通过将文本描述和图像进行向量化表示，实现了跨模态的语义匹配，用户可以通过输入文本描述来搜索与描述相符的图像。

1. 功能特点

- 多模态学习：使用 CLIP 模型，同时训练图像编码器和文本编码器，使其能够理解文本和图像之间的语义关联。
- 图像搜索：用户可以输入文本描述，系统将返回与描述相匹配的图像，并显示相似度分数。

- Web 应用程序：提供了基于 Streamlit 搭建的简单易用的 Web 用户界面，使用户能够方便地进行交互式搜索。

2. 技术细节
- CLIP 模型：使用 CLIP 模型进行文本描述和图像的语义匹配，该模型同时训练了一个图像编码器和一个文本编码器。
- 数据集准备：使用多个孟加拉语图像—文本数据集进行模型训练和图像搜索。
- 图像处理：使用 OpenCV 和 Albumentations 库对图像进行预处理，以便输入到 CLIP 模型中。
- Web 应用程序：使用 Streamlit 构建了用户友好的 Web 应用程序，实现了文本输入和图像展示功能。

3. 项目结构
- CLIP_model.py：定义了 CLIP 模型的架构，并实现了图像和文本的编码功能。
- dataset.py：数据集处理模块，用于加载和处理训练数据。
- prepare_datafiles.py：数据预处理脚本，用于准备训练所需的数据文件。
- image_search.py：图像搜索功能的实现，包括加载模型、图像处理和搜索功能。
- app.py：Web 应用程序的主要代码，提供了用户界面和交互功能。
- bang_prepare_dataset.py：准备孟加拉语数据集的脚本。

4. 技术栈
- PyTorch：深度学习框架，用于构建和训练 CLIP 模型。
- Streamlit：用于构建 Web 应用程序的 Python 库。
- OpenCV：用于图像处理和加载的库。
- Albumentations：用于图像增强的库。

5. 应用场景
- 图像搜索引擎：用户可以通过输入自然语言描述来搜索图像，适用于图片检索、电子商务等领域。
- 跨模态语义匹配：可以用于多模态内容的语义匹配，如图像标注、推荐系统等。

综上所述，通过该项目，用户可以体验到基于最新多模态大模型的图像搜索技术，实现更加智能和直观的图像检索。

5.5.2 CLIP 模型的配置参数

文件 config.py 定义了训练 Bangla CLIP 模型所需的配置参数，包括数据路径、训练超参数、模型和设备配置等。通过修改这些参数，可以调整模型的训练行为，例如批量大小、学习率、模型结构等。这些配置确保训练过程中的各种需求和设置得到满足，使模型能够有效地进行训练和验证。

```
import torch

debug = True    # 调试模式开关
image_path = ""    # 图像路径
captions_path = ""    # 标注路径

dataset_root = "/data/comps"    # 数据集根目录
train_json = "/data/comps/train.json"    # 训练集 JSON 文件路径
val_json = "/data/comps/val.json"    # 验证集 JSON 文件路径

batch_size = 200    # 批量大小
num_workers = 5    # 数据加载的工作线程数
head_lr = 1e-3    # 头部学习率
image_encoder_lr = 1e-4    # 图像编码器学习率
```

5.5 多模态图像搜索引擎

```
text_encoder_lr = 1e-5    # 文本编码器学习率
weight_decay = 1e-3    # 权重衰减
patience = 2    # 早停的耐心值
factor = 0.8    # 学习率衰减因子
epochs = 350    # 训练周期数

gpu = 1    # 使用的 GPU 编号
device = torch.device(f"cuda:{gpu}" if torch.cuda.is_available() else "cpu")    # 设备设置

# model_name = 'efficientnet_b2'
# image_embedding = 1408

model_name = 'resnet50'    # 图像编码器模型名称
image_embedding = 1000    # 图像嵌入维度

text_encoder_model = "neuralspace-reverie/indic-transformers-bn-bert"    # 文本编码器模型名称
text_tokenizer = "neuralspace-reverie/indic-transformers-bn-bert"    # 文本分词器
max_length = 100    # 文本最大长度

model_tag = f"{model_name}_{text_encoder_model.replace('/', '_')}_aug"    # 模型标签
log_tag = model_tag    # 日志标签

pretrained = True    # 是否使用预训练模型（图像和文本编码器）
trainable = True    # 是否训练模型（图像和文本编码器）
temperature = 1.0    # 温度参数

# 图像尺寸
size = 224

# 投影头
num_projection_layers = 1    # 投影层数
projection_dim = 256    # 投影维度
dropout = 0.1    # dropout 率
```

5.5.3 数据集处理

本项目使用了一个包含图像和文本对的数据集，这是一个专门为训练 Bangla CLIP 模型准备的孟加拉语图像-文本对数据集。该数据集包含了多个图像文件夹，每个图像都有一个对应的文本描述，元数据通过 JSON 文件进行管理。

（1）文件 bang_prepare_dataset.py 用于准备 Bangla CLIP 模型的训练和验证数据集，从多个来源的数据集中读取图像和对应的文本描述。这些数据集组成训练和验证数据，并保存为 CSV 文件。

```
import json
from pathlib import Path
from tqdm import tqdm
import pandas as pd

train_dataframe = {}
train_dataframe['caption'] = []
train_dataframe['image'] = []

valid_dataframe = {}
valid_dataframe['caption'] = []
valid_dataframe['image'] = []

# 从 Mendeley 获取的数据集
with open('/data/captions.json', encoding='utf-8') as fh:
    data = json.load(fh)

trn_split = int(0.8 * len(data))
```

```python
for sample in tqdm(data[:trn_split]):
    fn = sample['filename']
    cp = sample['caption']
    my_file = Path(f"/data/images/{fn}")
    if my_file.is_file():
        for tc in cp:
            tc = tc.replace(',', ' ')
            train_dataframe['caption'].append(tc)
            train_dataframe['image'].append(f"/data/images/{fn}")

for sample in tqdm(data[trn_split:]):
    fn = sample['filename']
    cp = sample['caption']
    my_file = Path(f"/data/images/{fn}")
    if my_file.is_file():
        for vc in cp:
            vc = vc.replace(',', ' ')
            valid_dataframe['caption'].append(vc)
            valid_dataframe['image'].append(f"/data/images/{fn}")

# 从 Kaggle 获取的 BNature 数据集
lines = open("/data/caption/caption.txt", "r").readlines()

trn_split = int(0.8 * len(lines))
for line in tqdm(lines[:trn_split]):
    fn = [x.strip() for x in line.split()][0]
    cp = ' '.join([x.strip() for x in line.split()][1:])
    my_file = Path(f"/data/Pictures/{fn}")
    if my_file.is_file():
        cp = cp.replace(',', ' ')
        train_dataframe['caption'].append(cp)
        train_dataframe['image'].append(f"/data/Pictures/{fn}")

for line in tqdm(lines[trn_split:]):
    fn = [x.strip() for x in line.split()][0]
    cp = ' '.join([x.strip() for x in line.split()][1:])
    my_file = Path(f"/data/Pictures/{fn}")
    if my_file.is_file():
        cp = cp.replace(',', ' ')
        valid_dataframe['caption'].append(cp)
        valid_dataframe['image'].append(f"/data/Pictures/{fn}")

# flickr8k 数据集的孟加拉语翻译
ban_caps = pd.read_csv("./../BAN-Cap_captiondata.csv")
cap_ids = list(ban_caps['caption_id'])
ban_trans = list(ban_caps['bengali_caption'])
trn_split = int(0.8 * len(cap_ids))
for j in tqdm(range(len(cap_ids[:trn_split]))):
    ci = cap_ids[j].split("#")[0]
    bt = ban_trans[j]
    fn = ci
    cp = bt
    my_file = Path(f"/data/flickr8k_images/{fn}")
    if my_file.is_file():
        cp = cp.replace(',', ' ')
        train_dataframe['caption'].append(cp)
        train_dataframe['image'].append(f"/data/flickr8k_images/{fn}")

for j in tqdm(range(len(cap_ids[trn_split:]))):
    ci = cap_ids[j].split("#")[0]
    bt = ban_trans[j]
    fn = ci
```

```
        cp = bt
        my_file = Path(f"/data/flickr8k_images/{fn}")
        if my_file.is_file():
            cp = cp.replace(',', ' ')
            valid_dataframe['caption'].append(cp)
            valid_dataframe['image'].append(f"/data/flickr8k_images/{fn}")

# 将数据转换为训练集和验证集并保存为 CSV 文件
train_dataframe = pd.DataFrame(train_dataframe)
valid_dataframe = pd.DataFrame(valid_dataframe)

print(train_dataframe.head())
print(valid_dataframe.head())

train_dataframe.to_csv('train_df_bang.csv', index=None)
valid_dataframe.to_csv('valid_df_bang.csv', index=None)
```

本项目的数据集来源如下。
- 从 Mendeley 获取的数据集，其中包含图像文件和对应的 JSON 格式的标注。
- 从 Kaggle 获取的 BNature 数据集，其中包含图像文件和对应的文本描述。
- flickr8k 数据集的孟加拉语翻译，数据保存在 BAN-Cap_captiondata.csv 文件中。

将数据集分为训练集和验证集，并保存为 CSV 文件以供后续使用。

（2）文件 prepare_datafiles.py 用于处理和组织 Bangla CLIP 模型的训练和验证数据集。它从指定的文件夹中读取图像文件及其对应的文本描述，将它们匹配并存储在 Pandas 数据帧中，最后保存为 CSV 文件。这样，数据集可以方便地用于模型的训练和验证。

```
import json
from glob import glob
from tqdm import tqdm
import config as CFG
import os
import pandas as pd

train_folders = ["train_text_img_pairs_0_compressed", "train_text_img_pairs_1_compressed",
"train_text_img_pairs_2_compressed", "train_text_img_pairs_3_compressed", "train_text_img_pairs_4_compressed",
"train_text_img_pairs_5_compressed", "train_text_img_pairs_6_compressed", "train_text_img_pairs_7_compressed",
"train_text_img_pairs_8_compressed"]
caption_labels = {}
train_json = json.load(open(CFG.train_json))

for tj in train_json:
    cap = tj['caption']    # 标注
    pro = tj['product']    # 产品
    caption_labels[pro] = cap

train_dataframe = {}
train_dataframe['caption'] = []
train_dataframe['image'] = []
print("Processing training images.")    # 处理训练图像
for tf in train_folders:
    print(tf)
    images_tf = list(glob(f"{CFG.dataset_root}/{tf}/*"))
    print("loading done")    # 加载完成
    for img_p in tqdm(images_tf):
        cap = caption_labels[os.path.basename(img_p)]
```

```python
        train_dataframe['caption'].append(cap)
        train_dataframe['image'].append(img_p)


val_folders = ["val_imgs"]
caption_labels = {}
val_json = json.load(open(CFG.val_json))

for tj in val_json:
    cap = tj['caption']    # 标注
    pro = tj['product']    # 产品
    caption_labels[pro] = cap
valid_dataframe = {}

valid_dataframe['caption'] = []
valid_dataframe['image'] = []
for tf in val_folders:
    print(tf)
    images_tf = list(glob(f"{CFG.dataset_root}/{tf}/*"))
    print("loading done")    # 加载完成
    for img_p in tqdm(images_tf):
        cap = caption_labels[os.path.basename(img_p)]

        valid_dataframe['caption'].append(cap)
        valid_dataframe['image'].append(img_p)


print(valid_dataframe)

train_dataframe = pd.DataFrame(train_dataframe)
valid_dataframe = pd.DataFrame(valid_dataframe)

print(train_dataframe.head())
print(valid_dataframe.head())

train_dataframe.to_csv('train_df.csv', index = None)   # 保存训练数据
valid_dataframe.to_csv('valid_df.csv', index = None)   # 保存验证数据
```

（3）文件 dataset.py 首先定义了一个自定义数据集类 CLIPDataset，用于加载和预处理图像及其对应的文本描述；然后定义了一个图像变换函数 get_transforms，用于在训练和验证过程中应用数据增强和归一化操作，提升模型的泛化能力和性能。

```python
import cv2
import torch
import albumentations as A

import config as CFG
from normalizer import normalize

class CLIPDataset(torch.utils.data.Dataset):
    def __init__(self, image_filenames, captions, tokenizer, transforms):
        """
        Bangla CLIP 数据集
        """
        self.image_filenames = image_filenames
        self.captions = [normalize(cap_sen) for cap_sen in list(captions)]   # 标准化文本描述

        self.transforms = transforms

    def __getitem__(self, idx):
        image = cv2.imread(f"{self.image_filenames[idx]}")   # 读取图像
        image = cv2.cvtColor(image, cv2.COLOR_BGR2RGB)        # 转换颜色空间
```

```python
            image = self.transforms(image=image)['image']  # 应用变换
            image = torch.tensor(image).permute(2, 0, 1).float()  # 转换为张量并调整维度
            caption = self.captions[idx]

            return image, caption

    def __len__(self):
        return len(self.captions)  # 返回数据集大小

def get_transforms(mode="train"):
    if mode == "train":
        config = {
            'aug_prob': 0.2  # 增强操作的概率
        }
        return A.Compose(
            [
                A.HueSaturationValue(hue_shift_limit=0.2, sat_shift_limit=0.2, val_
                    shift_limit=0.2, p=config['aug_prob']),
                A.RandomBrightnessContrast(brightness_limit=(-0.1, 0.1), contrast_
                    limit=(-0.1, 0.1), p=config['aug_prob']),
                A.CoarseDropout(p=config['aug_prob']),
                A.GaussNoise(p=config['aug_prob']),
                A.ZoomBlur(p=config['aug_prob']),
                A.RandomFog(p=config['aug_prob']),
                A.Rotate((-20., 20.), p=0.5),
                A.MotionBlur(p=config['aug_prob']),
                A.Resize(CFG.size, CFG.size, always_apply=True),  # 调整图像大小
                A.Normalize(max_pixel_value=255.0, always_apply=True),  # 归一化
            ]
        )
    else:
        return A.Compose(
            [
                A.Resize(CFG.size, CFG.size, always_apply=True),  # 调整图像大小
                A.Normalize(max_pixel_value=255.0, always_apply=True),  # 归一化
            ]
        )
```

5.5.4 实现 Bangla CLIP 模型

文件 CLIP_model.py 定义了一个用于实现 Bangla CLIP 模型的 CLIPModel 类，其中包含图像编码器和文本编码器。通过前向传播方法 forward 获取图像和文本的特征向量输出，同时提供了获取图像和文本特征向量的辅助方法，并包含了用于测试模型的代码。

```python
import torch
from torch import nn
import torch.nn.functional as F
from torchvision import models, transforms
from transformers import AutoTokenizer, AutoModel
import config as CFG
import cv2

class CLIPModel(nn.Module):
    """CLIP model for Bangla"""
    def __init__(self):
        super(CLIPModel, self).__init__()
        # 图像编码器
        self.image_encoder = models.efficientnet_b2(weights="EfficientNet_B2_Weights.DEFAULT")
         # 使用 EfficientNet-B2 模型作为图像编码器
        self.image_encoder.fc = nn.Identity()  # 移除最后一层全连接层
```

```python
        # 图像输出层
        self.image_out = nn.Sequential(
            nn.Linear(CFG.image_embedding, 256),  # 线性层将图像特征维度映射到 256 维
            nn.ReLU(),  # ReLU 激活函数
            nn.Linear(256, 256)  # 再次映射到 256 维
        )

        # 文本编码器
        self.text_encoder = AutoModel.from_pretrained(CFG.text_encoder_model)
        # 使用预训练的文本编码器模型
        self.target_token_idx = 0  # 目标 token 的索引

        # 文本输出层
        self.text_out = nn.Sequential(
            nn.Linear(768, 256),  # 线性层将文本特征维度映射到 256 维
            nn.ReLU(),  # ReLU 激活函数
            nn.Linear(256, 256)  # 再次映射到 256 维
        )

    def forward(self, image, text, mask):
        # 图像编码
        image_vec = self.image_encoder(image)  # 获取图像特征向量
        image_vec = self.image_out(image_vec)  # 图像特征向量经过输出层处理

        # 文本编码
        text_out = self.text_encoder(text, mask)  # 获取文本特征向量
        last_hidden_states = text_out.last_hidden_state
        last_hidden_states = last_hidden_states[:, self.target_token_idx, :]
        # 取出目标 token 的特征
        text_vec = self.text_out(last_hidden_states.view(-1, 768))  # 文本特征向量经过输出层处理

        return image_vec, text_vec

    def get_image_embeddings(self, image):
        # 获取图像特征向量
        image_vec = self.image_encoder(image)
        image_vec = self.image_out(image_vec)

        return image_vec

    def get_text_embeddings(self, text, mask):
        # 获取文本特征向量
        text_out = self.text_encoder(text, mask)
        last_hidden_states = text_out.last_hidden_state
        last_hidden_states = last_hidden_states[:, self.target_token_idx, :]
        text_vec = self.text_out(last_hidden_states.view(-1, 768))

        return text_vec

# 测试模型
if __name__ == '__main__':
    device = torch.device("cuda" if torch.cuda.is_available() else "cpu")

    images = torch.randn(40, 3, 224, 224).to(device)  # 40 张随机图像
    input_ids = torch.randint(5, 300, size=(40, 200)).to(device)
    # 40 个样本,每个样本长度为 200 的随机文本序列
    attention_mask = torch.ones(40, 200).to(device)  # 掩码,全为 1

    print("Building CLIP")
    clip_model = CLIPModel().to(device)  # 构建 CLIP 模型
    print(clip_model)

    img_vec, text_vec = clip_model(images, input_ids, attention_mask)  # 前向传播
```

```
        print(img_vec.shape)    # 输出图像特征向量形状
        print(text_vec.shape)   # 输出文本特征向量形状
```

上述代码的实现流程如下。

（1）图像编码器：使用 EfficientNet-B2 模型进行图像编码，并通过线性层将图像特征映射到 256 维。

（2）文本编码器：使用预训练的文本编码器模型，将文本序列编码为文本特征向量，并通过线性层将其映射到 256 维。

（3）前向传播方法：将图像和文本输入模型，获取它们的特征向量输出。

（4）辅助方法：提供获取图像和文本特征向量的方法。

（5）测试模块：在 if __name__ == '__main__': 中进行模型的测试，输出图像和文本特征向量的形状。

5.5.5　基于文本的图像搜索

文件 image_search.py 实现了图像搜索功能，通过给定的文本描述，在图像库中搜索与之相匹配的图像。主要包括如下功能。

- 函数 load_model 加载了预训练的 CLIP 模型和相应的分词器。
- 函数 process_image 用于处理图像，包括调整大小和归一化等操作。
- 函数 process_text 用于处理文本，包括分词和填充。
- 函数 search_images 实现了图像搜索功能，根据给定的文本描述在图像库中搜索与之相匹配的图像，并返回最相似的图像文件名和相似度分数。

文件 image_search.py 的具体实现代码如下所示。

```python
from transformers import AutoTokenizer
import config as CFG
from CLIP_model import CLIPModel
import cv2
import os
import torch
from glob import glob
import albumentations as A
import torch.nn.functional as F

def load_model(device, model_path):
    """加载模型和分词器"""
    model = CLIPModel().to(device)  # 加载 CLIP 模型
    model.load_state_dict(torch.load(model_path, map_location=device))  # 加载预训练的模型参数

    tokenizer = AutoTokenizer.from_pretrained(CFG.text_tokenizer)  # 加载文本分词器
    return model, tokenizer

def process_image(img_path):
    """处理图像"""
    imgs = []
    for ip in img_path:
        transforms_infer = A.Compose(
            [
                A.Resize(CFG.size, CFG.size, always_apply=True),  # 调整图像大小
                A.Normalize(max_pixel_value=255.0, always_apply=True),  # 归一化
            ])
        image = cv2.imread(ip)
        image = cv2.cvtColor(image, cv2.COLOR_BGR2RGB)
        image = transforms_infer(image=image)['image']
        image = torch.tensor(image).permute(2, 0, 1).float()  # 转换为 PyTorch 张量，并进行维度转换
        imgs.append(image)
```

```
        imgs = torch.stack(imgs)    # 将多个图像张量堆叠成一个张量
        return imgs

def process_text(caption, tokenizer):
    """处理文本"""
    caption = tokenizer(caption, padding=True)    # 使用分词器对文本进行分词,并进行填充
    e_text = torch.Tensor(caption["input_ids"]).long()    # 获取输入文本的 token ID
    mask = torch.Tensor(caption["attention_mask"]).long()    # 获取文本的注意力掩码
    return e_text, mask

def search_images(search_text, image_path, k=1):
    """在图像库中搜索与给定文本相匹配的图像"""
    device = torch.device("cuda" if torch.cuda.is_available() else "cpu")    # 检测可用设备
    model, tokenizer = load_model(device, model_path="models/clip_bangla.pt")    # 加载模型和分词器
    image_filenames = glob(image_path + "/*.jpg") + glob(image_path + "/*.JPEG") + \
                      glob(image_path + "/*.JPG") + glob(image_path + "/*.png") + glob(image_path + "/*.bmp")
                        # 获取图像文件名列表
    imgs = process_image(image_filenames)    # 处理图像

    if type(search_text) != list:
        search_text = [search_text]
    e_text, mask = process_text(search_text, tokenizer)    # 处理搜索文本

    with torch.no_grad():
        imgs = imgs.to(device)
        e_text = e_text.to(device)
        mask = mask.to(device)
        img_embeddings = model.get_image_embeddings(imgs)    # 获取图像特征向量
        text_embeddings = model.get_text_embeddings(e_text, mask)    # 获取文本特征向量
        image_embeddings_n = F.normalize(img_embeddings, p=2, dim=-1)    # 对图像特征向量进行 L2 归一化
        text_embeddings_n = F.normalize(text_embeddings, p=2, dim=-1)    # 对文本特征向量进行 L2 归一化
        dot_similarity = text_embeddings_n @ image_embeddings_n.T    # 计算文本与图像之间的相似度
    top_k_vals, top_k_indices = torch.topk(dot_similarity.detach().cpu(), min(k, len(image_filenames)))
    # 获取前 k 个最相似的图像
    top_k_vals = top_k_vals.flatten()
    top_k_indices = top_k_indices.flatten()

    images_ret = []
    scores_ret = []
    for i in range(len(top_k_indices)):
        images_ret.append(image_filenames[int(top_k_indices[i])])    # 获取最相似的图像文件名
        scores_ret.append(float(top_k_vals[i]))    # 获取相似度分数

    return images_ret, scores_ret, top_k_vals, top_k_indices
```

5.5.6 基于 Streamlit 的 Web 客户端

文件 app.py 是一个基于 Streamlit 的应用程序,允许用户输入文本描述,并在图像库中搜索与描述相匹配的图像。主要功能如下。

- 提供一个输入框,用户可以输入希望搜索的图像描述。
- 用户可以通过滑块选择要显示的搜索结果数量。
- 根据用户输入的文本描述,在图像库中搜索相似的图像。
- 显示搜索结果及其相似度分数。

```
import streamlit as st
from image_search import load_model, process_image, process_text, search_images

# 设置页面配置
```

```
st.set_page_config(
    page_title="Bangla CLIP Search",
    page_icon="chart_with_upwards_trend"
)

# 自定义页面样式
st.markdown(
    """
<style>
#introduction {
    padding: 10px 20px 10px 20px;
    background-color: #aad9fe;
    border-radius: 10px;
}

#introduction p {
    font-size: 1.1rem;
    color: #050e14;
}

img {
    padding: 5px;
}
</style>
""",
    unsafe_allow_html=True,
)

# 隐藏 Streamlit 默认样式
hide_streamlit_style = """
<style>
#MainMenu {visibility: hidden;}
footer {visibility: hidden;}
</style>
"""
st.markdown(hide_streamlit_style, unsafe_allow_html=True)

# 页面标题和介绍
st.markdown("# বাংলা CLIP সার্চ ইঞ্জিন ")  # 孟加拉语，翻译成中文是：孟加拉语 CLIP 搜索引擎
st.markdown("""---""")
st.markdown(
    """
<div id="introduction">
<p>
</p>
</div>
""",
    unsafe_allow_html=True,
)
st.markdown("""---""")

# 输入搜索文本
text_query = st.text_input(":mag_right: Search Images /

st.markdown("""---""")

# 选择结果数目
number_of_results = st.slider("Number of results ", 1, 100, 10)

st.markdown("""---""")

# 在图像库中搜索并显示结果
ret_imgs, ret_scores, _, _ = search_images(text_query, "demo_images/", k=number_of_results)
```

```
st.markdown("<div style='align: center; display: flex'>", unsafe_allow_html=True)
st.image([str(result) for result in ret_imgs], caption=["Score: " + str(r_s) for r_s in ret_sco
res],width=230)
st.markdown("</div>", unsafe_allow_html=True)
```

执行代码后可以输入文本搜索指定的图像,例如搜索老虎的执行结果如图 5-1 所示。

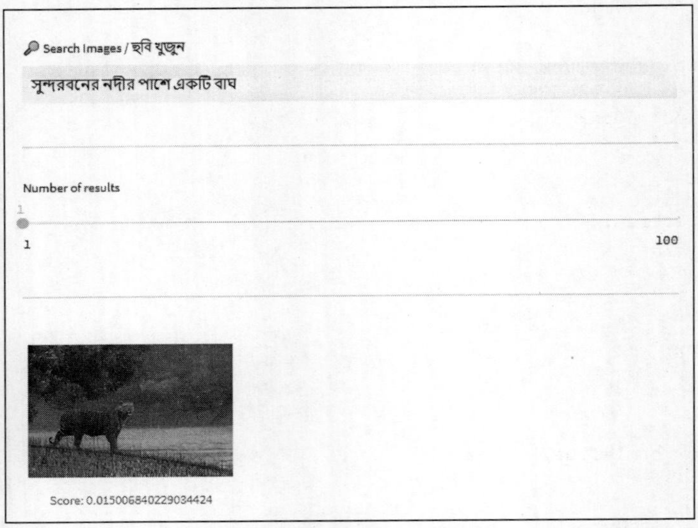

图 5-1　搜索老虎的执行结果

第 6 章　多模态大模型的训练

多模态大模型的训练是一种机器学习方法，旨在结合和处理来自不同模态（例如文本、图像、音频、视频等）的数据。这种方法通过整合多种数据源，可以提高模型的理解和表现能力。多模态模型在图像描述生成、视频内容分析和多模态情感识别等应用中表现出色。本章将详细讲解多模态模型训练的知识和用法。

6.1 模型训练的过程

在多模态大模型应用中，模型训练涉及将来自不同模态的数据（如文本、图像、音频、视频等）进行融合，以构建能够理解和处理多种信息源的统一模型。多模态大模型的训练过程如下。

1. 数据预处理与准备
（1）数据收集和预处理。
- 文本：进行分词、去停用词、词嵌入等处理。
- 图像：进行尺寸调整、归一化、数据增强等处理。
- 音频：进行降噪、特征提取（如 MFCC、梅尔频谱）等处理。
- 视频：进行关键帧提取、降采样、特征提取（如 3D 卷积神经网络）等处理。

（2）数据对齐：确保不同模态的数据在时间或语义上的同步和配对。

2. 特征提取
通过使用专门的模型，对不同模态的数据进行特征提取。
- 文本：使用 BERT、GPT 等模型提取文本特征。
- 图像：使用卷积神经网络如 ResNet、EfficientNet 提取图像特征。
- 音频：使用卷积神经网络或递归神经网络提取音频特征。
- 视频：使用时空卷积网络或长短期记忆网络提取视频特征。

3. 特征融合
将不同模态的特征进行融合，常用的方法如下。
- 拼接：将不同模态的特征向量直接拼接在一起。
- 加权平均：对不同模态的特征进行加权平均。
- 注意力机制：使用注意力机制对不同模态的特征进行加权和选择。
- 图神经网络：在多模态特征之间构建图结构进行融合。

4. 模型架构设计
设计能够处理多模态输入的模型架构，可用的架构方式如下。
- 单塔结构（Single Tower Architecture）：将所有模态的输入通过一个统一的网络进行处理。
- 双塔结构（Dual Tower Architecture）：不同模态的数据通过各自的子网络进行处理，然后在高层进行融合。
- 多塔结构（Multi-Tower Architecture）：每种模态的数据通过各自的网络进行处理，最后在高层进行综合。

5. 模型训练

（1）设计损失函数。
- 根据具体任务选择适当的损失函数，如分类任务中的交叉熵损失、回归任务中的均方误差等。
- 对于多任务学习，可以采用多任务损失函数，将多个任务的损失进行加权求和。

（2）优化算法。
- 使用优化算法如 Adam、SGD 等对模型参数进行优化。
- 调整学习率、批量大小等超参数，以提高训练效果。

（3）正则化与防过拟合。使用 Dropout、Batch Normalization 等技术防止过拟合，也可以使用数据增强和早停（Early Stopping）技术防过拟合策略。

6. 模型评估与调优
- 评估指标：根据具体任务选择适当的评估指标，如准确率、精确率、召回率、F1 分数等。
- 模型调优：通过交叉验证和超参数搜索（如网格搜索、随机搜索）对模型进行调优。结合验证集的表现，调整模型架构和训练策略。

7. 部署与推理
- 部署：将训练好的多模态模型部署到适当的环境中，如云服务、本地服务器、移动设备等。
- 推理：使用训练好的模型进行推理，处理多模态输入并输出预测结果。

通过上述步骤，多模态大模型可以有效地融合和处理来自不同模态的数据，提高模型理解和处理复杂信息的能力，从而在各种任务中表现出色。

6.2 训练策略

在多模态大模型的训练过程中，采用适当的训练策略至关重要，这些策略有助于提高模型的性能、稳定性和泛化能力。通过训练策略可以有效地提升多模态大模型的性能，使其能够更好地处理复杂的多模态任务。

6.2.1 预训练与微调

预训练（Pre-training）是指在大规模单一模态数据上进行预训练，这种方法利用大规模数据的优势，为模型提供了良好的初始参数。例如在大量的图像上预训练一个卷积神经网络，或者在大量的文本上预训练一个语言模型（如 BERT、GPT）。

微调（Fine-tuning）是指在多模态数据上对预训练的模型进行微调，以适应特定的任务。微调过程通常使用较小的学习率，以避免模型参数的剧烈变化。

下面的实例展示了一个多模态模型的训练过程。该模型结合了图像和文本信息，使用预训练的 ResNet 模型提取图像特征，使用预训练的 BERT 模型提取文本特征，并将两者融合后进行分类任务。实例 6-1 对图像模型进行了微调，而文本模型则保持预训练参数不变。

实例 6-1 实现对模型的预训练和微调操作

源码路径：codes/6/ yuwei.py

实例文件 yuwei.py 的具体实现代码如下所示。

```
import torch
```

```python
import torch.nn as nn
import torch.optim as optim
from torchvision import models, transforms
from transformers import BertTokenizer, BertModel
from torch.utils.data import Dataset, DataLoader
from PIL import Image

# 定义数据集
class MultimodalDataset(Dataset):
    def __init__(self, image_paths, texts, labels, transform=None):
        self.image_paths = image_paths
        self.texts = texts
        self.labels = labels
        self.transform = transform
        self.tokenizer = BertTokenizer.from_pretrained('bert-base-uncased')

    def __len__(self):
        return len(self.labels)

    def __getitem__(self, idx):
        image = Image.open(self.image_paths[idx]).convert("RGB")
        if self.transform:
            image = self.transform(image)

        text = self.texts[idx]
        encoding = self.tokenizer.encode_plus(
            text,
            add_special_tokens=True,
            max_length=128,
            return_token_type_ids=False,
            padding='max_length',
            return_attention_mask=True,
            return_tensors='pt',
        )
        input_ids = encoding['input_ids'].flatten()
        attention_mask = encoding['attention_mask'].flatten()

        label = self.labels[idx]

        return {
            'image': image,
            'input_ids': input_ids,
            'attention_mask': attention_mask,
            'label': torch.tensor(label, dtype=torch.long)
        }

# 定义多模态模型
class MultimodalModel(nn.Module):
    def __init__(self):
        super(MultimodalModel, self).__init__()
        # 图像模型
        self.image_model = models.resnet50(pretrained=True)
        # 获取图像模型最后一层的输入特征维度
        num_ftrs = self.image_model.fc.in_features
        # 替换图像模型最后一层为 Identity
        self.image_model.fc = nn.Identity()

        # 文本模型
        self.text_model = BertModel.from_pretrained('bert-base-uncased')

        # 分类层
        self.classifier = nn.Linear(num_ftrs + self.text_model.config.hidden_size, 2)
```

```python
    def forward(self, image, input_ids, attention_mask):
        # 提取图像特征
        image_features = self.image_model(image)

        # 提取文本特征
        text_outputs = self.text_model(input_ids=input_ids, attention_mask=attention_mask)
        text_features = text_outputs.last_hidden_state[:, 0, :]   # 取[CLS]标记的输出

        # 融合特征
        combined_features = torch.cat((image_features, text_features), dim=1)

        # 分类
        output = self.classifier(combined_features)

        return output

# 数据预处理和加载
transform = transforms.Compose([
    transforms.Resize((224, 224)),
    transforms.ToTensor(),
    transforms.Normalize(mean=[0.485, 0.456, 0.406], std=[0.229, 0.224, 0.225]),
])

# 示例数据
image_paths = ['path/to/image1.jpg', 'path/to/image2.jpg']   # 替换为实际路径
texts = ['This is a sample text 1', 'This is a sample text 2']
labels = [0, 1]

dataset = MultimodalDataset(image_paths, texts, labels, transform=transform)
dataloader = DataLoader(dataset, batch_size=2, shuffle=True)

# 初始化模型、损失函数和优化器
model = MultimodalModel()

# 对图像模型进行微调
params_to_update = []
for name, param in model.named_parameters():
    if 'image_model' in name:
        params_to_update.append(param)

optimizer = optim.Adam([
    {'params': params_to_update},
    {'params': model.classifier.parameters()}
], lr=1e-4)

criterion = nn.CrossEntropyLoss()

# 训练模型
model.train()
for epoch in range(5):   # 训练5个epoch
    for batch in dataloader:
        optimizer.zero_grad()
        outputs = model(batch['image'], batch['input_ids'], batch['attention_mask'])
        loss = criterion(outputs, batch['label'])
        loss.backward()
        optimizer.step()
        print(f'Epoch [{epoch+1}/5], Loss: {loss.item():.4f}')
```

上述代码的实现流程如下。

（1）定义一个多模态数据集类 MultimodalDataset，用于加载图像和文本数据，并进行预处理。

（2）创建多模态模型类 MultimodalModel，其中包括预训练的 ResNet 模型，用于图像特征提

取,以及预训练的 BERT 模型,用于文本特征提取。图像模型的全连接层被替换为 Identity,并获取其输入特征维度。文本模型的参数被冻结以保持预训练状态,然后定义一个用于分类的线性层。

(3)在训练过程中对图像模型进行微调,同时保持文本模型参数不变。使用交叉熵损失函数和 Adam 优化器进行模型训练。

执行代码后会输出:

```
Epoch [1/5], Loss: 0.8164
Epoch [2/5], Loss: 0.1799
Epoch [3/5], Loss: 0.0706
Epoch [4/5], Loss: 0.0362
Epoch [5/5], Loss: 0.0215
```

上面的输出显示了模型在训练过程中每个 epoch 的损失值,损失值逐渐减小,表明模型在训练过程中逐渐收敛。

6.2.2 多任务学习

多任务学习是指通过同时训练模型完成多个相关任务,提高模型的泛化能力。例如,可以同时训练一个模型进行图像分类和文本生成,从而使模型能够更好地理解图像和文本之间的关系。实例 6-2 演示了在多模态模型训练中使用预训练模型实现多任务学习的过程。在这个例子中,使用预训练的 ResNet 模型提取图像特征,并使用预训练的 BERT 模型提取文本特征,然后将这些特征用于两个不同的任务:图像分类和文本分类。

实例 6-2 使用预训练模型实现多任务学习
源码路径:codes/6/duo.py

实例文件 duo.py 的具体实现代码如下所示。

```python
import torch
import torch.nn as nn
import torch.optim as optim
from torchvision import models, transforms
from transformers import BertTokenizer, BertModel
from torch.utils.data import Dataset, DataLoader
from PIL import Image

# 定义多模态数据集类
class MultimodalDataset(Dataset):
    def __init__(self, image_paths, texts, image_labels, text_labels, transform=None):
        self.image_paths = image_paths
        self.texts = texts
        self.image_labels = image_labels
        self.text_labels = text_labels
        self.transform = transform
        self.tokenizer = BertTokenizer.from_pretrained('bert-base-uncased')

    def __len__(self):
        return len(self.image_labels)

    def __getitem__(self, idx):
        image = Image.open(self.image_paths[idx]).convert("RGB")
        if self.transform:
            image = self.transform(image)

        text = self.texts[idx]
        encoding = self.tokenizer.encode_plus(
            text,
```

```python
            add_special_tokens=True,
            max_length=128,
            return_token_type_ids=False,
            padding='max_length',
            return_attention_mask=True,
            return_tensors='pt',
        )
        input_ids = encoding['input_ids'].flatten()
        attention_mask = encoding['attention_mask'].flatten()

        image_label = self.image_labels[idx]
        text_label = self.text_labels[idx]

        return {
            'image': image,
            'input_ids': input_ids,
            'attention_mask': attention_mask,
            'image_label': torch.tensor(image_label, dtype=torch.long),
            'text_label': torch.tensor(text_label, dtype=torch.long)
        }

# 定义多模态模型
class MultimodalModel(nn.Module):
    def __init__(self, num_classes_image, num_classes_text):
        super(MultimodalModel, self).__init__()
        # 图像模型
        self.image_model = models.resnet50(pretrained=True)
        num_ftrs = self.image_model.fc.in_features
        self.image_model.fc = nn.Identity()

        # 文本模型
        self.text_model = BertModel.from_pretrained('bert-base-uncased')

        # 分类层
        self.image_classifier = nn.Linear(num_ftrs, num_classes_image)
        self.text_classifier = nn.Linear(self.text_model.config.hidden_size, num_classes_text)

    def forward(self, image, input_ids, attention_mask):
        # 提取图像特征
        image_features = self.image_model(image)

        # 提取文本特征
        text_outputs = self.text_model(input_ids=input_ids, attention_mask=attention_mask)
        text_features = text_outputs.last_hidden_state[:, 0, :]

        # 图像分类
        image_output = self.image_classifier(image_features)

        # 文本分类
        text_output = self.text_classifier(text_features)

        return image_output, text_output

# 数据预处理和加载
transform = transforms.Compose([
    transforms.Resize((224, 224)),
    transforms.ToTensor(),
    transforms.Normalize(mean=[0.485, 0.456, 0.406], std=[0.229, 0.224, 0.225]),
])

# 示例数据
image_paths = ['image1.jpg', 'image2.jpg']
texts = ['This is fuwuqu', 'This is Navigation Map']
```

```
image_labels = [0, 1]    # 图像分类标签
text_labels = [0, 1]     # 文本分类标签

dataset = MultimodalDataset(image_paths, texts, image_labels, text_labels, transform=
transform)
dataloader = DataLoader(dataset, batch_size=2, shuffle=True)

# 初始化模型、损失函数和优化器
model = MultimodalModel(num_classes_image=2, num_classes_text=2)
criterion_image = nn.CrossEntropyLoss()
criterion_text = nn.CrossEntropyLoss()
optimizer = optim.Adam(model.parameters(), lr=1e-4)

# 训练模型
model.train()
for epoch in range(5):    # 训练 5 个 epoch
    for batch in dataloader:
        optimizer.zero_grad()
        image_outputs, text_outputs = model(batch['image'], batch['input_ids'],
        batch['attention_mask'])
        loss_image = criterion_image(image_outputs, batch['image_label'])
        loss_text = criterion_text(text_outputs, batch['text_label'])
        total_loss = loss_image + loss_text
        total_loss.backward()
        optimizer.step()
        print(f'Epoch [{epoch+1}/5], Image Loss: {loss_image.item():.4f}, Text Loss:
        {loss_text.item():.4f}')
```

上述代码的实现流程如下。

（1）定义一个多模态数据集类 MultimodalDataset，用于加载图像和文本数据，并进行预处理。在数据集中，将图像路径、文本内容以及它们对应的标签传入。

（2）创建多模态模型类 MultimodalModel，该模型包括一个预训练的 ResNet 模型用于提取图像特征，一个预训练的 BERT 模型用于提取文本特征，并分别添加了用于图像分类和文本分类的线性层。

（3）进行数据预处理和加载操作，包括对图像进行 resize 和标准化处理，并创建数据加载器。

（4）分别初始化模型、损失函数和优化器。损失函数使用交叉熵损失函数，优化器选用 Adam 优化器。

（5）进入模型的训练循环中，在每个 epoch 中遍历数据加载器，获取图像、文本和它们对应的标签，将它们传入模型进行前向传播，计算图像分类和文本分类的损失，并进行反向传播更新模型参数。在训练过程中打印输出每个 epoch 的图像分类损失和文本分类损失。执行代码后会输出：

```
Epoch [1/5], Image Loss: 0.7888, Text Loss: 0.6868
Epoch [2/5], Image Loss: 0.1296, Text Loss: 0.4997
Epoch [3/5], Image Loss: 0.0436, Text Loss: 0.1485
Epoch [4/5], Image Loss: 0.0212, Text Loss: 0.0314
Epoch [5/5], Image Loss: 0.0120, Text Loss: 0.0158
```

6.2.3 全量微调

在多模态大模型的训练中，全量微调（Full Fine-Tuning）是一种常用技术，它涉及对整个模型的参数进行微调，以便将预训练的模型调整到特定任务或数据集上，而不是仅对部分参数进行调整。这种方法通常在预训练模型的基础上，通过在目标任务的数据集上进行训练来优化模型的表现。全量微调可以使模型更好地适应特定的任务或数据集，从而提高其性能。

实现全量微调的基本流程如下所示。

（1）预训练模型选择：选择一个在大规模数据集上进行预训练的多模态模型，例如 BERT、CLIP、ViT 等模型，它们在多个任务和数据集上进行了广泛的预训练。

（2）数据准备：准备与目标任务相关的数据集。这些数据集可以是针对特定应用场景的数据集，如图像分类、文本生成、视觉问答等。

（3）模型调整：将预训练模型加载到微调环境中，并将目标任务的数据集分为训练集、验证集和测试集。

（4）训练：在目标任务的数据集上进行全量微调，通常使用标准的优化算法（如 Adam）和适当的学习率来调整模型的所有参数。

（5）评估：在验证集和测试集上评估微调后的模型性能，确保其在目标任务上表现良好。

实例 6-3 演示了如何在多模态大模型训练中使用全量微调技术。在这个例子中，使用 transformers 库对一个预训练的多模态模型（如 CLIP 模型）进行全量微调，演示了在"图像-文本"匹配任务上进行微调的过程。

实例 6-3　对 CLIP 模型进行全量微调
源码路径：codes/6/zhuyi.py

实例文件 zhuyi.py 的具体实现代码如下所示。

```python
import torch
from transformers import CLIPProcessor, CLIPModel, AdamW
from torch.utils.data import DataLoader, Dataset
from torchvision import transforms
from PIL import Image

# 定义一个自定义数据集
class MultimodalDataset(Dataset):
    def __init__(self, image_paths, texts, processor):
        self.image_paths = image_paths
        self.texts = texts
        self.processor = processor

    def __len__(self):
        return len(self.image_paths)

    def __getitem__(self, idx):
        image = Image.open(self.image_paths[idx])
        text = self.texts[idx]
        inputs = self.processor(images=image, text=text, return_tensors="pt", padding=True, truncation=True)
        return inputs

# 加载预训练的 CLIP 模型和处理器
model = CLIPModel.from_pretrained("openai/clip-vit-base-patch32")
processor = CLIPProcessor.from_pretrained("openai/clip-vit-base-patch32")

# 准备数据
image_paths = ["image1.jpg", "image2.jpg"]  # 示例图片路径
texts = ["rest area of yishui", "Shudu Lake Boardwalk"]  # 对应文本
dataset = MultimodalDataset(image_paths, texts, processor)
dataloader = DataLoader(dataset, batch_size=2, shuffle=True)
# 配置优化器
optimizer = AdamW(model.parameters(), lr=5e-5)

# 微调模型
model.train()
for epoch in range(3):  # 训练 3 个 epoch
    for batch in dataloader:
```

```
        inputs = {k: v.squeeze(1).to(torch.device("cuda" if torch.cuda.is_available()
        else "cpu")) for k, v in batch.items()}
        outputs = model(**inputs)
        loss = outputs.loss
        
        optimizer.zero_grad()
        loss.backward()
        optimizer.step()
        
    print(f"Epoch {epoch+1} - Loss: {loss.item()}")

# 保存微调后的模型
model.save_pretrained("fine-tuned-model")
```

上述代码的实现流程如下所示。

（1）准备数据集：MultimodalDataset 类用于加载图像和文本数据，并使用 CLIP 处理器进行处理。image_paths 和 texts 列表提供了训练数据的路径和标签。

（2）模型和处理器：使用 CLIPModel 和 CLIPProcessor 从 Hugging Face 的 transformers 库加载预训练的 CLIP 模型和处理器。

（3）数据加载：使用 DataLoader 将数据集分批处理，以便进行训练。

（4）优化器：使用 AdamW 优化器对模型进行训练。

（5）训练过程：在训练过程中计算模型的损失，执行反向传播，并更新模型的参数。

（6）模型保存：训练完成后，将微调后的模型保存到"fine-tuned-model"路径。

执行后会输出以下结果，模型的输出形状为[2, 2]，表示两个样本分别对应两个类别的预测结果。

```
Model outputs shape: torch.Size([2, 2])
```

6.2.4 对比学习

对比学习通过构造正负样本对，让模型学习到不同模态之间的相似性和差异性。例如，在图像-文本匹配任务中，可以使用对比学习方法让模型区分匹配和不匹配的图像-文本对。实例 6-4 演示了使用对比学习方法训练模型学习图像-文本之间的相似性和差异性的过程。

实例 6-4　使用对比学习方法训练模型
源码路径：codes/6/duixue.py

实例文件 duixue.py 的具体实现代码如下所示。

```python
import torch
import torch.nn as nn
import torch.optim as optim
import torch.nn.functional as F

# 定义对比学习模型
class ContrastiveModel(nn.Module):
    def __init__(self, image_feature_dim, text_feature_dim, hidden_dim=512):
        super(ContrastiveModel, self).__init__()
        self.image_feature_dim = image_feature_dim
        self.text_feature_dim = text_feature_dim
        self.hidden_dim = hidden_dim
        
        # 图像特征处理
        self.image_linear = nn.Linear(image_feature_dim, hidden_dim)
        self.image_norm = nn.LayerNorm(hidden_dim)
```

```python
        # 文本特征处理
        self.text_linear = nn.Linear(text_feature_dim, hidden_dim)
        self.text_norm = nn.LayerNorm(hidden_dim)

        # 输出层
        self.output_layer = nn.Linear(hidden_dim, 1)

    def forward(self, image, text):
        # 图像特征处理
        image_features = F.normalize(self.image_linear(image), p=2, dim=1)
        image_features = self.image_norm(image_features)

        # 文本特征处理
        text_features = F.normalize(self.text_linear(text), p=2, dim=1)
        text_features = self.text_norm(text_features)

        # 计算图像-文本之间的相似度得分
        similarity_scores=torch.cosine_similarity(image_features,text_features,dim=1)
        return similarity_scores

# 创建正负样本对
def create_contrastive_pairs(image_features, text_features, labels, margin=0.5):
    # 计算正样本对的相似度得分
    positive_scores = torch.cosine_similarity(image_features, text_features, dim=1)

    # 打乱文本特征的顺序，构造负样本对
    text_features_shuffled = text_features[torch.randperm(text_features.size(0))]

    # 计算负样本对的相似度得分
    negative_scores = torch.cosine_similarity(image_features,text_features_shuffled,dim=1)

    # 计算对比损失
    losses = F.relu(margin - positive_scores + negative_scores)

    return losses.mean()

# 创建示例数据
image_feature_dim = 512
text_feature_dim = 512
batch_size = 4

# 随机生成图像特征和文本特征作为示例数据
image_features = torch.randn(batch_size, image_feature_dim, requires_grad=True)
# 设置 requires_grad 为 True
text_features = torch.randn(batch_size, text_feature_dim, requires_grad=True)
# 设置 requires_grad 为 True
labels = torch.randint(0, 2, (batch_size,))    # 随机生成标签，0 表示不匹配，1 表示匹配

# 创建对比学习模型实例
model = ContrastiveModel(image_feature_dim, text_feature_dim)

# 定义优化器
optimizer = optim.Adam(model.parameters(), lr=0.001)

# 训练模型
num_epochs = 10
for epoch in range(num_epochs):
    # 前向传播
    similarity_scores = model(image_features, text_features)

    # 计算对比损失
    loss = create_contrastive_pairs(image_features, text_features, labels)
```

```
    # 反向传播与优化
    optimizer.zero_grad()
    loss.backward()
    optimizer.step()

    print(f"Epoch [{epoch+1}/{num_epochs}], Loss: {loss.item():.4f}")
# 输出模型输出的形状
print("Model outputs shape:", similarity_scores.shape)
```

上述代码的实现流程如下。

（1）定义一个对比学习模型 ContrastiveModel，用于接收图像特征和文本特征，并计算它们之间的相似度得分。

（2）编写函数 create_contrastive_pairs，用于创建正负样本对，并计算对比损失。

（3）生成示例数据，包括图像特征、文本特征和标签。

（4）创建对比学习模型实例，并定义优化器。

（5）在训练循环中分别实现模型的前向传播、计算损失、反向传播和优化步骤。

（6）打印输出模型输出的形状。执行代码后会输出：

```
Epoch [1/10], Loss: 0.4508
Epoch [2/10], Loss: 0.5037
Epoch [3/10], Loss: 0.4508
Epoch [4/10], Loss: 0.4566
Epoch [5/10], Loss: 0.5000
Epoch [6/10], Loss: 0.4855
Epoch [7/10], Loss: 0.4998
Epoch [8/10], Loss: 0.5193
Epoch [9/10], Loss: 0.4597
Epoch [10/10], Loss: 0.5105
Model outputs shape: torch.Size([4])
```

上面的输出结果表明程序已经成功执行，并且模型输出的形状是 torch.Size([4])。

6.2.5 参数高效微调

在多模态大模型训练中，参数高效微调（Parameter-Efficient Fine-Tuning, PEFT）技术是一种用于优化大模型的训练方法。这些技术旨在保持预训练模型性能，同时通过对模型进行微调来减少计算资源消耗和训练时间。实际应用中常用的参数高效微调技术如下。

1. LoRA

LoRA 是一种通过引入低秩矩阵来优化大模型的微调技术，它将模型的参数矩阵分解为两个低秩矩阵，从而减少需要微调的参数数量。LoRA 的优势是能够显著减少模型微调所需的计算和存储资源，同时保持模型的原有性能。LoRA 被广泛应用于 NLP 和图像处理任务的微调，在大规模预训练模型，如 GPT 和 BERT 上效果显著。

2. Adapter Layers

Adapter 层是小型的可训练模块，可以插入预训练模型的中间层，通过仅微调这些适配器层而不是整个模型来实现高效微调。Adapter Layers 技术的优势是减少了需要训练的参数数量，且能够快速适应新任务，而不会干扰预训练的知识。该技术适用于各种任务的迁移学习，如机器翻译、文本分类等。

3. Prompt Tuning

Prompt Tuning 通过在输入数据中添加任务特定的提示，来引导预训练模型生成适合新任务的输出。这种方法不需要对模型权重进行修改,只需调整提示的参数。Prompt Tuning 技术的优势是降低了训练成本,因为只需调整少量参数(提示),而不必改变模型的内部结构。该技术适用于自然语言生成、文本分类等任务。

4. BitFit

BitFit 是一种高效微调技术,仅微调模型的偏置项(biases),而不调整其他参数。BitFit 技术的优势是极大地减少了需要调整的参数量,从而减少了计算和存储开销。BitFit 技术适用于 NLP 任务,如文本生成和文本分类。

5. Differentiable Search

Differentiable Search 技术通过在训练过程中使用可微分的搜索算法优化模型结构,从而提高微调效率。Differentiable Search 技术的优势是能够自动选择最优的微调策略,减少了人工调整的需求,此技术特别适用于需要结构调整的复杂模型训练任务。

在实际应用中,参数高效微调技术通过减少需要调整的模型参数和计算开销,使得大规模预训练模型在特定任务上进行微调变得更加高效。它们在多模态大模型训练中发挥了重要作用,帮助研究人员和工程师更好地利用现有资源,同时提高了模型的适应性和性能。下面的代码演示了配置和训练一个多模态大模型,利用 LoRA 进行高效微调以提高模型在特定任务上的性能的过程。

```python
import torch
from transformers import BertTokenizer, BertForSequenceClassification
from peft import LoraConfig, get_peft_model
from torch.utils.data import DataLoader, Dataset
from transformers import AdamW

# 定义一个简单的数据集
class SimpleDataset(Dataset):
    def __init__(self, texts, labels, tokenizer, max_length):
        self.texts = texts
        self.labels = labels
        self.tokenizer = tokenizer
        self.max_length = max_length

    def __len__(self):
        return len(self.texts)

    def __getitem__(self, idx):
        text = self.texts[idx]
        label = self.labels[idx]
        encoding = self.tokenizer(
            text,
            truncation=True,
            padding='max_length',
            max_length=self.max_length,
            return_tensors='pt'
        )
        return {
            'input_ids': encoding['input_ids'].squeeze(),
            'attention_mask': encoding['attention_mask'].squeeze(),
            'labels': torch.tensor(label, dtype=torch.long)
        }

# 初始化模型和分词器
model_name = 'bert-base-uncased'
tokenizer = BertTokenizer.from_pretrained(model_name)
model = BertForSequenceClassification.from_pretrained(model_name)

# 配置 LoRA
lora_config = LoraConfig(
```

```
        r=8,  # 低秩矩阵的秩
        lora_alpha=32,  # LoRA 的 alpha 超参数
        lora_dropout=0.1,  # Dropout 比率
        target_modules=['encoder.layer.*.attention', 'encoder.layer.*.intermediate']  # 选择
LoRA 应用的目标模块
)

# 准备模型进行 LoRA 微调
model = get_peft_model(model, lora_config)

# 创建一个简单的数据集
texts = ["Hello world!", "Hugging Face Transformers are awesome!", "LoRA is a great
technique."]
labels = [0, 1, 0]  # 示例标签
dataset = SimpleDataset(texts, labels, tokenizer, max_length=32)
dataloader = DataLoader(dataset, batch_size=2, shuffle=True)

# 定义优化器
optimizer = AdamW(model.parameters(), lr=5e-5)

# 训练循环
model.train()
for epoch in range(3):  # 训练 3 个 epoch
    for batch in dataloader:
        optimizer.zero_grad()

        inputs = {
            'input_ids': batch['input_ids'],
            'attention_mask': batch['attention_mask'],
            'labels': batch['labels']
        }

        outputs = model(**inputs)
        loss = outputs.loss

        loss.backward()
        optimizer.step()

    print(f"Epoch {epoch + 1} finished with loss: {loss.item()}")

# 保存微调后的模型
model.save_pretrained('finetuned_lora_model')
```

在上述代码中，首先加载预训练的多模态模型，并配置 LoRA 以实现高效的微调。然后配置数据加载器以读取和处理训练数据，确保数据能够适配模型输入要求。接着设置优化器和训练参数，准备好训练所需的环境和资源。最后，运行训练过程，逐步优化模型参数，并保存微调后的模型，以便后续使用和评估。

6.2.6 迁移学习

迁移学习（Transfer Learning）是指将从一个任务或领域中学到的知识应用到另一个相关任务或领域。例如，从自然图像分类任务中学到的特征可以迁移到医学图像分析任务中。实例 6-5 演示了利用迁移学习在多模态模型训练中使用预训练的自然图像分类模型的过程。

实例 6-5	利用迁移学习使用预训练的自然图像分类模型
	源码路径：codes/6/qian.py

实例文件 qian.py 的具体实现代码如下所示。

```python
import torch
import torch.nn as nn
import torchvision
from torchvision import transforms, models
from torch.utils.data import DataLoader, Dataset
from PIL import Image
import numpy as np

# 加载自然图像分类预训练模型,这里以 ResNet-18 为例
pretrained_model = models.resnet18(pretrained=True)
# 冻结预训练模型的参数
for param in pretrained_model.parameters():
    param.requires_grad = False

# 替换预训练模型的最后一层,适应新的任务(医学图像分析),这里以二分类为例
num_ftrs = pretrained_model.fc.in_features
pretrained_model.fc = nn.Linear(num_ftrs, 2)

# 定义图像数据预处理
transform = transforms.Compose([
    transforms.Resize((224, 224)),   # 将图像大小调整为预训练模型的输入尺寸
    transforms.ToTensor(),     # 将图像转换为 Tensor
    transforms.Normalize(mean=[0.485, 0.456, 0.406], std=[0.229, 0.224, 0.225])   # 标准化
])

# 假设这里有自然图像分类的数据集,用 MedicalDataset 代替
class MedicalDataset(Dataset):
    def __init__(self, transform=None):
        self.data = []     # 存放图像数据
        self.targets = []     # 存放图像对应的标签
        self.transform = transform
        # 生成一些示例数据(随机生成)
        for _ in range(100):
            self.data.append(np.random.randint(0, 256, size=(224, 224, 3), dtype=np.uint8))
            self.targets.append(np.random.randint(0, 2))

    def __len__(self):
        return len(self.data)

    def __getitem__(self, idx):
        image = Image.fromarray(self.data[idx])
        target = self.targets[idx]
        if self.transform:
            image = self.transform(image)
        return image, target

# 创建自然图像分类数据集的 DataLoader,这里用 MedicalDataset 代替
medical_dataset = MedicalDataset(transform=transform)
medical_dataloader = DataLoader(medical_dataset, batch_size=32, shuffle=True)

# 定义损失函数和优化器
criterion = nn.CrossEntropyLoss()
optimizer = torch.optim.SGD(pretrained_model.parameters(), lr=0.001, momentum=0.9)

# 训练模型
num_epochs = 5
for epoch in range(num_epochs):
    running_loss = 0.0
    for images, labels in medical_dataloader:
        optimizer.zero_grad()
        outputs = pretrained_model(images)
        loss = criterion(outputs, labels)
        loss.backward()
```

```
        optimizer.step()
        running_loss += loss.item()
    print(f"Epoch [{epoch+1}/{num_epochs}], Loss: {running_loss/len(medical_dataloader):
.4f}")

# 保存模型
torch.save(pretrained_model.state_dict(), 'medical_model.pth')
print("Model trained and saved.")
```

上述代码的实现流程如下。

（1）加载预训练的 ResNet-18 模型并替换最后一层以适应新的任务，这里的任务是二分类的医学图像分析。

（2）定义一个简单的医学图像数据集类（这里使用随机生成的示例数据），进行数据预处理并创建数据加载器。

（3）分别定义损失函数和优化器，并进行模型训练。

（4）打印输出模型训练过程中每个 epoch 的损失值，并保存训练好的模型。执行代码后会输出：

```
Epoch [1/5], Loss: 0.7988
Epoch [2/5], Loss: 0.7165
Epoch [3/5], Loss: 0.7389
Epoch [4/5], Loss: 0.6620
Epoch [5/5], Loss: 0.7141
Model trained and saved.
```

6.2.7 人类反馈强化学习

人类反馈强化学习（Reinforcement Learning from Human Feedback，RLHF）是一种结合了强化学习（RL）和人类反馈（HF）的训练技术，主要用于提升多模态大模型的性能。人类反馈强化学习结合强化学习和人类反馈，旨在通过人类提供的反馈来引导和优化模型的行为。其核心思想是利用人类对模型输出的评价作为奖励信号，从而提升模型在特定任务上的表现。通过 RLHF 这种技术，多模态大模型能够更好地理解和处理复杂的多模态数据，从而提升其在实际应用中的表现。实例 6-6 是一个使用 RLHF 的例子，模拟了一个简单的强化学习任务，其中人类反馈用于优化模型的行为。我们假设有一个模型生成文本，并通过人类反馈来优化模型生成的文本质量。

> **实例 6-6** 使用人类反馈强化学习优化模型生成的文本质量
> 源码路径：codes/6/zheng.py

实例文件 zheng.py 的具体实现代码如下所示。

```python
import torch
from transformers import GPT2LMHeadModel, GPT2Tokenizer
from datasets import load_dataset
import numpy as np

# 初始化模型和分词器
model_name = 'gpt2'
tokenizer = GPT2Tokenizer.from_pretrained(model_name)
model = GPT2LMHeadModel.from_pretrained(model_name)

# 人类反馈数据（假设我们有这样的反馈数据）
feedback_scores = {
    'example1': 0.8,
    'example2': 0.6,
    'example3': 0.9
}
```

```python
# 模型生成文本
def generate_text(prompt):
    inputs = tokenizer(prompt, return_tensors='pt')
    outputs = model.generate(inputs['input_ids'], max_length=50, pad_token_id=tokenizer.eos_token_id)
    return tokenizer.decode(outputs[0], skip_special_tokens=True)

# 计算奖励（基于人类反馈评分）
def compute_reward(text):
    return feedback_scores.get(text, 0.5)   # 默认奖励为 0.5

# 强化学习更新
def reinforce_update(prompt, model, optimizer):
    model.train()
    optimizer.zero_grad()

    generated_text = generate_text(prompt)

    # 将生成的文本进行分词，以便与输入进行比较
    inputs = tokenizer(prompt, return_tensors='pt')
    labels = tokenizer(generated_text, return_tensors='pt')['input_ids']

    # 确保标签的长度与输入相同
    if labels.size(1) < inputs['input_ids'].size(1):
        labels = torch.cat([labels, torch.full((1, inputs['input_ids'].size(1) - labels.
            size(1)), tokenizer.pad_token_id)], dim=1)
    elif labels.size(1) > inputs['input_ids'].size(1):
        labels = labels[:, :inputs['input_ids'].size(1)]

    # 计算损失
    outputs = model(input_ids=inputs['input_ids'], labels=labels)
    loss = -outputs.loss * compute_reward(generated_text)   # 使用负的损失和奖励来进行优化

    loss.backward()
    optimizer.step()

    return loss.item()

# 设置优化器
optimizer = torch.optim.AdamW(model.parameters(), lr=1e-5)

# 训练循环
prompts = ["Tell me a story about a robot.", "How does reinforcement learning work?"]
for epoch in range(3):   # 简化为 3 个 epoch
    for prompt in prompts:
        loss = reinforce_update(prompt, model, optimizer)
        print(f"Epoch {epoch}, Prompt: {prompt}, Loss: {loss}")

# 保存微调后的模型
model.save_pretrained('./fine-tuned-model')
tokenizer.save_pretrained('./fine-tuned-model')
```

上述代码的实现流程如下所示。

（1）初始化模型和分词器：使用 GPT-2 模型和分词器来生成文本。

（2）人类反馈数据：模拟一些人类反馈评分，用于优化模型生成的文本。

（3）生成文本：根据提示生成文本。

（4）计算奖励：根据人类反馈评分计算奖励。

（5）强化学习更新：用反馈和奖励更新模型参数，结合损失函数与奖励来优化模型。

（6）训练循环：对多个提示进行训练，进行强化学习更新。

（7）保存模型：保存微调后的模型以便后续使用。

执行后会输出下面的结果，可以看到在经过 3 个 epoch 的训练后，模型在每个 prompt 上的损失值不断变化，损失值总体上逐渐减少。这表明模型在强化学习更新过程中生成的文本可能越来越符合奖励标准。

```
Epoch 0, Prompt: Tell me a story about a robot., Loss: -1.985111951828003
Epoch 0, Prompt: How does reinforcement learning work?, Loss: -2.492488384246826
Epoch 1, Prompt: Tell me a story about a robot., Loss: -1.84115469455719
Epoch 1, Prompt: How does reinforcement learning work?, Loss: -2.8126277923583984
Epoch 2, Prompt: Tell me a story about a robot., Loss: -2.58296537399292
Epoch 2, Prompt: How does reinforcement learning work?, Loss: -3.103867292404175
```

6.2.8　动态学习率调整

在训练过程中可以调整动态学习率（Dynamic Learning Rate Adjustment），例如使用学习率衰减、余弦退火等方法。动态学习率可以在训练的不同阶段提供适当的学习率，从而提高训练效果。下面的例子演示了在模型训练中使用学习率调度器（Learning Rate Scheduler）动态调整学习率的过程。实例 6-7 中将使用 PyTorch 中的 CIFAR-10 数据集和一个卷积神经网络模型进行演示。

> **实例 6-7**　在模型训练过程中调整学习率
> 源码路径：codes/6/tiao.py

实例文件 tiao.py 的具体实现代码如下所示。

```python
import torch
import torch.nn as nn
import torch.optim as optim
import torchvision
import torchvision.transforms as transforms

# 定义简单的卷积神经网络模型
class SimpleCNN(nn.Module):
    def __init__(self):
        super(SimpleCNN, self).__init__()
        self.conv1 = nn.Conv2d(3, 16, 3, 1, padding=1)
        self.conv2 = nn.Conv2d(16, 32, 3, 1, padding=1)
        self.fc1 = nn.Linear(32*8*8, 128)
        self.fc2 = nn.Linear(128, 10)

    def forward(self, x):
        x = torch.relu(self.conv1(x))
        x = torch.max_pool2d(x, 2, 2)
        x = torch.relu(self.conv2(x))
        x = torch.max_pool2d(x, 2, 2)
        x = x.view(-1, 32*8*8)
        x = torch.relu(self.fc1(x))
        x = self.fc2(x)
        return x

# 数据预处理
transform = transforms.Compose([
    transforms.ToTensor(),
    transforms.Normalize((0.5, 0.5, 0.5), (0.5, 0.5, 0.5))
])

# 加载 CIFAR-10 数据集
trainset = torchvision.datasets.CIFAR10(root='./data', train=True, download=True,
transform=transform)
trainloader = torch.utils.data.DataLoader(trainset, batch_size=32, shuffle=True)
```

```python
# 初始化模型、损失函数和优化器
model = SimpleCNN()
criterion = nn.CrossEntropyLoss()
optimizer = optim.SGD(model.parameters(), lr=0.1)  # 初始学习率设为 0.1

# 定义学习率调度器
scheduler = optim.lr_scheduler.StepLR(optimizer, step_size=20, gamma=0.1)  # 每 20 个 epoch
将学习率乘以 0.1

# 训练模型
num_epochs = 50
for epoch in range(num_epochs):
    running_loss = 0.0
    for i, data in enumerate(trainloader, 0):
        inputs, labels = data

        optimizer.zero_grad()

        outputs = model(inputs)
        loss = criterion(outputs, labels)
        loss.backward()
        optimizer.step()

        running_loss += loss.item()
        if i % 200 == 199:   # 每 200 mini-batches 输出一次损失
            print('[%d, %5d] loss: %.3f' %
                  (epoch + 1, i + 1, running_loss / 200))
            running_loss = 0.0

    # 更新学习率
    scheduler.step()

print('Finished Training')
```

上述代码使用了 StepLR 学习率调度器，每 20 个 epochs 将学习率乘以 0.1。执行代码后会输出下面的结果，每 200 个 mini-batches 输出一次损失值，训练完成后会打印"Finished Training"。

```
[1,   200] loss: 2.057
[1,   400] loss: 1.728
[1,   600] loss: 1.573
[1,   800] loss: 1.467
[1,  1000] loss: 1.377
...
[50,   200] loss: 0.367
[50,   400] loss: 0.363
[50,   600] loss: 0.373
Finished Training
```

6.2.9　SFT 微调

在多模态大模型训练中，SFT（Supervised Fine-Tuning，监督微调）是一种常用的微调方法，主要用于提升预训练大模型在特定任务上的表现。SFT 是指在预训练模型基础上，通过监督学习的方式进行微调。这个过程通常包括使用带标签的数据集对模型进行进一步训练，以使模型更好地适应特定任务或数据分布。下面的例子演示了使用 SFT 技术的过程，通过微调预训练的图像编码器和文本编码器来适应特定的多模态任务。在实例 6-8 中，将冻结一部分模型参数，仅对最后几层或添加的全连接层进行微调。

6.2 训练策略

实例 6-8　使用 SFT 微调模型
源码路径：codes/6/hun.py

实例文件 hun.py 的具体实现代码如下所示。

```python
import torch
from torch.utils.data import Dataset, DataLoader
from PIL import Image
from transformers import BertTokenizer
import torch.optim as optim
import torch.nn as nn
import torchvision.models as models
import torchvision.transforms as transforms
from transformers import BertModel

# 自定义数据集类
class ImageTextDataset(Dataset):
    def __init__(self, image_paths, texts, tokenizer, transform=None):
        self.image_paths = image_paths
        self.texts = texts
        self.tokenizer = tokenizer
        self.transform = transform

    def __len__(self):
        return len(self.texts)

    def __getitem__(self, idx):
        # 加载图像并转换为 Tensor
        image = Image.open(self.image_paths[idx]).convert('RGB')
        if self.transform:
            image = self.transform(image)

        # 处理文本
        text = self.texts[idx]
        encoding = self.tokenizer(text, return_tensors='pt', padding='max_length', truncation=True, max_length=128)

        return image, encoding['input_ids'].squeeze(0), encoding['attention_mask'].squeeze(0)

# 图像转换器，将 PIL 图像转换为 Tensor
transform = transforms.Compose([
    transforms.Resize((224, 224)),
    transforms.ToTensor(),
    transforms.Normalize(mean=[0.485, 0.456, 0.406], std=[0.229, 0.224, 0.225])
])

# 加载数据
image_paths = ['image1.jpeg', 'image2.jpeg']
texts = ['A flower', 'A man']
tokenizer = BertTokenizer.from_pretrained('bert-base-uncased')
dataset = ImageTextDataset(image_paths, texts, tokenizer, transform=transform)
dataloader = DataLoader(dataset, batch_size=2, shuffle=True)

# 定义模型
class MultiModalModel(nn.Module):
    def __init__(self):
        super(MultiModalModel, self).__init__()
        # 图像编码器
        self.image_encoder = models.resnet50(pretrained=True)
        # 冻结 ResNet 的前几层参数，只微调后面的层
        for param in list(self.image_encoder.parameters())[:-10]:
            param.requires_grad = False
```

```python
        self.image_encoder.fc = nn.Identity()  # 移除 ResNet 的最后一个全连接层

        # 文本编码器
        self.text_encoder = BertModel.from_pretrained('bert-base-uncased')
        # 冻结 BERT 的前几层参数，只微调后面的层
        for param in list(self.text_encoder.parameters())[:-10]:
            param.requires_grad = False

        # 合并后的全连接层
        self.fc = nn.Linear(2048 + 768, 2)  # 假设有 2 个分类

    def forward(self, images, input_ids, attention_mask):
        # 图像编码
        image_features = self.image_encoder(images)
        # 文本编码
        text_features = self.text_encoder(input_ids=input_ids, attention_mask=
attention_mask).last_hidden_state[:, 0, :]
        # 合并特征
        combined_features = torch.cat((image_features, text_features), dim=1)
        # 分类输出
        outputs = self.fc(combined_features)
        return outputs

# 初始化模型
model = MultiModalModel()

# 损失函数和优化器
criterion = nn.CrossEntropyLoss()
optimizer = optim.Adam(filter(lambda p: p.requires_grad, model.parameters()), lr=1e-4)

# 训练循环
for epoch in range(5):  # 假设训练 5 个 epoch
    for images, input_ids, attention_masks in dataloader:
        # 清空梯度
        optimizer.zero_grad()
        # 前向传播
        outputs = model(images, input_ids, attention_masks)
        labels = torch.tensor([0, 1])  # 假设我们有标签
        loss = criterion(outputs, labels)
        # 反向传播和优化
        loss.backward()
        optimizer.step()
    print(f'Epoch {epoch + 1}, Loss: {loss.item()}')
```

在上述代码中，为了实现 SFT，我们在模型的图像编码器和文本编码器中冻结了前几层的参数，只微调后几层的参数。这确保了大部分预训练知识得以保留，同时适应特定任务的需求。执行后会输出下面的结果。

```
Epoch 1, Loss: 0.7242298126220703
Epoch 2, Loss: 0.34537628293037415
Epoch 3, Loss: 0.16260889172554016
Epoch 4, Loss: 2.5474729537963867
Epoch 5, Loss: 0.09362819045782089
```

从上面输出的训练过程中的损失值可以看出，模型在前几轮的训练中损失逐渐下降，说明模型正在学习。但是在第 4 轮时，损失突然增大，然后在第 5 轮又下降，造成这种情况的主要原因是当前数据集较小（仅 2 个样本），在实际应用中，建议大家使用更多样本以获得更可靠的结果。

6.3 CLIP 模型训练与微调

本项目提供了基于 PyTorch Lightning 的 CLIP 模型训练解决方案，支持从头开始训练和数据高效微调两种方式（见实例 6-9）。用户可以轻松地训练自己的 CLIP 模型，同时支持使用自定义数据集和预训练模型进行微调，实现图像和文本的多模态学习任务。项目旨在提供简单易用的训练流程，以实现对图像和文本之间关系的学习，为多模态任务的研究和应用提供了便利。

实例 6-9 多模态模型 CLIP 的训练与微调
源码路径：codes/6/train-CLIP-main

6.3.1 项目介绍

随着人工智能领域的发展，多模态学习成为一个备受关注的研究方向，涉及图像、文本、语音等多种数据模态的融合与学习。在多模态学习中，模型需要同时理解和处理不同模态的信息，从而更好地理解世界、进行推理和决策。CLIP 是由 OpenAI 提出的一种基于自监督学习的多模态模型，它通过对图像和文本之间的对比学习，实现了强大的视觉和语言理解能力，成了多个任务的基础模型。

本项目提供了基于 PyTorch Lightning 的 CLIP 模型训练解决方案，旨在帮助用户轻松训练自己的 CLIP 模型以及进行数据高效微调。通过该项目，用户可以从头开始训练 CLIP 模型，也可以利用预训练模型进行数据微调，实现对图像和文本之间关系的学习。同时，项目支持用户使用自定义数据集进行训练，为各种多模态任务（如图像分类、文本检索等）提供了灵活而高效的解决方案。无论是从头开始训练还是微调预训练模型，本项目都提供了简单易用的训练流程和接口，为多模态学习任务的研究和实践提供了便利。

本项目基于 PyTorch Lightning 框架实现，具体功能模块如下。

1. **模型的训练与微调**
- 提供了对 CLIP 模型的训练与微调功能，用户可以选择从头开始训练或者对预训练模型进行微调。
- 训练过程采用自监督学习的方法，通过对图像和文本之间的对比学习来学习模型的表示。
- 支持多种 CLIP 模型，用户可以根据需求选择不同的模型结构和参数进行训练。

2. **数据准备与加载**
- 提供了数据模块（DataModule），用于准备和加载图像与文本数据。
- 可以从指定文件夹加载数据，支持自定义数据集的加载，同时支持各种数据预处理操作。

3. **多模态模型的构建**
- 实现了 CLIP 模型的包装器（CLIPWrapper），用于训练原始 CLIP 模型。
- 实现了自定义的 CLIP 模型包装器（CustomCLIPWrapper），支持微调预训练的图像编码器和文本编码器。

4. **训练流程的管理**
- 使用 PyTorch Lightning 的 Trainer 来管理训练过程，支持分布式训练、混合精度训练等功能。
- 在训练过程中，根据用户指定的参数和模型配置，进行图像和文本的编码、对比学习等操作。

5. **命令行工具**
- 提供了命令行接口，用户可以通过命令行指定模型名称、数据文件夹、批处理大小等参数来启动训练过程。

- 支持从头训练和微调两种模式，使用户能够轻松地使用该项目进行训练。

本项目的实现流程遵循 CLIP 模型的自监督学习原理，通过最大化图像和文本之间的相似性来学习模型的表示，从而实现对图像和文本之间语义关系的理解和学习。

6.3.2 创建文本和图像配对数据集

文件 text_image_dm.py 实现了一个自定义的 PyTorch 数据集和数据模块，这是一个用于创建文本和图像配对数据集的工具，并为训练过程提供数据加载功能。

```python
class TextImageDataset(Dataset):
    def __init__(self,
                 folder: str,
                 image_size=224,
                 resize_ratio=0.75,
                 shuffle=False,
                 custom_tokenizer=False
                 ):
        """从包含文本和图像文件的目录中创建一个文本图像数据集。

        参数：
            folder (str)：包含图像和文本文件的文件夹，它们通过各自路径的 "stem" 匹配。
            image_size (int, optional)：输出图像的大小。默认为 224。
            resize_ratio (float, optional)：裁剪时包含的最小图像比例。默认为 0.75。
            shuffle (bool, optional)：是否在采样过程中进行打乱。默认为 False。
            custom_tokenizer (bool, optional)：是否有自定义分词器。默认为 False。
        """
        super().__init__()
        self.shuffle = shuffle
        path = Path(folder)

        text_files = [*path.glob('**/*.txt')]
        image_files = [
            *path.glob('**/*.png'), *path.glob('**/*.jpg'),
            *path.glob('**/*.jpeg'), *path.glob('**/*.bmp')
        ]

        text_files = {text_file.stem: text_file for text_file in text_files}
        image_files = {image_file.stem: image_file for image_file in image_files}

        keys = (image_files.keys() & text_files.keys())

        self.keys = list(keys)
        self.text_files = {k: v for k, v in text_files.items() if k in keys}
        self.image_files = {k: v for k, v in image_files.items() if k in keys}
        self.resize_ratio = resize_ratio
        self.image_transform = T.Compose([
            T.Lambda(self.fix_img),
            T.RandomResizedCrop(image_size,
                                scale=(self.resize_ratio, 1.),
                                ratio=(1., 1.)),
            T.ToTensor(),
            T.Normalize((0.48145466, 0.4578275, 0.40821073), (0.26862954, 0.26130258, 0.27577711))
        ])
        self.custom_tokenizer = custom_tokenizer

    def __len__(self):
        return len(self.keys)

    def fix_img(self, img):
        return img.convert('RGB') if img.mode != 'RGB' else img
```

```python
    def random_sample(self):
        return self.__getitem__(randint(0, self.__len__() - 1))

    def sequential_sample(self, ind):
        if ind >= self.__len__() - 1:
            return self.__getitem__(0)
        return self.__getitem__(ind + 1)

    def skip_sample(self, ind):
        if self.shuffle:
            return self.random_sample()
        return self.sequential_sample(ind=ind)

    def __getitem__(self, ind):
        key = self.keys[ind]

        text_file = self.text_files[key]
        image_file = self.image_files[key]

        descriptions = text_file.read_text().split('\n')
        descriptions = list(filter(lambda t: len(t) > 0, descriptions))
        try:
            description = choice(descriptions)
        except IndexError as zero_captions_in_file_ex:
            print(f"加载文件 {text_file} 时发生异常。")
            print(f"跳过索引 {ind}")
            return self.skip_sample(ind)

        tokenized_text = description if self.custom_tokenizer else clip.tokenize
        (description)[0]

        try:
            image_tensor = self.image_transform(PIL.Image.open(image_file))
        except (PIL.UnidentifiedImageError, OSError) as corrupt_image_exceptions:
            print(f"加载文件 {image_file} 时发生异常。")
            print(f"跳过索引 {ind}")
            return self.skip_sample(ind)

        # 成功
        return image_tensor, tokenized_text

class TextImageDataModule(LightningDataModule):
    def __init__(self,
                 folder: str,
                 batch_size: int,
                 num_workers=0,
                 image_size=224,
                 resize_ratio=0.75,
                 shuffle=False,
                 custom_tokenizer=None
                 ):
        """从包含文本和图像文件的目录中创建一个文本图像数据模块。

        参数:
            folder (str): 包含图像和文本文件的文件夹,它们通过各自路径的 "stem" 匹配。
            batch_size (int): 每个数据加载器的批处理大小。
            num_workers (int, optional): DataLoader 中的工作线程数。默认为 0。
            image_size (int, optional): 输出图像的大小。默认为 224。
            resize_ratio (float, optional): 裁剪时包含的最小图像比例。默认为 0.75。
            shuffle (bool, optional): 是否在采样过程中进行打乱。默认为 False。
            custom_tokenizer (transformers.AutoTokenizer, optional): 用于文本的分词器。默认为 None。
        """
```

```python
        super().__init__()
        self.folder =folder
        self.batch_size = batch_size
        self.num_workers = num_workers
        self.image_size = image_size
        self.resize_ratio = resize_ratio
        self.shuffle = shuffle
        self.custom_tokenizer = custom_tokenizer

    @staticmethod
    def add_argparse_args(parent_parser):
        parser = argparse.ArgumentParser(父级解析器=[parent_parser], 添加帮助=False)
        parser.add_argument('--folder', type=str, required=True, help='你的训练文件夹的目录')
        parser.add_argument('--batch_size', type=int, help='批处理大小')
        parser.add_argument('--num_workers', type=int, default=0, help='数据加载器的工作线程数')
        parser.add_argument('--image_size', type=int, default=224, help='图像的大小')
        parser.add_argument('--resize_ratio', type=float, default=0.75, help='随机裁剪时图像的最小尺寸')
        parser.add_argument('--shuffle', type=bool, default=False, help='采样时是否打乱顺序')
        return parser

    def setup(self, stage=None):
        self.dataset = TextImageDataset(self.folder, image_size=self.image_size, resize_ratio=
        self.resize_ratio, shuffle=self.shuffle, custom_tokenizer=not self.custom_tokenizer
        is None)

    def train_dataloader(self):
        return DataLoader(self.dataset, batch_size=self.batch_size, shuffle=self.shuffle,
        num_workers=self.num_workers, drop_last=True , collate_fn=self.dl_collate_fn)

    def dl_collate_fn(self, batch):
        if self.custom_tokenizer is None:
            return torch.stack([row[0] for row in batch]), torch.stack([row[1] for row in batch])
        else:
            return torch.stack([row[0] for row in batch]), self.custom_tokenizer([row[1]
            for row in batch], padding=True, truncation=True, return_tensors="pt")
```

对上述代码的具体说明如下。

（1）类 TextImageDataset 的功能是从指定的文件夹中加载和处理成对的图像和文本数据，该类通过匹配文件名来生成图像-文本对，并进行必要的预处理（如图像转换和文本分词），以便供模型训练使用。

（2）类 TextImageDataset 中__init__方法的功能是初始化数据集实例，加载指定文件夹中的图像和文本文件，并设置图像大小、裁剪比例、是否打乱顺序等参数。

（3）类 TextImageDataset 中__len__方法的功能是返回数据集中数据对的数量。

（4）类 TextImageDataset 中 fix_img 方法的功能是确保加载的图像为 RGB 格式，以保证图像处理的一致性。

（5）类 TextImageDataset 中 random_sample 方法的功能是随机从数据集中采样一个数据对。

（6）类 TextImageDataset 中 sequential_sample 方法的功能是按顺序采样下一个数据对，如果到达末尾，则返回第一个数据对，以实现循环采样。

（7）类 TextImageDataset 中 skip_sample 方法的功能是根据是否设置了打乱顺序，决定是随机采样一个数据对还是按顺序采样一个数据对。

（8）类 TextImageDataset 中__getitem__方法的功能是根据索引获取图像-文本数据对，并进行预处理。该方法确保图像转换为张量并对文本进行分词处理，然后返回预处理后的图像和文本张量。

（9）类 TextImageDataModule 的功能是创建一个方便管理数据加载的模块，利用 TextImageDataset 类，设置数据加载器的相关参数，并为模型训练提供数据。

（10）类 TextImageDataModule 中 __init__ 方法的功能是初始化数据模块实例，设置数据文件夹路径、批处理大小、工作线程数、图像大小、裁剪比例、是否打乱顺序以及自定义分词器等参数。

（11）类 TextImageDataModule 中 add_argparse_args 静态方法的功能是为命令行参数解析添加功能，使用户可以通过命令行传入数据模块的参数，如文件夹路径、批处理大小等。

（12）类 TextImageDataModule 中 setup 方法的功能是初始化 TextImageDataset 数据集实例，准备数据集以供训练使用。

（13）类 TextImageDataModule 中 train_dataloader 方法的功能是返回训练数据加载器，该加载器使用 TextImageDataset 类实例，并根据设置的参数（如批处理大小、是否打乱顺序等）进行配置。

（14）类 TextImageDataModule 中 dl_collate_fn 方法的功能是定义批处理函数，用于将单个样本组成批次。该函数根据是否使用自定义分词器，堆叠图像和文本数据，以便在训练过程中进行批处理。

6.3.3 构建多模态模型

本项目的 models 目录，包含了对 CLIP 模型和其自定义包装器进行详细定义和配置的代码文件。其中，CLIPWrapper 类是对 CLIP 模型的 Lightning 包装器，用于训练和验证 CLIP 模型；而 CustomCLIPWrapper 类则是对 CLIPWrapper 的定制，引入了自我蒸馏和其他自定义功能以增强模型性能。此外，models 目录还包括用于配置不同模型参数的 YAML 文件，如 RN.yaml 和 ViT.yaml。这些文件共同构成了一个完整的模型训练和配置环境，用于实现对 CLIP 模型的训练和评估。

1. 实现 CLIP 模型

文件 model.py 定义了一个 CLIP 模型的实现，包含了视觉和文本处理模块。其中视觉部分使用了改进的 ResNet 或视觉 Transformer，文本部分使用了自定义的 Transformer。文件 model.py 实现了多个类，如 Bottleneck、AttentionPool2d、ModifiedResNet、LayerNorm、QuickGELU、ResidualAttentionBlock、Transformer 和 VisualTransformer，它们分别负责特征提取、注意力机制、层归一化和非线性激活等操作。CLIP 类整合了这些组件，能将图像和文本编码为高维特征向量，并通过余弦相似度计算图像和文本的相似性，用于多模态任务。

文件 model.py 的具体实现流程如下。

（1）类 Bottleneck 定义了一个用于残差网络的瓶颈结构模块，它通过 3 个卷积层进行特征提取，并在需要下采样时添加了一个平均池化层以调整输入特征图的尺寸。该模块采用了残差连接方式，通过跳跃连接来缓解深度网络中的梯度消失问题，从而有效地进行特征学习。

```
class Bottleneck(nn.Module):
    expansion = 4

    def __init__(self, inplanes, planes, stride=1):
        super().__init__()

        #所有卷积层的步幅都是1，当步幅大于1时，在第二次卷积后进行平均池化
        self.conv1 = nn.Conv2d(inplanes, planes, 1, bias=False)
        self.bn1 = nn.BatchNorm2d(planes)

        self.conv2 = nn.Conv2d(planes, planes, 3, padding=1, bias=False)
        self.bn2 = nn.BatchNorm2d(planes)

        self.avgpool = nn.AvgPool2d(stride) if stride > 1 else nn.Identity()
```

```
            self.conv3 = nn.Conv2d(planes, planes * self.expansion, 1, bias=False)
            self.bn3 = nn.BatchNorm2d(planes * self.expansion)

            self.relu = nn.ReLU(inplace=True)
            self.downsample = None
            self.stride = stride

            if stride > 1 or inplanes != planes * Bottleneck.expansion:
                # 下采样层前置一个平均池化层，随后的卷积步幅为 1
                self.downsample = nn.Sequential(OrderedDict([
                    ("-1", nn.AvgPool2d(stride)),
                    ("0", nn.Conv2d(inplanes, planes * self.expansion, 1, stride=1, bias=False)),
                    ("1", nn.BatchNorm2d(planes * self.expansion))
                ]))

        def forward(self, x: torch.Tensor):
            identity = x

            out = self.relu(self.bn1(self.conv1(x)))
            out = self.relu(self.bn2(self.conv2(out)))
            out = self.avgpool(out)
            out = self.bn3(self.conv3(out))

            if self.downsample is not None:
                identity = self.downsample(x)

            out += identity
            out = self.relu(out)
            return out
```

（2）类 AttentionPool2d 实现了一种基于多头自注意力机制的二维池化操作。首先，对输入进行重塑和排列，然后通过在输入上添加位置嵌入来增强其空间信息。接着，利用多头自注意力机制计算查询、键和值之间的注意力分布，并通过线性变换得到最终输出。类 AttentionPool2d 可以用来对二维特征图进行全局的信息汇聚，从而在视觉任务中捕获全局特征。

```
class AttentionPool2d(nn.Module):
    def __init__(self, spacial_dim: int, embed_dim: int, num_heads: int, output_dim:int = None):
        super().__init__()
        self.positional_embedding = nn.Parameter(torch.randn(spacial_dim ** 2 +1, embed_dim) / embed_dim ** 0.5)
        self.k_proj = nn.Linear(embed_dim, embed_dim)
        self.q_proj = nn.Linear(embed_dim, embed_dim)
        self.v_proj = nn.Linear(embed_dim, embed_dim)
        self.c_proj = nn.Linear(embed_dim, output_dim or embed_dim)
        self.num_heads = num_heads

    def forward(self, x):
        # NCHW -> (HW)NC
        x = x.reshape(x.shape[0], x.shape[1], x.shape[2] * x.shape[3]).permute(2, 0, 1)
        # (HW+1)NC
        x = torch.cat([x.mean(dim=0, keepdim=True), x], dim=0)
        # (HW+1)NC
        x = x + self.positional_embedding[:, None, :].to(x.dtype)
        x, _ = F.multi_head_attention_forward(
            query=x, key=x, value=x,
            embed_dim_to_check=x.shape[-1],
            num_heads=self.num_heads,
            q_proj_weight=self.q_proj.weight,
            k_proj_weight=self.k_proj.weight,
```

```
                v_proj_weight=self.v_proj.weight,
                in_proj_weight=None,
                in_proj_bias=torch.cat([self.q_proj.bias, self.k_proj.bias, self.v_
                    proj.bias]),
                bias_k=None,
                bias_v=None,
                add_zero_attn=False,
                dropout_p=0,
                out_proj_weight=self.c_proj.weight,
                out_proj_bias=self.c_proj.bias,
                use_separate_proj_weight=True,
                training=self.training,
                need_weights=False
            )
        
        return x[0]
```

（3）定义类 ModifiedResNet，这是对 ResNet 的一种改进实现。相比于传统的 ResNet，这个版本在以下方面进行了修改。

- 使用了 3 层卷积代替单层卷积作为网络的 "stem"，并用平均池化代替最大池化。
- 在步幅大于 1 的卷积前添加平均池化实现反混叠处理。
- 在池化层中使用 QKV 注意力机制代替传统的平均池化，从而增强了模型对全局信息的捕捉能力。

```
class ModifiedResNet(nn.Module):
    """
    一个与 torchvision 中的 ResNet 类似的 ResNet 类，但包含以下更改:
    - 现在有 3 个 "stem 卷积，而不是 1 个，并且使用平均池化代替最大池化
    - 执行反混叠的步幅卷积，其中一个 avgpool 被添加到步幅大于 1 的卷积前
    - 最后的池化层是 QKV 注意力，而不是平均池化
    """
    
    def __init__(self, layers, output_dim, heads, input_resolution=224, width=64):
        super().__init__()
        self.output_dim = output_dim
        self.input_resolution = input_resolution
        
        # 3 层 stem
        self.conv1 = nn.Conv2d(3, width // 2, kernel_size=3, stride=2, padding=1, bias=False)
        self.bn1 = nn.BatchNorm2d(width // 2)
        self.conv2 = nn.Conv2d(width // 2, width // 2, kernel_size=3, padding=1, bias=False)
        self.bn2 = nn.BatchNorm2d(width // 2)
        self.conv3 = nn.Conv2d(width // 2, width, kernel_size=3, padding=1, bias=False)
        self.bn3 = nn.BatchNorm2d(width)
        self.avgpool = nn.AvgPool2d(2)
        self.relu = nn.ReLU(inplace=True)
        
        # 残差层
        self._inplanes = width  # 这是一个在构造过程中使用的 "可变" 变量
        self.layer1 = self._make_layer(width, layers[0])
        self.layer2 = self._make_layer(width * 2, layers[1], stride=2)
        self.layer3 = self._make_layer(width * 4, layers[2], stride=2)
        self.layer4 = self._make_layer(width * 8, layers[3], stride=2)
        
        embed_dim = width * 32  # ResNet 特征维度
        self.attnpool = AttentionPool2d(input_resolution // 32, embed_dim, heads, output_dim)
    
    def _make_layer(self, planes, blocks, stride=1):
        layers = [Bottleneck(self._inplanes, planes, stride)]
```

```
        self._inplanes = planes * Bottleneck.expansion
        for _ in range(1, blocks):
            layers.append(Bottleneck(self._inplanes, planes))

        return nn.Sequential(*layers)

    def forward(self, x):
        def stem(x):
            for conv, bn in [(self.conv1, self.bn1), (self.conv2, self.bn2), (self.conv3,
            self.bn3)]:
                x = self.relu(bn(conv(x)))
            x = self.avgpool(x)
            return x

        x = x.type(self.conv1.weight.dtype)
        x = stem(x)
        x = self.layer1(x)
        x = self.layer2(x)
        x = self.layer3(x)
        x = self.layer4(x)
        x = self.attnpool(x)

        return x
```

（4）类 LayerNorm 对 PyTorch 中的类 LayerNorm 实现了子类化处理，用于处理 fp16 数据。类 QuickGELU 定义了一个快速的 GELU（Gaussian Error Linear Unit）激活函数的模块。

```
class LayerNorm(nn.LayerNorm):
    """子类化 torch 的 LayerNorm 以处理 fp16。"""

    def forward(self, x: torch.Tensor):
        orig_type = x.dtype
        ret = super().forward(x.type(torch.float32))
        return ret.type(orig_type)


class QuickGELU(nn.Module):
    def forward(self, x: torch.Tensor):
        return x * torch.sigmoid(1.702 * x)
```

（5）类 ResidualAttentionBlock 实现了一个残差注意力块，其中包含了多头注意力机制和多层感知机。在前向传播过程中，输入通过注意力层和多层感知机，然后将结果与输入进行残差连接，并经过 LayerNorm 处理。

```
class ResidualAttentionBlock(nn.Module):
    def __init__(self, d_model: int, n_head: int, attn_mask: torch.Tensor = None):
        super().__init__()

        self.attn = nn.MultiheadAttention(d_model, n_head)
        self.ln_1 = LayerNorm(d_model)    # 第一个 LayerNorm 模块
        self.mlp = nn.Sequential(OrderedDict([
            ("c_fc", nn.Linear(d_model, d_model * 4)),    # 多层感知机的全连接层
            ("gelu", QuickGELU()),    # 快速 GELU 激活函数
            ("c_proj", nn.Linear(d_model * 4, d_model))    # 全连接层
        ]))
        self.ln_2 = LayerNorm(d_model)    # 第二个 LayerNorm 模块
        self.attn_mask = attn_mask    # 注意力屏蔽矩阵

    def attention(self, x: torch.Tensor):
        """执行多头注意力机制"""
        self.attn_mask = self.attn_mask.to(dtype=x.dtype, device=x.device) if self.
```

```
        attn_mask is not None else None
        return self.attn(x, x, x, need_weights=False, attn_mask=self.attn_mask)[0]

    def forward(self, x: torch.Tensor):
        """前向传播过程"""
        x = x + self.attention(self.ln_1(x))  # 加上注意力机制的残差连接和 LayerNorm
        x = x + self.mlp(self.ln_2(x))        # 加上多层感知机的残差连接和 LayerNorm
        return x
```

（6）类 Transformer 实现了一个简单的 Transformer 模型，其中包含多个残差注意力块并将其作为 Transformer 的层。在前向传播过程中，输入通过多个残差注意力块进行处理，然后返回输出。

```
class Transformer(nn.Module):
    def __init__(self, width: int, layers: int, heads: int, attn_mask: torch.Tensor = None):
        super().__init__()
        self.width = width    # 模型宽度
        self.layers = layers  # Transformer 层的数量
        self.resblocks = nn.Sequential(*[ResidualAttentionBlock(width, heads, attn_mask)
        for _ in range(layers)])

    def forward(self, x: torch.Tensor):
        """Transformer 的前向传播过程"""
        return self.resblocks(x)
```

（7）类 VisualTransformer 实现了一个视觉 Transformer 模型，用于处理图像数据。该模型首先使用卷积提取特征，然后将特征送入 Transformer 模型进行特征编码，最后将编码后的特征投影到指定维度空间。

```
class VisualTransformer(nn.Module):
    def __init__(self, input_resolution: int, patch_size: int, width: int, layers: int,
    heads: int, output_dim: int):
        super().__init__()
        self.input_resolution = input_resolution  # 输入图像的分辨率
        self.output_dim = output_dim   # 输出维度
        self.conv1 = nn.Conv2d(in_channels=3, out_channels=width, kernel_size=patch_size,
        stride=patch_size, bias=False)

        scale = width ** -0.5
        self.class_embedding = nn.Parameter(scale * torch.randn(width))    # 类别嵌入向量
        self.positional_embedding = nn.Parameter(scale * torch.randn((input_resolution
        //patch_size) ** 2 + 1, width))
        # 位置嵌入向量
        self.ln_pre = LayerNorm(width)  # 输入嵌入的 Layer Normalization

        self.transformer = Transformer(width, layers, heads)
        # 使用 Transformer 模型进行特征提取和编码

        self.ln_post = LayerNorm(width)  # 输出嵌入的 Layer Normalization
        self.proj = nn.Parameter(scale * torch.randn(width, output_dim))
        # 投影矩阵，将特征映射到指定维度空间

    def forward(self, x: torch.Tensor):
        x = self.conv1(x)   # 卷积提取特征
        x = x.reshape(x.shape[0], x.shape[1], -1)  # 将特征图展平
        x = x.permute(0, 2, 1)   # 调整维度顺序以适应 Transformer 模型输入
        x = torch.cat([self.class_embedding.to(x.dtype) + torch.zeros(x.shape[0], 1, x.
        shape[-1], dtype=x.dtype, device=x.device), x], dim=1)
            # 添加类别嵌入特征向量中
        x = x + self.positional_embedding.to(x.dtype)   # 加上位置嵌入向量
        x = self.ln_pre(x)   # 应用 Layer Normalization
```

```
            x = x.permute(1, 0, 2)  # 调整维度顺序以适应 Transformer 模型输入格式
            x = self.transformer(x)  # 通过 Transformer 进行特征编码
            x = x.permute(1, 0, 2)  # 调整维度顺序以适应输出格式

            x = self.ln_post(x[:, 0, :])  # 应用 Layer Normalization，并选择序列的第一个位置作为输出

            if self.proj is not None:
                x = x @ self.proj  # 使用投影矩阵将特征映射到指定维度空间

            return x
```

（8）类 CLIP 实现了 CLIP 模型，结合文本和视觉信息，使用 Transformer 编码文本和视觉特征，并计算它们之间的余弦相似度作为对数概率。其中，视觉特征提取器可以是 ModifiedResNet 或 VisualTransformer。

```
class CLIP(nn.Module):
    def __init__(self,
                 embed_dim: int,
                 # 视觉
                 image_resolution: int,
                 vision_layers: Union[Tuple[int, int, int, int], int],
                 vision_width: int,
                 vision_patch_size: int,
                 # 文本
                 context_length: int,
                 vocab_size: int,
                 transformer_width: int,
                 transformer_heads: int,
                 transformer_layers: int
                 ):
        super().__init__()

        self.context_length = context_length  # 上下文长度

        # 根据输入确定使用 ModifiedResNet 还是 VisualTransformer 进行视觉特征提取和编码
        if isinstance(vision_layers, (tuple, list)):
            vision_heads = vision_width * 32 // 64
            self.visual = ModifiedResNet(
                layers=vision_layers,
                output_dim=embed_dim,
                heads=vision_heads,
                input_resolution=image_resolution,
                width=vision_width
            )
        else:
            vision_heads = vision_width // 64
            self.visual = VisualTransformer(
                input_resolution=image_resolution,
                patch_size=vision_patch_size,
                width=vision_width,
                layers=vision_layers,
                heads=vision_heads,
                output_dim=embed_dim
            )

        # 文本编码器使用 Transformer
        self.transformer = Transformer(
            width=transformer_width,
            layers=transformer_layers,
            heads=transformer_heads,
            attn_mask=self.build_attention_mask()
        )
```

```python
        self.vocab_size = vocab_size
        self.token_embedding = nn.Embedding(vocab_size, transformer_width)  # 文本词嵌入
        self.positional_embedding = nn.Parameter(torch.empty(self.context_length,
        transformer_width))  # 位置嵌入
        self.ln_final = LayerNorm(transformer_width)  # 最终的 Layer Normalization

        self.text_projection = nn.Parameter(torch.empty(transformer_width, embed_dim))
        # 文本特征投影矩阵
        self.logit_scale = nn.Parameter(torch.ones([]) * np.log(1 / 0.07))  # 对数尺度参数

        self.initialize_parameters()

    def initialize_parameters(self):
        # 初始化参数
        nn.init.normal_(self.token_embedding.weight, std=0.02)
        nn.init.normal_(self.positional_embedding, std=0.01)

        # 初始化视觉编码器和 Transformer 模型中的参数
        if isinstance(self.visual, ModifiedResNet):
            if self.visual.attnpool is not None:
                std = self.visual.attnpool.c_proj.in_features ** -0.5
                nn.init.normal_(self.visual.attnpool.q_proj.weight, std=std)
                nn.init.normal_(self.visual.attnpool.k_proj.weight, std=std)
                nn.init.normal_(self.visual.attnpool.v_proj.weight, std=std)
                nn.init.normal_(self.visual.attnpool.c_proj.weight, std=std)

            for resnet_block in [self.visual.layer1, self.visual.layer2, self.visual.
            layer3, self.visual.layer4]:
                for name, param in resnet_block.named_parameters():
                    if name.endswith("bn3.weight"):
                        nn.init.zeros_(param)

        proj_std = (self.transformer.width ** -0.5) * ((2 * self.transformer.layers) **-0.5)
        attn_std = self.transformer.width ** -0.5
        fc_std = (2 * self.transformer.width) ** -0.5
        for block in self.transformer.resblocks:
            nn.init.normal_(block.attn.in_proj_weight, std=attn_std)
            nn.init.normal_(block.attn.out_proj.weight, std=proj_std)
            nn.init.normal_(block.mlp.c_fc.weight, std=fc_std)
            nn.init.normal_(block.mlp.c_proj.weight, std=proj_std)

        if self.text_projection is not None:
            nn.init.normal_(self.text_projection, std=self.transformer.width ** -0.5)

    def build_attention_mask(self):
        # 惰性创建自回归注意力掩码，视觉和文本序列之间的全局注意力
        # PyTorch 使用加性注意力掩码；填充为 -inf
        mask = torch.empty(self.context_length, self.context_length)
        mask.fill_(float("-inf"))
        mask.triu_(1)  # 将下三角区域置零
        return mask

    @property
    def dtype(self):
        return self.visual.conv1.weight.dtype

    def encode_image(self, image):
        return self.visual(image.type(self.dtype))

    def encode_text(self, text):
        x = self.token_embedding(text).type(self.dtype)  # 获取文本嵌入向量
    [batch_size, n_ctx, d_model]
```

```
        x = x + self.positional_embedding.type(self.dtype)
        x = x.permute(1, 0, 2)  # 调整维度顺序以适应 Transformer 模型输入
        x = self.transformer(x)
        x = x.permute(1, 0, 2)  # 调整维度顺序
        x = self.ln_final(x).type(self.dtype)

        # x.shape = [batch_size, n_ctx, transformer.width]
        # 提取来自 EOT 嵌入的特征 ( EOT 标记是每个序列中的最高数值 )
        x = x[torch.arange(x.shape[0]), text.argmax(dim=-1)] @ self.text_projection

        return x

    def forward(self, image, text):
        image_features = self.encode_image(image)
        text_features = self.encode_text(text)

        # 归一化特征向量
        image_features = image_features / image_features.norm(dim=-1, keepdim=True)
        text_features = text_features / text_features.norm(dim=-1, keepdim=True)

        # 计算余弦相似度作为对数概率
        logit_scale = self.logit_scale.exp()
        logits_per_image = logit_scale * image_features @ text_features.t()
        logits_per_text = logit_scale * text_features @ image_features.t()

        # shape = [global_batch_size, global_batch_size]
        return logits_per_image, logits_per_text
```

对上述代码的具体说明如下。

- __init__ 方法初始化了 CLIP 模型，接收多个参数，包括嵌入维度、图像分辨率、视觉编码器的层数、宽度、补丁大小等，以及文本编码器的相关参数。根据参数设置，初始化视觉编码器为 ModifiedResNet 或 VisualTransformer，初始化文本编码器为 Transformer，同时初始化了文本词嵌入、位置嵌入、LayerNorm 等参数。
- initialize_parameters 方法用于初始化模型参数，包括文本词嵌入、位置嵌入以及视觉编码器和 Transformer 模型中的参数。
- build_attention_mask 方法惰性创建自回归注意力掩码，用于 Transformer 中的注意力机制，以实现全局注意力。
- encode_image 方法用于对图像进行编码，调用视觉编码器将图像转换为视觉特征。
- encode_text 方法用于对文本进行编码，包括文本词嵌入、位置嵌入以及 Transformer 编码器，最后提取文本特征。
- forward 方法定义了 CLIP 模型的前向传播过程，接收图像和文本输入，并计算它们之间的余弦相似度，将其作为对数概率输出。

（9）函数 convert_weights 用于将适用的模型参数转换为半精度浮点数（fp16）格式。它遍历模型的所有层，对满足条件的参数执行相应的转换操作，包括卷积层、线性层、多头注意力层以及文本投影和输出投影。

```
def convert_weights(model: nn.Module):
    """将适用的模型参数转换为 fp16 格式"""

    def _convert_weights_to_fp16(l):
        # 如果是卷积层、线性层，则将权重和偏置转换为 fp16 格式
        if isinstance(l, (nn.Conv1d, nn.Conv2d, nn.Linear)):
            l.weight.data = l.weight.data.half()
            if l.bias is not None:
```

6.3　CLIP 模型训练与微调

```
            l.bias.data = l.bias.data.half()

        # 如果是多头注意力层，则将相关参数转换为 fp16 格式
        if isinstance(l, nn.MultiheadAttention):
            for attr in [*[f"{s}_proj_weight" for s in ["in", "q", "k", "v"]], "in_proj_
                bias", "bias_k", "bias_v"]:
                tensor = getattr(l, attr)
                if tensor is not None:
                    tensor.data = tensor.data.half()

        # 如果存在文本投影或输出投影，则将其参数转换为 fp16 格式
        for name in ["text_projection", "proj"]:
            if hasattr(l, name):
                attr = getattr(l, name)
                if attr is not None:
                    attr.data = attr.data.half()

    # 对模型应用参数转换函数
    model.apply(_convert_weights_to_fp16)
```

（10）函数 build_model 用于构建模型并加载预训练权重。它解析了预训练权重中各个部分的参数，然后根据这些参数构建了适用于 CLIP 模型的实例，并加载了预训练权重。同时，函数 build_model 也负责将加载的权重转换为半精度浮点数（fp16）格式。

```
def build_model(state_dict: dict):
    """构建模型并加载预训练权重"""

    # 检查预训练权重是否适用于 VisualTransformer
    vit = "visual.proj" in state_dict

    # 解析视觉部分的参数
    if vit:
        vision_width = state_dict["visual.conv1.weight"].shape[0]
        vision_layers = len([k for k in state_dict.keys() if k.startswith("visual.") and
        k.endswith(".attn.in_proj_weight")])
        vision_patch_size = state_dict["visual.conv1.weight"].shape[-1]
        grid_size = round((state_dict["visual.positional_embedding"].shape[0] - 1) ** 0.5)
        image_resolution = vision_patch_size * grid_size
    else:
        counts: list = [len(set(k.split(".")[2] for k in state_dict if k.startswith
        (f"visual.layer{b}"))) for b in [1, 2, 3, 4]]
        vision_layers = tuple(counts)
        vision_width = state_dict["visual.layer1.0.conv1.weight"].shape[0]
        output_width = round((state_dict["visual.attnpool.positional_embedding"].shape
        [0] - 1) ** 0.5)
        vision_patch_size = None
        assert output_width ** 2 + 1 == state_dict["visual.attnpool.positional_embedding"].
        shape[0]
        image_resolution = output_width * 32

    # 解析文本部分的参数
    embed_dim = state_dict["text_projection"].shape[1]
    context_length = state_dict["positional_embedding"].shape[0]
    vocab_size = state_dict["token_embedding.weight"].shape[0]
    transformer_width = state_dict["ln_final.weight"].shape[0]
    transformer_heads = transformer_width // 64
    transformer_layers = len(set(k.split(".")[2] for k in state_dict if k.startswith
    (f"transformer.resblocks")))

    # 创建 CLIP 模型实例
    model = CLIP(
        embed_dim,
```

```
            image_resolution, vision_layers, vision_width, vision_patch_size,
            context_length, vocab_size, transformer_width, transformer_heads, transformer_layers
    )

    # 删除 state_dict 中的不必要键
    for key in ["input_resolution", "context_length", "vocab_size"]:
        if key in state_dict:
            del state_dict[key]

    # 将模型参数转换为 fp16 格式并加载预训练权重
    convert_weights(model)
    model.load_state_dict(state_dict)

    return model.eval()
```

2. Lightning 包装器

文件 wrapper.py 实现了对 CLIP 模型的 Lightning 包装器，提供了训练、验证和优化器配置等功能。Lightning 包装器是指将一个模型或训练过程包装在 PyTorch Lightning 框架中的类。PyTorch Lightning 是一个用于深度学习研究和生产的高级训练框架，它通过提供预定义的训练循环和一些实用功能来简化模型训练过程。在这种情况下，CLIPWrapper 和 CustomCLIPWrapper 就是将 CLIP 模型与 PyTorch Lightning 框架结合起来的包装器。这些包装器让使用 PyTorch Lightning 的训练、验证和优化功能变得更加简单和高效。

文件 wrapper.py 的具体实现流程如下。

（1）下面代码定义了一个用于 CLIP 模型的 Lightning 包装器，包含训练、验证和优化器配置等功能。__init__ 方法初始化模型和参数，num_training_steps 方法计算总训练步骤，training_step 方法定义了训练过程，validation_step 方法定义了验证过程，configure_optimizers 方法配置了优化器和学习率调度器。

```
class CLIPWrapper(pl.LightningModule):
    def __init__(self,
                 model_name: str,
                 config: dict,
                 minibatch_size: int
                 ):
        """一个适用于 CLIP 模型的 Lightning 包装器，如论文中所述。

        Args:
            model_name (str): 区分大小写的视觉模型名称。
            config (dict): 包含 CLIP 实例化参数的字典。
        """
        super().__init__()
        self.model_name = model_name
        self.model = CLIP(**config)
        self.minibatch_size = minibatch_size
        self.isViT = 'ViT' in self.model_name
        self.automatic_optimization = False

    @property
    def num_training_steps(self) -> int:
        """从数据模块和设备推断出的总训练步骤。"""
        dataset = self.train_dataloader()
        if self.trainer.max_steps:
            return self.trainer.max_steps

        dataset_size = len(dataset)

        num_devices = max(1, self.trainer.num_gpus, self.trainer.num_processes)
```

6.3 CLIP 模型训练与微调

```python
        if self.trainer.tpu_cores:
            num_devices = max(num_devices, self.trainer.tpu_cores)

        effective_batch_size = dataset.batch_size * self.trainer.accumulate_grad_batches
        * num_devices
        return (dataset_size // effective_batch_size) * self.trainer.max_epochs

    def training_step(self, train_batch, idx):
        """训练步骤。

        Args:
            train_batch: 一个批次的训练数据。
            idx: 步骤索引。
        """
        # 获取优化器和调度器
        optimizer = self.optimizers()

        image, text = train_batch
        n = math.ceil(len(image) // self.minibatch_size)
        image_mbs = torch.chunk(image, n)
        text_mbs = torch.chunk(text, n)

        # 计算原始统计数据
        with torch.no_grad():
            ims = [F.normalize(self.model.encode_image(im), dim=1) for im in image_mbs]
            txt = [F.normalize(self.model.encode_text(t), dim=1) for t in text_mbs]
            # 从所有的 GPU 收集数据
            ims = self.all_gather(torch.cat(ims))
            txt = self.all_gather(torch.cat(txt))

            if len(ims.shape) == 3:
                ims = list(ims)
                txt = list(txt)
            else:
                ims = [ims]
                txt = [txt]

            image_logits = torch.cat(ims) @ torch.cat(txt).t() * self.model.logit_scale.exp()
            ground_truth = torch.arange(len(image_logits)).long().to(image_logits.device)
            loss = (F.cross_entropy(image_logits, ground_truth) + F.cross_entropy(image_
            logits.t(), ground_truth)).div(2)
            acc_i = (torch.argmax(image_logits, 1) == ground_truth).sum()
            acc_t = (torch.argmax(image_logits, 0) == ground_truth).sum()
            self.log_dict({'loss': loss / len(ims), 'acc': (acc_i + acc_t) / 2 / len
            (image) / len(ims)}, prog_bar=True)

        if isinstance(optimizer, list):
            optimizer = optimizer[0]
        optimizer.zero_grad()

        # 图像损失
        for j, mb in enumerate(image_mbs):
            images_tmp = copy.deepcopy(ims)
            images_tmp[self.global_rank][j*self.minibatch_size:(j+1)*self.minibatch_size]
            = F.normalize(self.model.encode_image(mb), dim=1)
            image_logits = torch.cat(images_tmp) @ torch.cat(txt).t() * self.model.logit
            _scale.exp()
            ground_truth = torch.arange(len(image_logits)).long().to(image_logits.device)
            loss = (F.cross_entropy(image_logits, ground_truth) + F.cross_entropy(image_
            logits.t(), ground_truth))/2
            self.manual_backward(loss)

        # 文本损失
```

```python
        for j, mb in enumerate(text_mbs):
            text_tmp = copy.deepcopy(txt)
            text_tmp[self.global_rank][j*self.minibatch_size:(j+1)*self.
            minibatch_size] = F.normalize(self.model.encode_text(mb), dim=1)
            image_logits = torch.cat(ims) @ torch.cat(text_tmp).t() * self.model.logit_
            scale.exp()
            loss = (F.cross_entropy(image_logits, ground_truth) + F.cross_entropy(image_
            logits.t(), ground_truth))/2
            self.manual_backward(loss)

    optimizer.step()
    lr_scheduler = self.lr_schedulers()
    lr_scheduler.step()
    self.model.logit_scale.data.clamp_(-np.log(100), np.log(100))

def validation_step(self, val_batch, idx):
    """验证步骤。

    Args:
        val_batch: 一个批次的验证数据。
        idx: 步骤索引。
    """
    image, text = val_batch
    image_logits, text_logits = self.forward(image, text)
    ground_truth = torch.arange(len(image_logits))
    loss = (F.cross_entropy(image_logits, ground_truth) + F.cross_entropy(text_logits,
    ground_truth)).div(2)
    self.log('val_loss', loss)

def configure_optimizers(self):
    lr = {
        "RN50": 5e-4,
        "RN101": 5e-4,
        "RN50x4": 5e-4,
        "RN50x16": 4e-4,
        "RN50x64": 3.6e-4,
        "ViT-B/32": 5e-4,
        "ViT-B/16": 5e-4,
        "ViT-L/14": 4e-4,
        "ViT-L/14-336px": 2e-5
    }[self.model_name]

    optimizer = torch.optim.AdamW(
        self.model.parameters(),
        lr=lr,
        betas=(
            0.9,
            0.98 if self.isViT else 0.999
        ),
        eps=1e-6 if self.isViT else 1e-8,
        weight_decay=0.2
    )

    # 使用库 pytorch-cosine-annealing-with-warmup, 需要用 pip 安装此库
    lr_scheduler = CosineAnnealingWarmupRestarts(
        optimizer,
        first_cycle_steps=self.num_training_steps,
        cycle_mult=1.0,
        max_lr=lr,
        min_lr=0,
        warmup_steps=2000
    )
```

```
        return {'optimizer': optimizer, 'lr_scheduler': lr_scheduler}
```

对上述代码的具体说明如下。

- __init__ 方法用于初始化 CLIPWrapper 类，接收模型名称、配置参数和最小批处理大小作为参数，并创建 CLIP 模型的实例。
- num_training_steps 方法用于计算总的训练步数，根据数据模块和设备的情况推断。
- training_step 方法定义了训练步骤，包括计算损失、反向传播和更新优化器等操作。
- validation_step 方法定义了验证步骤，用于计算验证损失。
- configure_optimizers 方法配置优化器和学习率调度器，用于模型训练时的参数优化。

（2）下面代码定义了一个自定义的 CLIP 包装器类 CustomCLIPWrapper，该类继承自 CLIPWrapper，用于训练和微调 CLIP 模型。它允许替换默认的图像编码器和文本编码器，并支持自蒸馏（distillation）的训练方式。

```
class CustomCLIPWrapper(CLIPWrapper):
    def __init__(self,
                 image_encoder,
                 text_encoder,
                 minibatch_size,
                 learning_rate=3e-3,
                 kl_coeff=1.0,
                 avg_word_embs=False
                 ):
        with open('models/configs/RN.yaml') as fin:
            config = yaml.safe_load(fin)['RN50']
        super().__init__('RN50', config, minibatch_size)
        del self.model.visual
        del self.model.transformer
        self.model.visual = image_encoder
        self.model.transformer = text_encoder
        self.learning_rate = learning_rate
        self.avg_word_embs = avg_word_embs
        self.sink_temp = nn.Parameter(torch.ones([]) * np.log(1 / 0.07))

        self.teacher = copy.deepcopy(self.model)
        self.kl_coeff = kl_coeff

    def training_step(self, train_batch, idx):
        #获取优化器和调度器
        optimizer = self.optimizers()

        image, text = train_batch
        n = math.ceil(len(image) // self.minibatch_size)
        image_mbs = torch.chunk(image, n)
        text_mbs_ids = torch.chunk(torch.arange(len(image)), n)

        text_mbs = []
        for s in text_mbs_ids:
            d = {}
            for key in list(text.keys()):
                d[key] = text[key][s]
            text_mbs.append(d)

        with torch.no_grad():
            ims = [F.normalize(self.model.encode_image(im), dim=1) for im in image_mbs]
            txt = [F.normalize(self.encode_text(t), dim=1) for t in text_mbs]
            # 从所有的 GPU 收集数据
            ims = self.all_gather(torch.cat(ims))
            txt = self.all_gather(torch.cat(txt))
```

```python
            if len(ims.shape) == 3:
                ims = list(ims)
                txt = list(txt)
            else:
                ims = [ims]
                txt = [txt]

            image_logits_notemp = torch.cat(ims) @ torch.cat(txt).t()
            image_logits = image_logits_notemp * self.model.logit_scale.exp()
            ground_truth = torch.arange(len(image_logits)).long().to(image_logits.device)
            loss = (F.cross_entropy(image_logits, ground_truth) + F.cross_entropy(image_
            logits.t(), ground_truth)).div(2)
            acc_i = (torch.argmax(image_logits, 1) == ground_truth).sum()
            acc_t = (torch.argmax(image_logits, 0) == ground_truth).sum()
            teacher_ims = [F.normalize(self.teacher.encode_image(im), dim=1) for im in
            image_mbs]
            teacher_txt = [F.normalize(self.encode_text(t, teacher=True), dim=1) for t
            in text_mbs]

            teacher_ims = self.all_gather(torch.cat(teacher_ims))
            teacher_txt = self.all_gather(torch.cat(teacher_txt))

            if len(teacher_ims.shape) == 3:
                teacher_ims = list(teacher_ims)
                teacher_txt = list(teacher_txt)
            else:
                teacher_ims = [teacher_ims]
                teacher_txt = [teacher_txt]

            sim_ii, sim_tt, sim_it, sim_ti = self.compute_similarities(torch.cat
            (teacher_ims), torch.cat(teacher_txt))

            # 最优传输
            img_cost = - (sim_ii + sim_tt + sim_it)
            txt_cost = - (sim_ii + sim_tt + sim_ti)
            img_target = self.sinkhorn(img_cost)
            txt_target = self.sinkhorn(txt_cost)
            loss += (F.kl_div(F.log_softmax(image_logits_notemp * self.sink_temp, dim=-1),
            img_target,
            reduction='batchmean') + F.kl_div(F.log_softmax(image_logits_notemp.t() *
            self.sink_temp, dim=-1), txt_target,
            reduction='batchmean')) / 2 * self.kl_coeff
            self.log_dict({'loss': loss / len(ims), 'acc': (acc_i + acc_t) / 2 / len
            (image) / len(ims)}, prog_bar=True)

        if isinstance(optimizer, list):
            optimizer = optimizer[0]
        optimizer.zero_grad()

        for j, mb in enumerate(image_mbs):
            images_tmp = copy.deepcopy(ims)
            images_tmp[self.global_rank][j*self.minibatch_size:(j+1)*self.minibatch_size] =
            F.normalize(self.model.encode_image(mb), dim=1)
            image_logits_notemp = torch.cat(images_tmp) @ torch.cat(txt).t()
            image_logits = image_logits_notemp * self.model.logit_scale.exp()
            loss = (F.cross_entropy(image_logits, ground_truth) + F.cross_entropy(image_
            logits.t(), ground_truth))/2
            loss += (F.kl_div(F.log_softmax(image_logits_notemp * self.sink_temp, dim=-1),
            img_target, reduction='batchmean') + F.kl_div(F.log_softmax(image_logits_
            notemp.t() *
            self.sink_temp, dim=-1), txt_target, reduction='batchmean')) / 2 * self.kl_coeff
            self.manual_backward(loss)
```

```python
        for j, mb in enumerate(text_mbs):
            text_tmp = copy.deepcopy(txt)
            text_tmp[self.global_rank][j*self.minibatch_size:(j+1)*self. minibatch_size]
            = F.normalize(self.encode_text(mb), dim=1)
            image_logits_notemp = torch.cat(ims) @ torch.cat(text_tmp).t()
            image_logits = image_logits_notemp * self.model.logit_scale.exp()
            loss = (F.cross_entropy(image_logits, ground_truth) + F.cross_entropy(image_
            logits.t(), ground_truth))/2
            loss += (F.kl_div(F.log_softmax(image_logits_notemp * self.sink_temp, dim=-1),
            img_target, reduction='batchmean') + F.kl_div(F.log_softmax(image_logits_
            notemp.t() *
            self.sink_temp, dim=-1), txt_target, reduction='batchmean')) / 2 * self.kl_coeff
            self.manual_backward(loss)

    optimizer.step()
    lr_scheduler = self.lr_schedulers()
    lr_scheduler.step()
    self.model.logit_scale.data.clamp_(-np.log(100), np.log(100))
    self.sink_temp.data.clamp_(-np.log(100), np.log(100))
    self.update_teacher()

def encode_text(self, inputs, teacher=False):
    if self.avg_word_embs:
        sequence_output = self.teacher.transformer(**inputs)[0] if teacher
        else self.model.transformer(**inputs)[0]

        embeddings = torch.sum(
            sequence_output * inputs["attention_mask"].unsqueeze(-1), dim=1) / torch.
            clamp(torch.sum(inputs["attention_mask"], dim=1, keepdims=True), min=1e-9)

        return embeddings
    else:
        return self.teacher.transformer(**inputs)[1] if teacher else self.
        model.transformer(**inputs)[1]

def compute_similarities(self, I_emb, T_emb):
    sim_ii, sim_tt = I_emb @ I_emb.t(), T_emb @ T_emb.t()
    sim_it, sim_ti = I_emb @ T_emb.t(), T_emb @ I_emb.t()
    return sim_ii, sim_tt, sim_it, sim_ti

def update_teacher(self):
    for teacher, student in zip(self.teacher.parameters(), self.model.
    parameters()):
        teacher.data.copy_(self.ema(student.data, teacher.data))

def ema(self, s, t):
    return s * (1 - 0.999) + t * 0.999

def forward(self, images, text):
    logits = F.normalize(self.model.encode_image(images), dim=1) @ F.normalize
     (self.encode_text(text), dim=1).t() *
    self.model.logit_scale.exp()
    return logits, logits.t()

def sinkhorn(self, out):
    Q = torch.exp(out / 0.05).t()    # Q 的值等于 K 乘以 B
    B = Q.shape[1]     # 要分配的样本数量
    K = Q.shape[0]     # 有多少个原型

    sum_Q = torch.sum(Q)
    Q /= sum_Q
```

```python
        for it in range(3):
            #规范化每一行: 每个原型的总权重必须是 1/K
            sum_of_rows = torch.sum(Q, dim=1, keepdim=True)
            Q /= sum_of_rows
            Q /= K

            #规范化每一列: 每个样本的总权重必须是 1/B
            Q /= torch.sum(Q, dim=0, keepdim=True)
            Q /= B

        Q *= B  #各列之和必须为 1, 以便 Q 是一个分配矩阵
        return Q.t()

    def configure_optimizers(self):
        lr = self.learning_rate

        optimizer = torch.optim.SGD(
            self.parameters(),
            lr=lr,
            momentum=0.9
        )

        lr_scheduler = CosineAnnealingWarmupRestarts(
            optimizer,
            first_cycle_steps=self.num_training_steps,
            cycle_mult=1.0,
            max_lr=lr,
            min_lr=0,
            warmup_steps=2000
        )

        return {'optimizer': optimizer, 'lr_scheduler': lr_scheduler}
```

对上述代码的具体说明如下。

- __init__ 方法初始化自定义的包装器，加载配置文件，并替换模型的视觉和转换器部分为自定义的编码器，同时设置超参数。
- training_step 方法实现了训练步骤，包括损失计算、反向传播、参数更新等操作。它支持在多个 GPU 上进行分布式训练，并使用自蒸馏技术进行模型训练。
- encode_text 方法用于编码文本输入，它可以根据需要选择是否使用平均词嵌入（avg_word_embs）或直接获取 CLIP 模型的文本编码。
- compute_similarities 方法用于计算图像和文本之间的相似度，用于自蒸馏训练中的损失计算。
- update_teacher 方法用于更新教师模型的参数，实现自蒸馏过程中的参数更新。
- ema 方法实现了指数移动平均，用于更新教师模型的参数。
- forward 方法执行前向传播操作，计算图像和文本之间的相似度。
- sinkhorn 方法用于执行 Sinkhorn 操作，能够优化分布匹配问题，这在自蒸馏过程中用于计算损失。
- configure_optimizers 方法用于配置优化器和学习率调度器，能够实现模型训练的优化器设置。

3. ResNet 模型配置

配置文件 RN.yaml 定义了不同规模和分辨率的 CLIP 模型的架构和超参数，用于设置 ResNet 模型的配置信息，例如 RN50、RN101、RN50x4 等。每个模型配置包括嵌入维度（embed_dim）、图像分辨率（image_resolution）、视觉层参数（vision_layers、vision_width、vision_patch_size）、上下文长度（context_length）、词汇表大小（vocab_size）以及 Transformer 的参数（transformer_width、transformer_heads、transformer_layers）等。

4. Vision Transformer（ViT）模型配置

配置文件 ViT.yaml 定义了不同规模和分辨率的 ViT 模型的架构和超参数，用于设置 Vision Transformer（ViT）模型的配置信息，例如 ViT-B/32、ViT-B/16、ViT-L/14 和 ViT-L/14-336px。每个模型配置包括嵌入维度（embed_dim）、图像分辨率（image_resolution）、视觉层参数（vision_layers、vision_width、vision_patch_size）、上下文长度（context_length）、词汇表大小（vocab_size）以及 Transformer 的参数（transformer_width、transformer_heads、transformer_layers）等。

6.3.4 训练模型

文件 train.py 是训练 CLIP 模型的主程序，首先根据命令行参数指定的模型名称加载相应的配置文件，然后创建一个 CLIPWrapper 模型实例，并根据命令行参数初始化数据模块。接着，使用 PyTorch Lightning 的 Trainer 对象进行训练。

```python
import yaml
from argparse import ArgumentParser
from pytorch_lightning import Trainer
from data.text_image_dm import TextImageDataModule
from models import CLIPWrapper


def main(hparams):
    config_dir = 'models/configs/ViT.yaml' if 'ViT' in hparams.model_name else 'models/configs/RN.yaml'
    with open(config_dir) as fin:
        config = yaml.safe_load(fin)[hparams.model_name]

    if hparams.minibatch_size < 1:
        hparams.minibatch_size = hparams.batch_size

    model = CLIPWrapper(hparams.model_name, config, hparams.minibatch_size)
    del hparams.model_name
    dm = TextImageDataModule.from_argparse_args(hparams)
    trainer = Trainer.from_argparse_args(hparams, precision=16, max_epochs=32)
    trainer.fit(model, dm)


if __name__ == '__main__':
    parser = ArgumentParser()
    parser.add_argument('--model_name', type=str, required=True)
    parser.add_argument('--minibatch_size', type=int, default=0)
    parser = TextImageDataModule.add_argparse_args(parser)
    parser = Trainer.add_argparse_args(parser)
    args = parser.parse_args()

    main(args)
```

对上述代码的具体说明如下。
- 加载模型配置文件：根据模型名称确定加载 ViT.yaml 还是 RN.yaml 配置文件。
- 创建模型实例：使用 CLIPWrapper 类创建模型实例，传入模型名称、配置和最小批次大小。
- 初始化数据模块：使用 TextImageDataModule.from_argparse_args 根据命令行参数初始化数据模块。
- 设置训练器参数：使用 Trainer.from_argparse_args 根据命令行参数设置训练器，包括精度和最大训练周期。

- 开始训练：使用 trainer.fit 方法开始训练模型。

6.3.5　模型微调

文件 train_finetune.py 用于微调 CLIP 模型的主程序 train_finetune.py，首先加载预训练的图像编码器（ResNet-50）、文本编码器（declutr-sci-base）以及相应的 tokenizer，然后创建一个 CustomCLIPWrapper 模型实例进行微调训练。

```
import torch
from argparse import ArgumentParser
from pytorch_lightning import Trainer
from data.text_image_dm import TextImageDataModule
from models import CustomCLIPWrapper
from torchvision.models import resnet50
from transformers import AutoTokenizer, AutoModel

def main(hparams):
    img_encoder = resnet50(pretrained=True)
    img_encoder.fc = torch.nn.Linear(2048, 768)
    tokenizer = AutoTokenizer.from_pretrained("johngiorgi/declutr-sci-base")
    txt_encoder = AutoModel.from_pretrained("johngiorgi/declutr-sci-base")
    if hparams.minibatch_size < 1:
        hparams.minibatch_size = hparams.batch_size
    model = CustomCLIPWrapper(img_encoder, txt_encoder, hparams.minibatch_size, avg_word
    _embs=True)
    dm = TextImageDataModule.from_argparse_args(hparams, custom_tokenizer=tokenizer)
    trainer = Trainer.from_argparse_args(hparams, precision=16, max_epochs=32)
    trainer.fit(model, dm)


if __name__ == '__main__':
    parser = ArgumentParser()
    parser.add_argument('--minibatch_size', type=int, default=0)
    parser = TextImageDataModule.add_argparse_args(parser)
    parser = Trainer.add_argparse_args(parser)
    args = parser.parse_args()
    main(args)
```

对上述代码的具体说明如下。
- 加载预训练模型和 tokenizer：加载预训练的 ResNet-50 图像编码器和 declutr-sci-base 文本编码器，以及相应的 tokenizer。
- 修改图像编码器：将 ResNet-50 的全连接层替换为一个线性层，将输出维度调整为 768。
- 创建模型实例：使用 CustomCLIPWrapper 类创建模型实例，传入图像编码器、文本编码器和其他参数，如最小批次大小。
- 初始化数据模块：使用 TextImageDataModule.from_argparse_args 根据命令行参数初始化数据模块，同时传入自定义的 tokenizer。
- 设置训练器参数：使用 Trainer.from_argparse_args 根据命令行参数设置训练器，包括精度和最大训练周期。
- 开始微调训练：使用 trainer.fit 方法开始微调训练模型。

6.3.6　调试运行

根据自己的需要，大家可以按照如下 3 种方式训练多模态模型 CLIP。

1. 全新训练

在训练多模态模型 CLIP 时可以直接使用项目中的配置信息，只需提供一个训练目录或自己的数据集即可。在训练时需要指定模型名称，并告诉训练文件夹和批量大小，所有可能的模型都可以在 models/config 目录下的 yaml 文件中找到。例如运行命令如下。

```
python train.py --model_name RN50 --folder data_dir --batchsize 512
```

2. 微调训练

为了更高效地进行 CLIP 训练，可以使用类 CustomCLIPWrapper，这个类用于微调预训练的图像和语言模型，这样可以大大提高性能效率。要使用这个功能，只需修改 train_finetune.py 文件，传入一个图像编码器和 Hugging Face 文本编码器。

```
img_encoder = resnet50(pretrained=True)
img_encoder.fc = torch.nn.Linear(2048, 768)

tokenizer = AutoTokenizer.from_pretrained("johngiorgi/declutr-sci-base")
txt_encoder = AutoModel.from_pretrained("johngiorgi/declutr-sci-base")

model = CustomCLIPWrapper(img_encoder, txt_encoder, hparams.minibatch_size, avg_word_embs=True)
```

具体的命令行参数与之前一样，只是去掉了--model_name 标志。

```
python train_finetune.py --folder data_dir --batchsize 51
```

3. 使用自己的 DataModule 进行训练

此时需要每个图像-文本对具有相同的 stem 名称（即 coco_img1.png 和 coco_img1.txt），你只需在运行时指定文件夹即可。任何子文件夹结构都将被忽略，这意味着只要 foo/bar/image1.jpg 和 myster/folder/image1.txt 共享一个共同的父文件夹，foo/bar/image1.jpg 就始终能找到它的 myster/folder/image1.txt。所有图像后缀都可以使用，唯一的期望是标题由\n 分隔。

4. 使用自己的数据进行训练

如果你有不同的训练需求，可以插入自己的 DataLoader。首先注释掉项目中的 DataModule，并将自己的 DataModule 插入 trainer.fit（model，your_data）中，然后编辑 train.py 脚本以满足需求。唯一的期望是返回元组的第一项是图像批次，第二项是文本批次。

第 7 章 多模态大模型的评估与验证

多模态大模型的评估与验证是指通过多种数据类型（如文本、图像、音频等）进行模型性能的全面测试和验证。在评估过程中，采用精度、召回率、F1 值等指标来衡量模型在不同模态数据上的表现。多模态大模型的验证不仅关注单一模态的性能，还特别注重多模态数据融合后的效果，以确保模型在实际应用中的可靠性和有效性。本章将详细讲解多模态大模型评估与验证的知识。

7.1 模型评估

模型的评估与验证是机器学习和人工智能开发过程中的关键步骤，旨在确保模型的准确性、鲁棒性和可泛化性。通过系统的评估与验证，可以确保模型在实际应用中的可靠性和有效性，减少因模型缺陷带来的风险和损失。

7.1.1 模型评估的必要性

模型评估是机器学习和人工智能开发过程中至关重要的步骤。在多模态大模型应用中，模型评估是一个复杂且多层次的过程，需要考查模型处理和融合不同模态数据的能力。

1. **评估模型性能**

模型评估的首要目的是衡量模型在完成特定任务时的性能。通过评估，可以确定模型是否能够准确、高效地处理输入数据并产生期望的输出。

- 准确性：确保模型能够在大多数情况下做出正确的预测或分类。
- 精确率和召回率：帮助理解模型在处理正类和负类样本时的表现，特别是在不平衡数据集上。
- F1 值：综合考虑精度和召回率，提供一个衡量模型整体性能的单一指标。

2. **检测模型的泛化能力**

评估模型在不同数据集上的表现，特别是在验证集和测试集上的表现，可以帮助我们了解模型的泛化能力，即模型在处理未见过的数据时的表现。

- 防止过拟合（Overfitting）：通过评估，可以检测到模型是否过度拟合训练数据，导致在新数据上的性能下降。
- 防止欠拟合（Underfitting）：评估可以帮助识别模型是否过于简单，以致无法捕捉数据中的复杂模式。

3. **选择最佳模型**

在模型开发过程中，通常会训练多个模型或调整不同的超参数，这时模型评估可以帮助我们选择在特定任务中表现最佳的模型。

- 比较不同的模型：通过一致的评估指标比较不同算法或不同超参数配置下的模型，选择最优方案。

- 调优模型：评估结果可以指导超参数调优、特征选择和其他模型优化步骤。

4. **确保模型的可靠性和稳定性**

模型在实际应用中需要具备可靠性和稳定性，通过评估可以提前发现潜在的问题。
- 鲁棒性测试：通过在不同条件下（如噪声、干扰、数据分布变化等）评估模型，可以确保模型的稳定性表现。
- 一致性验证：评估模型在不同环境和设置下的一致性，确保其在各种使用场景中的可靠性。

5. **提供透明性和可解释性**

模型评估提供了关于模型性能的详细信息，有助于理解和解释模型的行为，特别是在关键领域（如医疗、金融）中需要模型的决策过程透明。
- 性能报告：详细的评估报告可以帮助用户和决策者理解模型的优缺点和适用范围。
- 解释性分析：通过评估指标和错误分析，可以解释模型的决策机制和潜在的改进方向。

6. **满足监管和合规要求**

在许多行业，特别是医疗、金融和法律领域，模型评估是满足监管和合规要求的重要步骤。
- 合规性验证：通过评估，确保模型符合行业标准和法律法规的要求。
- 风险评估：识别和评估模型在应用中的潜在风险，确保其安全可靠。

7. **指导后续改进**

评估结果提供了关于模型性能的反馈，有助于指导后续的开发和改进。
- 发现问题和不足：通过评估，可以发现模型存在的问题和不足，为改进提供方向。
- 持续优化：评估结果可以作为持续优化和迭代开发的依据，不断提升模型性能。

8. **用户和客户信任**

通过严格的评估，向用户和客户展示模型的可靠性和有效性，增强其信任和接受度。
- 性能证明：通过评估结果证明模型的性能和可靠性，增强用户信心。
- 透明沟通：评估报告和结果可以作为与用户和客户沟通的工具，解释模型的能力和局限性。

总之，模型评估是确保模型在实际应用中有效性、可靠性和稳定性的关键步骤，是机器学习和人工智能项目成功的基石。

7.1.2 评估指标

多模态大模型的评估需要使用多种指标来衡量其在不同模态和整体融合上的表现，常用的评估指标如下。
- 准确率（Accuracy）：衡量模型在所有模态数据上的整体正确率。
- 精确率（Precision）、召回率（Recall）、F1值（F1 Score）：这些指标用于评估模型在分类任务中的表现，分别反映了模型的精确性、覆盖率和综合性能。
- ROC曲线（ROC Curve）和AUC（Area Under Curve）：用于评估二分类或多分类任务中的分类性能，特别是在不同阈值下的表现。
- BLEU、ROUGE等：在处理文本生成任务时，评估生成文本的质量，特别是在自然语言处理任务中。
- 均方误差（Mean Squared Error，MSE）、平均绝对误差（Mean Absolute Error，MAE）：用于回归任务中的性能评估，衡量模型预测值与真实值之间的差异。

实例7-1演示了对多种任务（分类、回归、文本生成）的模型进行评估和可视化操作的过程，包括分类任务的指标计算与可视化、回归任务的指标计算与可视化，以及文本生成任务的指标计算与可视化。

实例 7-1　对多任务模型进行评估和可视化操作
源码路径：codes/7/zhibiao.py

实例文件 zhibiao.py 的具体实现代码如下所示。

```python
import numpy as np
import matplotlib.pyplot as plt
import seaborn as sns
import pandas as pd
from sklearn.metrics import accuracy_score, precision_score, recall_score, f1_score, roc_auc_score, mean_squared_error, mean_absolute_error, confusion_matrix, classification_report
from nltk.translate.bleu_score import sentence_bleu, SmoothingFunction
from rouge import Rouge

# 示例数据
true_labels_classification = np.array([1, 0, 1, 1, 0, 1, 0, 0, 1, 0])
pred_labels_classification = np.array([1, 0, 1, 0, 0, 1, 1, 0, 1, 0])

true_labels_regression = np.array([3.5, 2.1, 4.0, 5.6, 3.3])
pred_labels_regression = np.array([3.7, 2.0, 4.1, 5.4, 3.5])

reference_texts = [["this", "is", "a", "test"], ["another", "test", "sentence"]]
generated_texts = [["this", "is", "test"], ["another", "sentence"]]

# 计算分类指标
accuracy = accuracy_score(true_labels_classification, pred_labels_classification)
precision = precision_score(true_labels_classification, pred_labels_classification)
recall = recall_score(true_labels_classification, pred_labels_classification)
f1 = f1_score(true_labels_classification, pred_labels_classification)
roc_auc = roc_auc_score(true_labels_classification, pred_labels_classification)

# 打印分类指标
print("Classification Metrics:")
print(f"Accuracy: {accuracy:.2f}")
print(f"Precision: {precision:.2f}")
print(f"Recall: {recall:.2f}")
print(f"F1 Score: {f1:.2f}")
print(f"ROC AUC: {roc_auc:.2f}")

# 计算回归指标
mse = mean_squared_error(true_labels_regression, pred_labels_regression)
mae = mean_absolute_error(true_labels_regression, pred_labels_regression)

# 打印回归指标
print("\nRegression Metrics:")
print(f"MSE: {mse:.2f}")
print(f"MAE: {mae:.2f}")

# 计算文本生成指标
smoothing_function = SmoothingFunction().method1
bleu_scores = [sentence_bleu([ref], gen, smoothing_function=smoothing_function) for ref, gen in zip(reference_texts, generated_texts)]
average_bleu = np.mean(bleu_scores)

rouge = Rouge()
scores = rouge.get_scores([" ".join(gen) for gen in generated_texts], [" ".join(ref) for ref in reference_texts], avg=True)

# 打印文本生成指标
print("\nText Generation Metrics:")
print(f"Average BLEU Score: {average_bleu:.2f}")
print(f"ROUGE Scores: {scores}")
```

```python
# 可视化分类报告
conf_matrix = confusion_matrix(true_labels_classification, pred_labels_classification)plt.
figure(figsize=(6, 4))
sns.heatmap(conf_matrix, annot=True, fmt="d", cmap="Blues")
plt.title("Confusion Matrix")
plt.xlabel("Predicted Labels")
plt.ylabel("True Labels")
plt.show()

# 打印分类报告
class_report = classification_report(true_labels_classification, pred_labels_classification, output_dict=True)
plt.figure(figsize=(10, 6))
sns.heatmap(pd.DataFrame(class_report).iloc[:-1, :].T, annot=True, cmap="Blues")
plt.title("Classification Report")
plt.show()

# 可视化回归误差
plt.figure(figsize=(6, 4))
plt.scatter(true_labels_regression, pred_labels_regression)
plt.plot([min(true_labels_regression), max(true_labels_regression)], [min(true_labels_regression), max(true_labels_regression)], 'r--')
plt.xlabel("True Values")
plt.ylabel("Predicted Values")
plt.title("Regression Results")
plt.show()

# 可视化文本生成指标
metrics_names = ['BLEU', 'ROUGE-1', 'ROUGE-2', 'ROUGE-L']
metrics_values = [average_bleu, scores['rouge-1']['f'], scores['rouge-2']['f'], scores['rouge-l']['f']]

plt.figure(figsize=(8, 6))
sns.barplot(x=metrics_names, y=metrics_values, palette="viridis")
plt.title("Text Generation Metrics")
plt.ylabel("Scores")
plt.show()
```

在上述代码中，首先通过计算准确率、精确率、召回率、F1 值、ROC、AUC 等指标评估分类任务的性能，并通过混淆矩阵和分类报告可视化分类结果。然后，计算并可视化回归任务的均方误差和平均绝对误差。最后，计算并展示文本生成任务的 BLEU 和 ROUGE 分数，并进行可视化展示。这些可视化图表有助于更直观地理解多模态模型在不同任务上的表现。执行代码后会输出：

```
Classification Metrics:
Accuracy: 0.80
Precision: 0.80
Recall: 0.80
F1 Score: 0.80
ROC AUC: 0.80

Regression Metrics:
MSE: 0.03
MAE: 0.16

Text Generation Metrics:
Average BLEU Score: 0.15
ROUGE Scores: {'rouge-1': {'r': 0.7083333333333333, 'p': 1.0, 'f': 0.8285714237224491}, 'rouge-2': {'r': 0.16666666666666666, 'p': 0.25, 'f': 0.19999999760000003}, 'rouge-l': {'r': 0.7083333333333333, 'p': 1.0, 'f': 0.8285714237224491}}
```

执行代码后不但会输出上面的分类任务、回归任务和文本生成任务的评估结果，还会绘制

图 7-1 所示的 4 张可视化图，这 4 张图分别对应了分类任务的混淆矩阵和分类报告热图、回归任务的回归结果散点图，以及文本生成任务的文本生成指标条形图。具体说明如下。

- 混淆矩阵：显示分类模型的真实标签和预测标签之间的关系。
- 分类报告热图：显示分类模型的精确率、召回率、F1 值等指标。
- 回归结果散点图：显示回归模型的真实值和预测值的对比。
- 文本生成指标条形图：显示文本生成任务的 BLEU 和 ROUGE 分数。

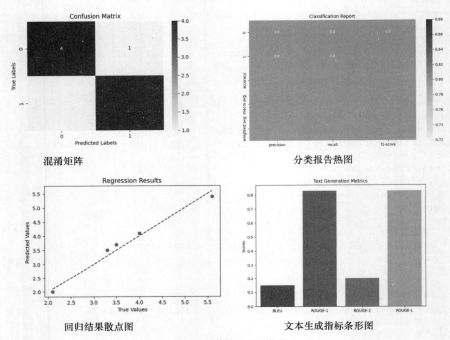

图 7-1　评估结果可视化图

7.1.3　单模态性能评估

在多模态大模型应用中评估每个单独模态的性能具有重要意义，这样做可以帮助我们深入了解每个模态对整体模型性能的贡献，发现模态之间的差异和潜在问题，并指导模型的进一步改进和优化，从而提高多模态大模型的整体性能和鲁棒性。单模态性能评估的信息如下。

- 文本：评估自然语言处理任务中的性能，如文本分类、情感分析、文本生成等。
- 图像：评估计算机视觉任务中的性能，如图像分类、物体检测、图像生成等。
- 音频：评估语音识别、情感识别等音频处理任务的性能。

实例 7-2 演示了使用 Python 相关库评估模型性能的方法。其中的文本数据是随机生成的，并使用随机生成的标签进行评估。本实例从指定路径读取了图像数据，计算了每个图像的预测结果，并可视化展示了图像信息。

实例 7-2　评估多模态中文本模型的性能
源码路径：codes/7/dan.py

实例文件 dan.py 的具体实现代码如下所示。

```
from sklearn.metrics import accuracy_score, classification_report
import numpy as np
import matplotlib.pyplot as plt
```

7.1 模型评估

```
from PIL import Image

# 模拟生成文本、图像数据
# 这里只是示例，实际应用中需要替换为真实数据
text_data = np.random.randint(0, 2, size=100)   # 示例文本数据
true_labels = np.random.randint(0, 2, size=100)  # 示例标签

# 图像数据路径
image_paths = ["image1.png", "image2.png", "image3.png", "image4.png", "image5.png"]image_data = []

# 读取图像数据并计算每个图像的预测结果
for path in image_paths:
    img = Image.open(path)
    img = img.resize((10, 10))   # 将图像调整为示例大小（10×10）
    img_array = np.array(img) / 255.0   # 将图像像素值缩放到 [0, 1] 范围内
    # 计算每个图像的预测结果（取每个通道的平均值）
    image_predictions = np.round(np.mean(img_array))
    image_data.append(image_predictions)

image_data = np.array(image_data)

# 模拟评估单个模态性能
text_accuracy = accuracy_score(true_labels, text_data)

# 打印单个模态性能评估结果
print("Text Modality Accuracy:", text_accuracy)

# 打印单个模态分类报告
print("\nText Modality Classification Report:")
print(classification_report(true_labels, text_data))

# 显示图像示例
plt.figure(figsize=(12, 3))
for i in range(len(image_paths)):
    plt.subplot(1, len(image_paths), i + 1)
    plt.imshow(np.array(Image.open(image_paths[i])))
    plt.title(f"Image {i+1}")
    plt.axis('off')
plt.show()
```

上述代码的实现流程如下。

（1）导入所需要的库，accuracy_score、classification_report 用于评估，numpy 用于数据处理，matplotlib.pyplot 用于绘图，PIL.Image 用于处理图像。

（2）模拟生成示例的文本和图像数据。文本数据使用 np.random.randint 随机生成，表示文本的分类标签，图像数据从指定路径读取，并将每张图像调整为示例大小（10×10），然后将像素值缩放到[0, 1]范围内。

（3）评估文本模态的性能。使用 accuracy_score 计算文本数据的准确率，并打印输出。

（4）使用 classification_report 打印文本模态的分类报告，包括精确率、召回率、F1 值等指标。打印输出下面的内容：

```
Text Modality Accuracy: 0.52

Text Modality Classification Report:
              precision    recall  f1-score   support

           0       0.45      0.51      0.48        43
           1       0.59      0.53      0.56        57

    accuracy                           0.52       100
```

```
    macro avg       0.52        0.52        0.52        100
 weighted avg       0.53        0.52        0.52        100
```

（5）使用 matplotlib.pyplot 展示图像模态的示例，将每张图像显示在子图中，并添加相应的标题，关闭坐标轴，如图 7-2 所示。

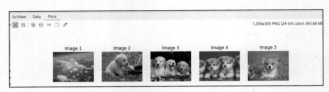

图 7-2　可视化展示图像模态数据

7.1.4　多模态融合性能评估

多模态大模型的核心优势在于能够融合不同模态的数据，因此评估融合性能至关重要。需要评估的指标包括以下几点。

- 融合策略的效果：评估早期融合、晚期融合和中期融合策略的效果，确定哪种策略能够更好地提升模型性能。
- 跨模态一致性：评估不同模态之间的一致性和互补性，确保融合后的模型能够充分利用各模态的信息。
- 信息增益：评估多模态融合是否带来了信息增益，即融合后的模型是否显著优于单模态模型。

下面的代码演示了使用 DummyClassifier 模拟多模态融合性能评估的过程。其中结合示例的文本数据和随机生成的图像特征，并使用虚拟的分类器进行训练和评估。最终，打印出多模态的准确率和分类报告。

```
import numpy as np
from sklearn.metrics import accuracy_score, classification_report
from sklearn.dummy import DummyClassifier

# 示例文本数据
text_data = ["This is a sample text for multimodal fusion.", "Another example text for evaluation."]

# 示例图像特征（随机生成）
image_features = np.random.rand(len(text_data), 10)    # 假设每个图像有 10 个特征

# 模拟多模态融合
multimodal_data = np.concatenate([np.array(text_data).reshape(-1, 1), image_features], axis=1)

# 假设这里有一些示例标签
true_labels = np.random.randint(0, 2, size=len(text_data))

# 使用 DummyClassifier 模拟分类器
clf = DummyClassifier(strategy="most_frequent")
clf.fit(multimodal_data, true_labels)

# 模拟多模态性能评估
multimodal_predictions = clf.predict(multimodal_data)

# 计算多模态准确率
multimodal_accuracy = accuracy_score(true_labels, multimodal_predictions)

# 打印多模态性能评估结果
print("Multimodal Accuracy:", multimodal_accuracy)
```

```
# 打印多模态分类报告
print("\nMultimodal Classification Report:")
print(classification_report(true_labels, multimodal_predictions))
```

执行代码后会输出：

```
Multimodal Accuracy: 0.5

Multimodal Classification Report:
              precision    recall  f1-score   support

           0       0.50      1.00      0.67         1
           1       0.00      0.00      0.00         1

    accuracy                           0.50         2
   macro avg       0.25      0.50      0.33         2
weighted avg       0.25      0.50      0.33         2
```

7.1.5　效率与资源使用

多模态大模型通常需要极高的计算资源，因此评估其效率和资源使用情况也是重要的一环。
- 计算效率：评估模型在训练和推理阶段的计算效率，考虑处理时间和资源消耗。
- 内存使用：评估模型在训练和推理过程中对内存的使用情况，确保其在资源受限的环境中也能有效运行。

实例 7-3 演示了使用随机森林分类器和 Dummy 分类器对多模态数据进行训练和推理的过程，并比较它们的效率。

实例 7-3　对比两种分类器的效率

源码路径：codes/7/xiao.py

实例文件 xiao.py 的具体实现代码如下所示。

```python
import time
import numpy as np
from sklearn.ensemble import RandomForestClassifier
from sklearn.dummy import DummyClassifier

# 模拟多模态数据：文本和图像特征
text_features = np.random.rand(1000, 1000)   # 假设文本特征维度为1000
image_features = np.random.rand(1000, 1000)  # 假设图像特征维度为1000

# 模拟标签
true_labels = np.random.randint(0, 2, size=1000)

# 方案1：随机森林分类器
print("Training and Inference with Random Forest Classifier:")
start_time = time.time()

# 模拟多模态融合
multimodal_data = np.concatenate([text_features, image_features], axis=1)

# 使用随机森林分类器进行训练
clf_rf = RandomForestClassifier(n_estimators=100, random_state=42)
clf_rf.fit(multimodal_data, true_labels)

train_time_rf = time.time() - start_time
print("Training Time (Random Forest):", train_time_rf, "seconds")

# 推理阶段
start_time = time.time()
```

```
# 进行推理
multimodal_predictions_rf = clf_rf.predict(multimodal_data)

inference_time_rf = time.time() - start_time
print("Inference Time (Random Forest):", inference_time_rf, "seconds")

# 方案2：Dummy 分类器（用于比较）
print("\nTraining and Inference with Dummy Classifier:")
start_time = time.time()

# 使用 Dummy 分类器进行训练
clf_dummy = DummyClassifier(strategy="most_frequent")
clf_dummy.fit(multimodal_data, true_labels)

train_time_dummy = time.time() - start_time
print("Training Time (Dummy Classifier):", train_time_dummy, "seconds")

# 推理阶段
start_time = time.time()

# 进行推理
multimodal_predictions_dummy = clf_dummy.predict(multimodal_data)

inference_time_dummy = time.time() - start_time
print("Inference Time (Dummy Classifier):", inference_time_dummy, "seconds")
```

上述代码的实现流程如下。

（1）准备多模态数据，并定义两种分类器——随机森林分类器和 Dummy 分类器。

（2）使用随机森林分类器对数据进行训练，并计算训练时间。然后对同样的数据进行推理，并计算推理时间。

（3）使用 Dummy 分类器对数据进行训练，并计算训练时间。同样地，对数据进行推理，并计算推理时间。

（4）打印输出训练和推理的时间，展示随机森林分类器和 Dummy 分类器的效率对比。执行代码后会输出：

```
Training and Inference with Random Forest Classifier:
Training Time (Random Forest): 2.3421850204467773 seconds
Inference Time (Random Forest): 0.014086008071899414 seconds

Training and Inference with Dummy Classifier:
Training Time (Dummy Classifier): 0.0009493827819824219 seconds
Inference Time (Dummy Classifier): 0.0 seconds
```

上面的输出结果表明在这个例子中，Dummy 分类器的训练和推理速度明显快于随机森林分类器，但是大家需要注意的是，在处理复杂数据和任务时，随机森林分类器的性能通常会更好。

7.1.6　定性评估和复杂场景评估

在实际应用中，定性评估也是多模态大模型评估的重要方面。常用的定性评估如下。

- 人工审查：通过人工审查生成的结果，评估其合理性和一致性。例如，检查生成的图像和对应的文本描述是否匹配。
- 用户反馈：收集用户对模型输出的反馈，评估其在实际使用中的效果和用户满意度。

通过这些评估方法，可以全面了解多模态大模型的性能，确定其优缺点，指导模型的改进和优化。另外，多模态大模型通常被应用于复杂场景中，因此需要针对实际应用场景进行如下评估。

- 任务特定评估：针对具体任务（如自动驾驶中的多模态感知、医疗诊断中的多模态数据分析）进行性能评估，确保模型在特定任务中的实用性。
- 环境鲁棒性：评估模型在不同环境和条件下的表现，如不同光照条件下的图像识别、不同背景噪声下的语音识别等。

7.1.7 语音命令识别系统

本项目是一个多模态命令识别系统，旨在识别和分类不同的语音命令（见实例 7-4）。通过建立和训练模型，能够准确地识别用户说出的命令，包括"no""yes""down""up""left""right""go""stop"等。除了音频数据，该项目还提供了一个大模型框架，可以轻松扩展到多模态数据，如结合文本、图像和音频数据。具体来说，本项目实现了如下功能。

- 交叉验证：使用交叉验证方法评估模型在不同数据子集上的性能表现，以确保模型的泛化能力。
- 验证集评估：独立的验证集评估模型在未见过的数据上的性能，以检查模型是否过拟合或欠拟合。
- 建立混淆矩阵：分析模型在每个类别上的预测结果，以查看模型在各个类别上的表现情况。
- 绘制准确率和损失曲线：可视化训练过程中的准确率和损失，以了解模型的训练情况。
- 模型保存与加载：保存训练好的模型，并重新加载以确保模型的保存和加载正常，验证模型的一致性。

通过以上评估与验证方法，可以全面了解本模型在不同数据集上的性能表现，并确保模型的可靠性和有效性。

实例 7-4　识别和分类不同的语音命令
源码路径：codes/7/audio_recognizing_keywords

实例文件 audio_recognizing_keywords 的具体实现代码如下。

（1）导入所需的库，设置环境，以进行语音指令识别项目的分析和建模，为后续的数据处理、模型构建和可视化作好准备。

```
import os
import pathlib

import matplotlib.pyplot as plt
import numpy as np
import seaborn as sns
import tensorflow as tf

from tensorflow.keras import layers
from tensorflow.keras import models
from IPython import display
```

（2）设置随机种子，确保每次运行代码时生成的随机数是一样的，这对于调试和结果复现是非常有帮助的。

```
seed = 42
tf.random.set_seed(seed)
np.random.seed(seed)
```

（3）下载名为"mini_speech_commands"的数据集，该数据集包含了用于语音指令识别的音频样本。如果指定路径下不存在数据集文件，则从给定的链接下载数据集，并将其解压到指定路径中。

```
DATASET_PATH = 'PATH'

if not os.path.exists(DATASET_PATH):
```

```
    os.makedirs(DATASET_PATH)

dataset_file = os.path.join(DATASET_PATH, 'mini_speech_commands.zip')

# 下载数据集

if not os.path.exists(dataset_file):
    tf.keras.utils.get_file(
        'mini_speech_commands.zip',
        origin="http://storage.googleapis.com/download.tensorflow.org/data/mini_speech_commands
        .zip", extract=True,
        cache_dir='.', cache_subdir='data')
```

执行代码后显示下载过程：

```
Downloading data from http://storage.googleapis.com/download.tensorflow.org/data/mini_
speech_commands.zip
182082353/182082353 [==============================] - 6s 0us/step
```

（4）读取已下载的语音指令数据集目录，并列出其中的指令类别。

```
data_dir = 'PATH/mini_speech_commands'
commands = np.array(tf.io.gfile.listdir(str(data_dir)))
commands = commands[(commands != 'README.md') & (commands != '.DS_Store')]
print('Commands:', commands)
```

对上述代码的具体说明如下。
- data_dir='PATH/mini_speech_commands'：指定数据集所在的目录路径。
- commands=np.array(tf.io.gfile.listdir(str(data_dir)))：使用 TensorFlow 的 tf.io.gfile.listdir()函数列出指定路径下的文件和目录，并将结果转换为 NumPy 数组。
- commands=commands[(commands!='README.md')&(commands!='.DS_Store')]：过滤掉文件列表中的 "README.md" 和 ".DS_Store"（如果存在），得到真正的指令类别列表。
- 最后通过 print()打印输出指令类别列表。

执行代码后会获取数据集中的所有指令类别，并将其打印出来：

```
Commands: ['right' 'go' 'no' 'left' 'stop' 'up' 'down' 'yes']
```

（5）从指定目录中创建训练集和验证集的数据管道，并获取训练集中的类别标签数组，为后续的模型训练和评估做准备。

```
train_ds, val_ds = tf.keras.utils.audio_dataset_from_directory(directory=data_dir,
                                                batch_size=64,
                                                validation_split=0.2,
                                                seed=0,
                                                output_sequence_length=16000,
                                                subset='both')

label_names = np.array(train_ds.class_names)
print()
print('label names: ', label_names)
```

在上述代码中，使用函数 tf.keras.utils.audio_dataset_from_directory()从指定目录中创建音频数据集，其主要功能如下。
- 指定目录 data_dir，其中包含音频文件。
- 将数据集分成训练集和验证集，验证集占总数据集的 20%。
- 设置随机种子以确保数据集分割的一致性。

- 输出的音频序列长度为 16000 个采样点。
- 返回训练集和验证集的数据管道 train_ds 和 val_ds。

通过 train_ds.class_names 获取训练集中的类别标签，并将其转换为 NumPy 数组 label_names。执行代码后会输出：

```
Found 8000 files belonging to 8 classes.
Using 6400 files for training.
Using 1600 files for validation.
label names:  ['down' 'go' 'left' 'no' 'right' 'stop' 'up' 'yes']
```

（6）通过如下代码返回训练集数据管道中每个元素的规格（specification），即每个样本的结构。

```
train_ds.element_spec
```

执行代码后会输出：

```
(TensorSpec(shape=(None, 16000, None), dtype=tf.float32, name=None),
 TensorSpec(shape=(None,), dtype=tf.int32, name=None))
```

在 TensorFlow 数据管道中，element_spec 属性描述了数据管道中每个元素的组成结构，包括数据类型、形状等信息。训练集数据管道中，每个元素的规格可能包括输入数据和标签数据，其结构通常是一个元组（input, label），其中 input 是输入数据，label 是对应的标签数据。

（7）定义函数 squeeze，功能是对训练集和验证集中的音频数据进行维度压缩，以便后续模型的输入处理。

```
def squeeze(audio, labels):
  audio = tf.squeeze(audio, axis=-1)
  return audio, labels

train_ds = train_ds.map(squeeze, tf.data.AUTOTUNE)
val_ds = val_ds.map(squeeze, tf.data.AUTOTUNE)
```

（8）将验证集 val_ds 分成两个子数据集，并返回其中一个作为测试集 test_ds，另一个作为新的验证集 val_ds。

```
test_ds = val_ds.shard(num_shards=2, index=0)
val_ds = val_ds.shard(num_shards=2, index=1)
```

在上述代码中，使用函数 shard() 进行数据集的分片操作，以便将数据集分成多个部分。具体说明如下。

- test_ds = val_ds.shard(num_shards=2, index=0)：将验证集 val_ds 分成两个子数据集，将其中的第一个子数据集作为测试集 test_ds。
- val_ds = val_ds.shard(num_shards=2, index=1)：将验证集 val_ds 分成两个子数据集，将其中的第二个子数据集作为新的验证集 val_ds。

（9）从训练集 train_ds 中取出一个批次的样本，并打印出每个样本的音频数据和标签的形状。

```
for example_audio, example_labels in train_ds.take(1):
  print(example_audio.shape)
  print(example_labels.shape)
```

执行代码后会打印输出每个样本音频数据的形状和标签的形状。

```
(64, 16000)
(64,)
```

（10）使用数组[1，1，3，0]作为索引，从标签名称数组 label_names 中获取对应索引位置的标签名称。例如，如果 label_names 数组中包含['down', 'go', 'left', 'no', 'right', 'stop', 'up', 'yes']，那么 label_names[[1，1，3，0]]将返回['go', 'go', 'no', 'down']，分别是索引为 1、1、3、0 的标签名称。

```
label_names[[1,1,3,0]]
```

执行代码后会输出：

```
array(['go', 'go', 'no', 'down'], dtype='<U5')
```

（11）使用 Matplotlib 绘制一个 3×3 的子图网格，每个子图展示一个音频样本的波形图像，并在标题中显示对应的标签名称。

```
fig, axs = plt.subplots(3, 3, figsize=(16, 10))

for i, ax in enumerate(axs.flat):
    if i < len(example_audio):
        audio_signal = example_audio[i]
        ax.plot(audio_signal)
        ax.set_title(label_names[example_labels[i]])
        ax.set_yticks(np.arange(-1.2, 1.2, 0.2))
        ax.set_ylim([-1.1, 1.1])
    else:
        ax.axis('off')

plt.tight_layout()
plt.show()
```

在上述代码中，首先创建了一个 3×3 的子图网格，以准备在每个子图中展示一个音频样本的波形图像。然后，使用 enumerate 函数遍历所有子图，并在每个子图中绘制相应的音频波形。在绘制波形时，根据索引获取对应的音频信号和标签，并设置子图的标题为对应的标签名称。最后，通过调整子图布局和显示绘制的图像，来呈现每个子图中的音频样本波形及其对应的标签。绘制的音频样本的波形图如图 7-3 所示。

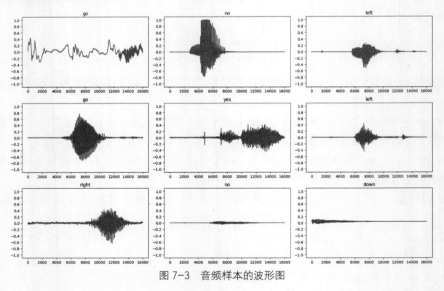

图 7-3　音频样本的波形图

（12）定义函数 get_spectrogram，用于将音频波形转换为频谱图像。首先，通过短时傅里叶变换（STFT）将波形转换为频谱图，然后获取其幅度。为了能够将频谱图像作为类似图像的输入数据传递给卷积层，添加了一个额外的维度。

```python
def get_spectrogram(waveform):
    # 将波形通过 STFT 转换为频谱图，显示随时间变化的频率，并可表示为二维图像
    spectrogram = tf.signal.stft(
        waveform, frame_length=255, frame_step=128)

    # 获取 STFT 的幅度
    spectrogram = tf.abs(spectrogram)

    # 添加一个 channels 维度，以便频谱图作为类似图像的输入数据传递给卷积层
    # 卷积层期望的形状为 ('batch_size', 'height', 'width', 'channels')
    spectrogram = spectrogram[..., tf.newaxis]
    return spectrogram
```

（13）下面的循环遍历前 3 个样本，对每个样本进行以下操作。
- 获取样本的标签名称和音频波形。
- 调用 get_spectrogram 函数将音频波形转换为频谱图。
- 打印标签名称、音频波形形状和频谱图形状。
- 使用 display.Audio 函数播放音频波形。

```python
for i in range(3):
    label = label_names[example_labels[i]]
    waveform = example_audio[i]
    spectrogram = get_spectrogram(waveform)

    print('Label:', label)
    print('Waveform shape:', waveform.shape)
    print('Spectrogram shape:', spectrogram.shape)
    print('Audio playback')
    display.display(display.Audio(waveform, rate=16000))
```

执行代码后会打印输出每个样本的标签名称、音频波形的形状和频谱图的形状，并且显示对应的播放界面，会播放每个样本的音频，如图 7-4 所示。

（14）定义函数 plot_spectrogram，用于可视化显示频谱图。首先，将频谱图转换为对数刻度，并将其转置以使时间表示在 x 轴上。然后，使用函数 pcolormesh 在坐标轴上绘制频谱图。

```python
# 用于显示频谱图的函数
def plot_spectrogram(spectrogram, ax):
    if len(spectrogram.shape) > 2:
        assert len(spectrogram.shape) == 3
        spectrogram = np.squeeze(spectrogram, axis=-1)
    # 将频率转换为对数刻度并转置，以便时间表示在 x 轴上（列）
    # 添加一个 epsilon 以避免取对数时出现零
    log_spec = np.log(spectrogram.T + np.finfo(float).eps)
    height = log_spec.shape[0]
    width = log_spec.shape[1]
    X = np.linspace(0, np.size(spectrogram), num=width, dtype=int)
    Y = range(height)
    ax.pcolormesh(X, Y, log_spec)
```

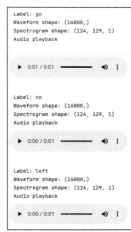

图 7-4　输出内容

（15）在单个图中同时显示一个音频样本的波形和频谱图，以便直观地展示音频数据的时域和频域特征。

```python
fig, axes = plt.subplots(2, figsize=(12, 8))
```

```
timescale = np.arange(waveform.shape[0])
axes[0].plot(timescale, waveform.numpy())
axes[0].set_title('Waveform')
axes[0].set_xlim([0, 16000])

plot_spectrogram(spectrogram.numpy(), axes[1])
axes[1].set_title('Spectrogram')
plt.suptitle(label.title())
plt.show()
```

执行效果如图 7-5 所示。

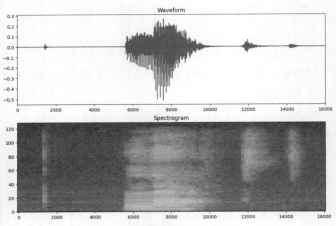

图 7-5 音频样本的波形和频谱图

（16）定义函数 make_spec_ds(ds)，此函数接收一个音频数据集 ds，并将其转换为一个新的数据集，其中每个样本都包含音频波形的频谱图像和对应的标签。

```
def make_spec_ds(ds):
  return ds.map(
      map_func=lambda audio,label: (get_spectrogram(audio), label),
      num_parallel_calls=tf.data.AUTOTUNE)

train_spectrogram_ds = make_spec_ds(train_ds)
val_spectrogram_ds = make_spec_ds(val_ds)
test_spectrogram_ds = make_spec_ds(test_ds)

for example_spectrograms, example_spect_labels in train_spectrogram_ds.take(1):
  break
```

上述代码的实现流程如下。

- 首先，定义一个名为 make_spec_ds 的函数，接收一个音频数据集 ds 作为输入。然后，使用 map 函数对输入的数据集中的每个样本进行映射操作。在映射函数中，对每个样本的音频数据应用 get_spectrogram 函数，将其转换为频谱图像，并保持标签不变，返回转换后的新数据集。
- 该函数分别应用于训练集、验证集和测试集，生成对应的频谱图像数据集 train_spectrogram_ds、val_spectrogram_ds 和 test_spectrogram_ds。
- 最后，通过取出训练集的第一个样本来检查转换后的数据集。

（17）对训练集、验证集和测试集的频谱图像数据集应用缓存、随机打乱和预取操作，这样处理后的数据集在训练、验证和测试过程中能够更高效地加载和使用数据。

```
train_spectrogram_ds = train_spectrogram_ds.cache().shuffle(10000).prefetch(tf.data.AUTOTUNE)
val_spectrogram_ds = val_spectrogram_ds.cache().prefetch(tf.data.AUTOTUNE)
```

```
test_spectrogram_ds = test_spectrogram_ds.cache().prefetch(tf.data.AUTOTUNE)
```

（18）下面这段代码的功能是构建一个卷积神经网络模型，用于对频谱图像进行分类。首先，获取频谱图像数据集中一个样本的输入形状，并计算标签的数量。然后，初始化一个归一化层，并根据训练集的数据来适应该归一化层。接着，构建一个包含卷积、池化、全连接等层的序贯模型（Sequential Model），用于对频谱图像进行特征提取和分类。

```
input_shape = example_spectrograms.shape[1:]
print('Input shape:', input_shape)
num_labels = len(label_names)

norm_layer = layers.Normalization()
norm_layer.adapt(data=train_spectrogram_ds.map(map_func=lambda spec, label: spec))

model = models.Sequential([
    layers.Input(shape=input_shape),
    layers.Resizing(32, 32),
    norm_layer,
    layers.Conv2D(32, 3, activation='relu'),
    layers.Conv2D(64, 3, activation='relu'),
    layers.MaxPooling2D(),
    layers.Dropout(0.25),
    layers.Flatten(),
    layers.Dense(128, activation='relu'),
    layers.Dropout(0.5),
    layers.Dense(num_labels),
])
```

（19）构建一个卷积神经网络模型用于频谱图像的分类任务，并打印输出模型的摘要信息，其中包括每一层的名称、类型、输出形状以及参数数量，以便于了解模型结构和参数量。

```
model.summary()
```

执行代码后会输出：

```
Model: "sequential"
_____
Layer (type)                 Output Shape              Param #
=================================================================
resizing (Resizing)          (None, 32, 32, 1)         0

normalization (Normalizatio  (None, 32, 32, 1)         3
n)

conv2d (Conv2D)              (None, 30, 30, 32)        320

conv2d_1 (Conv2D)            (None, 28, 28, 64)        18496

max_pooling2d (MaxPooling2D  (None, 14, 14, 64)        0
)

dropout (Dropout)            (None, 14, 14, 64)        0

flatten (Flatten)            (None, 12544)             0

dense (Dense)                (None, 128)               1605760

dropout_1 (Dropout)          (None, 128)               0

dense_1 (Dense)              (None, 8)                 1032
```

```
=================================================================
Total params: 1,625,611
Trainable params: 1,625,608
Non-trainable params: 3
```

（20）使用 Adam 优化器、稀疏分类交叉熵损失函数和准确率作为评估指标来编译模型，Adam 优化器用于优化模型参数，稀疏分类交叉熵损失函数用于多类别分类问题中的模型训练，而准确率则用于评估模型的性能。

```
model.compile(
    optimizer=tf.keras.optimizers.Adam(),
    loss=tf.keras.losses.SparseCategoricalCrossentropy(from_logits=True),
    metrics=['accuracy'],
)
```

（21）下面的这段代码用于训练模型，指定了训练的 epoch 数量为 10，并使用提前停止回调函数在验证集上监控模型性能，当性能不再提升时提前终止训练。训练过程中的训练损失、训练准确率、验证损失和验证准确率会保存在 history 对象中。

```
EPOCHS = 10
history = model.fit(
    train_spectrogram_ds,
    validation_data=val_spectrogram_ds,
    epochs=EPOCHS,
    callbacks=tf.keras.callbacks.EarlyStopping(verbose=1, patience=2),
)
```

执行代码后会输出训练过程：

```
100/100 [==============================] - 39s 370ms/step - loss: 1.7898 - accuracy: 0.3455 - val_loss: 1.3351 - val_accuracy: 0.5885
Epoch 2/10
100/100 [==============================] - 30s 297ms/step - loss: 1.2387 - accuracy: 0.5572 - val_loss: 0.9381 - val_accuracy: 0.7109
Epoch 3/10
100/100 [==============================] - 28s 280ms/step - loss: 0.9419 - accuracy: 0.6675 - val_loss: 0.7712 - val_accuracy: 0.7643
Epoch 4/10
100/100 [==============================] - 32s 323ms/step - loss: 0.7672 - accuracy: 0.7256 - val_loss: 0.6794 - val_accuracy: 0.8086
Epoch 5/10
100/100 [==============================] - 32s 315ms/step - loss: 0.6509 - accuracy: 0.7641 - val_loss: 0.6028 - val_accuracy: 0.8268
Epoch 6/10
100/100 [==============================] - 31s 312ms/step - loss: 0.5723 - accuracy: 0.8000 - val_loss: 0.5704 - val_accuracy: 0.8372
Epoch 7/10
100/100 [==============================] - 30s 303ms/step - loss: 0.4909 - accuracy: 0.8228 - val_loss: 0.5112 - val_accuracy: 0.8385
Epoch 8/10
100/100 [==============================] - 28s 276ms/step - loss: 0.4514 - accuracy: 0.8416 - val_loss: 0.5039 - val_accuracy: 0.8451
Epoch 9/10
100/100 [==============================] - 31s 314ms/step - loss: 0.4093 - accuracy: 0.8502 - val_loss: 0.4557 - val_accuracy: 0.8542
Epoch 10/10
100/100 [==============================] - 31s 305ms/step - loss: 0.3731 - accuracy: 0.8675 - val_loss: 0.4488 - val_accuracy: 0.8672
```

（22）下面的这段代码用于可视化模型训练过程中的损失和准确率变化。通过创建包含两个子图的图表，第一个子图显示训练损失和验证损失随 epoch 变化的曲线，第二个子图显示训练准确率和验证准确率随 epoch 变化的曲线，以直观地了解模型在训练过程中的性能表现。

```
metrics = history.history
plt.figure(figsize=(16,6))
plt.subplot(1,2,1)
plt.plot(history.epoch, metrics['loss'], metrics['val_loss'])
plt.legend(['loss', 'val_loss'])
plt.ylim([0, max(plt.ylim())])
plt.xlabel('Epoch')
plt.ylabel('Loss [CrossEntropy]')

plt.subplot(1,2,2)
plt.plot(history.epoch, 100*np.array(metrics['accuracy']), 100*np.array(metrics['val_accuracy']))
plt.legend(['accuracy', 'val_accuracy'])
plt.ylim([0, 100])
plt.xlabel('Epoch')
plt.ylabel('Accuracy [%]')
```

执行效果如图 7-6 所示。

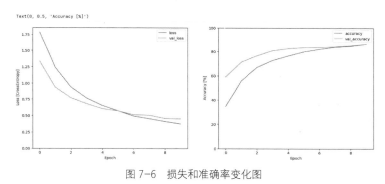

图 7-6 损失和准确率变化图

（23）将训练好的模型保存到名为"best_model.hdf5"的文件中，以便之后重新加载和使用该模型。

```
from keras.models import load_model
model.save('best_model.hdf5')
```

（24）评估模型在测试数据集上的性能，通过对"test_spectrogram_ds"进行评估，返回一个包含评估指标的字典，其中包括测试损失和准确率等。

```
model.evaluate(test_spectrogram_ds, return_dict=True)
```

执行代码后会输出：

```
13/13 [==============================] - 6s 395ms/step - loss: 0.4840 - accuracy: 0.8389
{'loss': 0.48401975631713867, 'accuracy': 0.838942289352417}
```

（25）读取一个指定的音频文件，将其转换为频谱图，并使用训练好的模型对其进行预测，随后可视化预测结果的概率分布，并播放该音频样本，以直观地展示模型对音频分类的结果。

```
x = 'PATH/mini_speech_commands/no/01bb6a2a_nohash_0.wav'
x = tf.io.read_file(str(x))
x, sample_rate = tf.audio.decode_wav(x, desired_channels=1, desired_samples=16000,)
x = tf.squeeze(x, axis=-1)
waveform = x
x = get_spectrogram(x)
```

```
x = x[tf.newaxis,...]

prediction = model(x)
x_labels = ['no', 'yes', 'down', 'go', 'left', 'up', 'right', 'stop']
plt.bar(x_labels, tf.nn.softmax(prediction[0]))
plt.title('No')
plt.show()

display.display(display.Audio(waveform, rate=16000))
```

预测结果的概率分布图和音频播放如图 7-7 所示。

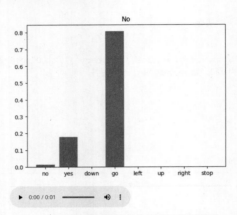

图 7-7　预测结果的概率分布图和音频播放

（26）定义类 ExportModel，用于将模型导出并进行推理。该类可以接收文件名字符串或音频波形作为输入，处理输入数据并进行预测，返回预测结果、类别 ID 和类别名称。

```
class ExportModel(tf.Module):
  def __init__(self, model):
    self.model = model

    # 接收字符串文件名或音频波形，可以为单个波形或不规则批次添加额外的签名
    self.__call__.get_concrete_function(
        x=tf.TensorSpec(shape=(), dtype=tf.string))
    self.__call__.get_concrete_function(
      x=tf.TensorSpec(shape=[None, 16000], dtype=tf.float32))


  @tf.function
  def __call__(self, x):
    # 如果传递的是字符串，则加载文件并解码
    if x.dtype == tf.string:
      x = tf.io.read_file(x)
      x, _ = tf.audio.decode_wav(x, desired_channels=1, desired_samples=16000,)
      x = tf.squeeze(x, axis=-1)
      x = x[tf.newaxis, :]

    x = get_spectrogram(x)
    result = self.model(x, training=False)

    class_ids = tf.argmax(result, axis=-1)
    class_names = tf.gather(label_names, class_ids)
    return {'predictions': result,
            'class_ids': class_ids,
            'class_names': class_names}
```

（27）通过类 ExportModel 的实例对指定的音频文件进行预测，返回预测结果、类别 ID 和类别名称。

```
export = ExportModel(model)
export(tf.constant(str('PATH/no/01bb6a2a_nohash_0.wav')))
```

执行代码后会输出：

```
{'predictions': <tf.Tensor: shape=(1, 8), dtype=float32, numpy=
array([[ 0.9276998,  3.4083233, -2.6086864,  4.9220834, -4.5850563,
        -3.0693464, -3.6959574, -3.6129196]], dtype=float32)>,
 'class_ids': <tf.Tensor: shape=(1,), dtype=int64, numpy=array([3])>,
 'class_names': <tf.Tensor: shape=(1,), dtype=string, numpy=array([b'no'], dtype=object)>}
```

（28）将 ExportModel 实例保存为 TensorFlow SavedModel 格式，并重新加载该模型进行预测，以验证模型保存和加载的正确性。

```
tf.saved_model.save(export, "saved")
imported = tf.saved_model.load("saved")
imported(waveform[tf.newaxis, :])
```

执行代码后会输出：

```
{'predictions': <tf.Tensor: shape=(1, 8), dtype=float32, numpy=
array([[ 0.9276998,  3.4083233, -2.6086864,  4.9220834, -4.5850563,
        -3.0693464, -3.6959574, -3.6129196]], dtype=float32)>,
 'class_ids': <tf.Tensor: shape=(1,), dtype=int64, numpy=array([3])>,
 'class_names': <tf.Tensor: shape=(1,), dtype=string, numpy=array([b'no'], dtype=object)>}
```

7.2 模型验证

在多模态大模型应用中，模型验证是确保模型在不同数据模态上的性能和泛化能力的关键步骤。验证的目的是检测模型是否发生过拟合、欠拟合，并评估其在实际应用中的表现。

7.2.1 模型验证的必要性

模型验证是确保机器学习模型可靠性和泛化能力的关键步骤，其主要功能如下。

- 评估模型性能：验证模型可以帮助我们了解模型在未见过的数据上的表现。仅依靠训练集上的表现不能完全反映模型的泛化能力，因此需要使用验证集来评估模型的性能。
- 检测过拟合和欠拟合：通过验证集可以检测模型是否出现过拟合（在训练集上表现良好但在验证集上表现差）或欠拟合（模型无法很好地拟合训练集数据）的问题。
- 调整超参数：通过验证集的表现可以调整模型的超参数，如学习率、正则化参数等，以提高模型的性能和泛化能力。
- 选择最佳模型：在多个模型中选择最佳模型时，需要使用验证集评估它们的性能，并选择性能最好的模型。
- 验证模型稳定性：模型验证可以帮助验证模型的稳定性和一致性，确保模型的预测结果是可靠的。
- 提高信任度：通过验证模型，可以提高对模型的信任度，使其在实际应用中更可靠。

总之，模型验证是确保机器学习模型有效性和可靠性的必要步骤，有助于确保模型能够在不同数据上表现良好并产生可信赖的预测结果。

7.2.2 数据准备和分割

在多模态大模型验证中,数据准备和分割是基础步骤,确保各模态数据在训练和验证过程中都能被合理利用。

- 训练集、验证集、测试集分割:将数据集分为训练集、验证集和测试集,确保每个数据集都包含足够的各模态数据。
- 保持模态同步:确保在分割数据时,保持不同模态数据的同步和对应关系。例如,在图像-文本对中,确保图像和相应的文本描述同时出现在同一个数据集中。

实例 7-5 使用 CIFAR-10 图像数据集和随机生成的假文本数据作为演示,展示了对多模态数据集的分割过程。

实例 7-5	分割多模态数据集 源码路径:codes/7/zhunfen.py

实例文件 zhunfen.py 的具体实现代码如下所示。

```python
import numpy as np
from sklearn.model_selection import train_test_split
from torchvision import models, datasets, transforms

# 加载预训练的 ResNet 模型(仅提取特征)
resnet_model = models.resnet18(pretrained=True)
resnet_model.fc = torch.nn.Identity()  # 去掉最后一层全连接层

# 加载图像数据集 CIFAR-10
transform = transforms.Compose([
    transforms.Resize(224),    # ResNet 需要输入大小为 224×224
    transforms.ToTensor(),
    transforms.Normalize((0.5, 0.5, 0.5), (0.5, 0.5, 0.5))  # 图像标准化
])

cifar_dataset = datasets.CIFAR10(root='./data', train=True, download=True, transform=transform)
cifar_data = cifar_dataset.data          # CIFAR-10 图像数据
cifar_labels = cifar_dataset.targets     # CIFAR-10 图像标签

# 使用随机生成的假数据作为文本特征
text_data = np.random.rand(len(cifar_labels), 300)   # 假设每个文本特征是长度为 300 的向量

# 划分数据集
image_train, image_remain, text_train, text_remain, label_train, label_remain = train_test_split(cifar_data, text_data, cifar_labels, test_size=0.2, random_state=42)

image_valid, image_test, text_valid, text_test, label_valid, label_test = train_test_split(
    image_remain, text_remain, label_remain, test_size=0.5, random_state=42)

# 打印数据集大小
print("训练集大小:", len(image_train))
print("验证集大小:", len(image_valid))
print("测试集大小:", len(image_test))

# 确保图像、文本和标签的对应关系
print("\n训练集示例:")
for img, txt, lbl in zip(image_train, text_train, label_train):
    print("Image:", img.shape, "- Text:", txt.shape, "- Label:", lbl)

print("\n验证集示例:")
for img, txt, lbl in zip(image_valid, text_valid, label_valid):
    print("Image:", img.shape, "- Text:", txt.shape, "- Label:", lbl)
```

```
print("\n测试集示例:")
for img, txt, lbl in zip(image_test, text_test, label_test):
    print("Image:", img.shape, "- Text:", txt.shape, "- Label:", lbl)
```

在上述代码中，首先，利用预训练的 ResNet 模型提取图像特征，然后使用随机生成的假文本数据作为文本特征。接着，将数据集分割为训练集、验证集和测试集，并确保图像、文本和标签之间的对应关系。最后打印输出如下数据集中的部分示例，包括图像、文本特征和标签信息。

```
Text: (300,) - Label: 8
Image: (32, 32, 3) - Text: (300,) - Label: 7
Image: (32, 32, 3) - Text: (300,) - Label: 7
Image: (32, 32, 3) - Text: (300,) - Label: 9
Image: (32, 32, 3) - Text: (300,) - Label: 3
Image: (32, 32, 3) - Text: (300,) - Label: 9
Image: (32, 32, 3) - Text: (300,) - Label: 8
Image: (32, 32, 3) - Text: (300,) - Label: 4
Image: (32, 32, 3) - Text: (300,) - Label: 5
Image: (32, 32, 3) - Text: (300,) - Label: 5
Image: (32, 32, 3) - Text: (300,) - Label: 4
Image: (32, 32, 3) - Text: (300,) - Label: 0
Image: (32, 32, 3) - Text: (300,) - Label: 8
Image: (32, 32, 3) - Text: (300,) - Label: 5
……
```

7.2.3 交叉验证

交叉验证是多模态大模型验证的常用方法，有助于获得模型在不同数据子集上的性能表现，减少偏差。

- k 折交叉验证（k-Fold Cross-Validation）：将数据集分成 k 个子集，轮流使用 $k-1$ 个子集进行训练，剩下的一个子集进行验证。重复 k 次，计算平均性能。
- 留一法交叉验证（Leave-One-Out Cross-Validation, LOOCV）：每次只留一个样本作为验证集，其余样本作为训练集，适用于小数据集。

实例 7-6 演示了在多模态大模型上使用 k 折交叉验证和留一法交叉验证的过程。在每个样本中包含了图像、文本和标签信息，确保了在交叉验证过程中每个 fold 或每个留出的样本都包含完整的多模态数据。

实例 7-6　使用 k 折交叉验证和留一法交叉验证
源码路径：codes/7/jiao.py

实例文件 jiao.py 的具体实现代码如下所示。

```
import numpy as np
from sklearn.model_selection import KFold, LeaveOneOut

# 假设有图像数据和文本描述数据以及标签
# 这里用两个简单的列表代替，实际情况下可以是图像文件路径和文本数据
image_data = ['image1.jpg', 'image2.jpg', 'image3.jpg', 'image4.jpg', 'image5.jpg']
text_data = ['Description 1', 'Description 2', 'Description 3', 'Description 4', 'Description 5']
labels = [0, 1, 0, 1, 1]    # 假设有标签数据

# 定义 k 折交叉验证
k_folds = 3
kf = KFold(n_splits=k_folds)

fold = 1
```

```python
for train_index, test_index in kf.split(image_data):
    print(f"Fold {fold}:")
    print("Train Index:", train_index)
    print("Test Index:", test_index)

    # 分割数据
    image_train, image_test = [image_data[i] for i in train_index], [image_data[i] for i in test_index]
    text_train, text_test = [text_data[i] for i in train_index], [text_data[i] for i in test_index]
    labels_train, labels_test = [labels[i] for i in train_index], [labels[i] for i in test_index]

    # 在这里进行多模态模型的训练和验证
    # 可以使用 image_train, text_train, labels_train 进行训练，然后使用 image_test, text_test,
    # labels_test 进行验证

    # 打印当前 fold 的数据情况
    print("Train Set:")
    for img, txt, lbl in zip(image_train, text_train, labels_train):
        print("Image:", img, "- Text:", txt, "- Label:", lbl)

    print("Test Set:")
    for img, txt, lbl in zip(image_test, text_test, labels_test):
        print("Image:", img, "- Text:", txt, "- Label:", lbl)

    print("\n-----------------------")
    fold += 1

# 定义留一法交叉验证
loo = LeaveOneOut()

for train_index, test_index in loo.split(image_data):
    print("Train Index:", train_index, "Test Index:", test_index)

    # 分割数据
    image_train, image_test = [image_data[i] for i in train_index], [image_data[i] for i in test_index]
    text_train, text_test = [text_data[i] for i in train_index], [text_data[i] for i in test_index]
    labels_train, labels_test = [labels[i] for i in train_index], [labels[i] for i in test_index]

    # 在这里进行多模态模型的训练和验证
    # 可以使用 image_train, text_train, labels_train 进行训练，然后使用 image_test, text_test,
    labels_test 进行验证

    # 打印当前留出的样本
    print("Test Sample:")
    print("Image:", image_test[0], "- Text:", text_test[0], "- Label:", labels_test[0])
    print("\n-----------------------")
```

上述代码的实现流程如下。

（1）定义图像数据、文本描述数据以及标签数据。

（2）使用 k 折交叉验证。

- 定义 k 的值，即将数据集分成几个子集。
- 使用 KFold 对象进行数据集的划分，获取每个 fold 的训练集和测试集的索引。
- 对于每个 fold，首先将数据集分割成训练集和测试集，分别包含图像、文本和标签信息。然后打印输出当前 fold 的索引、训练集和测试集的信息。

（3）使用留一法交叉验证。

- 使用 LeaveOneOut 对象进行数据集的划分，每次只留一个样本作为测试集，其余样本作为训练集。
- 每次分割的训练集和测试集分别包含图像、文本和标签信息。然后打印输出当前留出的样本的索引和训练集、测试集的信息。

（4）打印输出每个 fold 或每个留出的样本的图像、文本和标签信息，以及对应的训练集和测试集信息。执行代码后会输出：

```
Fold 1:
Train Index: [2 3 4]
Test Index: [0 1]
Train Set:
Image: image3.jpg - Text: Description 3 - Label: 0
Image: image4.jpg - Text: Description 4 - Label: 1
Image: image5.jpg - Text: Description 5 - Label: 1
Test Set:
Image: image1.jpg - Text: Description 1 - Label: 0
Image: image2.jpg - Text: Description 2 - Label: 1

----------------------
Fold 2:
Train Index: [0 1 4]
Test Index: [2 3]
Train Set:
Image: image1.jpg - Text: Description 1 - Label: 0
Image: image2.jpg - Text: Description 2 - Label: 1
Image: image5.jpg - Text: Description 5 - Label: 1
Test Set:
Image: image3.jpg - Text: Description 3 - Label: 0
Image: image4.jpg - Text: Description 4 - Label: 1

----------------------
Fold 3:
Train Index: [0 1 2 3]
Test Index: [4]
Train Set:
Image: image1.jpg - Text: Description 1 - Label: 0
Image: image2.jpg - Text: Description 2 - Label: 1
Image: image3.jpg - Text: Description 3 - Label: 0
Image: image4.jpg - Text: Description 4 - Label: 1
Test Set:
Image: image5.jpg - Text: Description 5 - Label: 1

----------------------
Train Index: [1 2 3 4] Test Index: [0]
Test Sample:
Image: image1.jpg - Text: Description 1 - Label: 0

----------------------
Train Index: [0 2 3 4] Test Index: [1]
Test Sample:
Image: image2.jpg - Text: Description 2 - Label: 1

----------------------
Train Index: [0 1 3 4] Test Index: [2]
Test Sample:
Image: image3.jpg - Text: Description 3 - Label: 0

----------------------
Train Index: [0 1 2 4] Test Index: [3]
Test Sample:
Image: image4.jpg - Text: Description 4 - Label: 1
```

```
------------------------
Train Index: [0 1 2 3] Test Index: [4]
Test Sample:
Image: image5.jpg - Text: Description 5 - Label: 1

------------------------
```

7.2.4 嵌套交叉验证

嵌套交叉验证用于同时调优模型的超参数和验证模型性能,特别适用于多模态大模型的复杂性。
- 外层交叉验证:用于评估模型的泛化性能。
- 内层交叉验证:用于超参数调优,确保选择的参数能够在验证集上表现良好。

例如,下面是一个使用嵌套交叉验证同时调优模型超参数并验证模型性能的例子,在例子中使用了一个简单的模拟数据集来进行演示。

```python
import numpy as np
from sklearn.model_selection import GridSearchCV, KFold
from sklearn.svm import SVC

# 模拟图像和文本数据
image_data = np.random.rand(100, 10)  # 假设有 100 个图像样本,每个样本有 10 个特征
text_data = np.random.rand(100, 20)   # 假设有 100 个文本样本,每个样本有 20 个特征
labels = np.random.randint(2, size=100)  # 随机生成 0 和 1 的标签,假设有 100 个样本

# 定义模型
svm = SVC()

# 定义超参数网格
param_grid = {
    'C': [0.1, 1, 10],
    'gamma': [0.01, 0.1, 1],
    'kernel': ['linear', 'rbf']
}

# 定义嵌套交叉验证
outer_cv = KFold(n_splits=5, shuffle=True, random_state=42)
inner_cv = KFold(n_splits=3, shuffle=True, random_state=42)

# 执行嵌套交叉验证及超参数调优
grid_search = GridSearchCV(estimator=svm, param_grid=param_grid, cv=inner_cv)
nested_score = []

for train_index, test_index in outer_cv.split(image_data):
    X_train_image, X_test_image = image_data[train_index], image_data[test_index]
    X_train_text, X_test_text = text_data[train_index], text_data[test_index]
    y_train, y_test = labels[train_index], labels[test_index]

    # 在内层交叉验证上进行超参数调优
    grid_search.fit(X_train_image, y_train)  # 使用图像数据进行内层交叉验证
    best_params = grid_search.best_params_

    # 在外层交叉验证上评估性能
    svm_best = SVC(**best_params)
    svm_best.fit(X_train_text, y_train)  # 使用文本数据训练模型
    score = svm_best.score(X_test_text, y_test)  # 使用文本数据验证模型性能
    nested_score.append(score)

# 输出嵌套交叉验证结果
print("嵌套交叉验证结果: ", nested_score)
print("平均准确率: ", np.mean(nested_score))
```

在上述代码的内层交叉验证中，使用了 GridSearchCV 来选择最佳超参数；在外层交叉验证中，评估了模型的性能。

7.2.5 模态间一致性验证

在多模态大模型应用中，模态间一致性验证是指确保不同模态（比如图像、文本、语音等）之间的信息是一致的或者是相互补充的，这种一致性验证对于确保模型的有效性和泛化能力非常重要。具体来说，在实际应用中使用模态间一致性验证的主要原因如下。

- 信息互补性：不同模态往往会提供样本不同方面的信息，比如图像可以提供视觉信息，文本可以提供语义信息，音频可以提供声音信息。模态间一致性验证可以确保这些信息是相互补充的，而不是相互矛盾的。
- 模态相关性：不同模态之间可能存在相关性，比如一张图片和相应的文本描述通常是相关的。模态间一致性验证可以确保模型能够充分利用这种相关性。

实现模态间一致性验证的主要方法如下。

- 特征级一致性验证：检查不同模态提取的特征之间的相似性或相关性。可以使用相关系数、协方差等方法进行衡量。
- 模型级一致性验证：在模型中引入跨模态的一致性约束，确保模型学习到的表示能够在不同模态之间保持一致性。比如，在多模态融合模型中，可以使用跨模态的损失函数来促使模型学习到一致的表示。
- 交叉模态验证：使用一个模态的数据来验证在另一个模态上训练的模型的性能。例如，使用图像数据来验证在文本数据上训练的模型，以及使用文本数据来验证在图像数据上训练的模型，以确保模型对不同模态的泛化能力。

在实际应用中，实现模态间一致性验证的主要步骤如下。
（1）特征提取：对每个模态的数据进行特征提取，确保得到可用于模型的表示。
（2）模型训练：分别使用每个模态的数据训练模型，或者使用多模态数据训练融合模型。
（3）实现模态间一致性验证，主要步骤如下。

- 检查特征级一致性：比较不同模态提取的特征之间的相似性。
- 引入模型级一致性约束：在模型中引入跨模态的一致性约束。
- 进行交叉模态验证：使用一个模态的数据来验证在另一个模态上训练的模型的性能。
- 调优和评估：根据验证结果调优模型，并最终评估模型的性能。

通过模态间一致性验证，可以确保多模态模型能够充分利用不同模态之间的信息，提高模型的鲁棒性和泛化能力。实例 7-7 展示了在多模态大模型应用中实现模态间一致性验证的过程。

实例 7-7 在多模态大模型中实现模态间一致性验证
源码路径：codes/7/mo.py

实例文件 mo.py 的具体实现代码如下所示。

```
import numpy as np
from sklearn.metrics.pairwise import cosine_similarity
from sklearn.model_selection import cross_val_score, KFold
from sklearn.svm import SVC

# 模拟图像和文本数据
image_data = np.random.rand(100, 10)   # 假设有 100 个图像样本,每个样本有 10 个特征
text_data = np.random.rand(100, 10)    # 假设有 100 个文本样本,每个样本有 10 个特征
labels = np.random.randint(2, size=100)   # 随机生成 0 和 1 的标签,假设有 100 个样本
```

```python
# 特征提取
image_features = image_data.mean(axis=1, keepdims=True)   # 图像特征为特征均值
text_features = text_data.mean(axis=1, keepdims=True)     # 文本特征为特征均值

# 模型训练
svm_image = SVC()
svm_text = SVC()

# 实现模态间一致性验证
# 检查特征级一致性: 计算图像和文本特征之间的相似性（这里使用余弦相似度）
feature_similarity = cosine_similarity(image_features, text_features)
print("特征级一致性（余弦相似度）:\n", feature_similarity)

# 引入模型级一致性约束: 在模型中引入跨模态的一致性约束（这里简单列举使用两个模型的预测结果的平均）
kf = KFold(n_splits=5, shuffle=True, random_state=42)
for train_index, test_index in kf.split(image_data):
    X_train_image, X_test_image = image_features[train_index], image_features[test_index]
    X_train_text, X_test_text = text_features[train_index], text_features[test_index]
    y_train, y_test = labels[train_index], labels[test_index]

    # 在图像数据上训练模型
    svm_image.fit(X_train_image, y_train)
    # 在文本数据上训练模型
    svm_text.fit(X_train_text, y_train)

    # 模型级一致性约束: 使用两个模型的预测结果的平均作为最终的预测结果
    pred_image = svm_image.predict(X_test_image)
    pred_text = svm_text.predict(X_test_text)
    pred_combined = (pred_image + pred_text) / 2    # 取平均
    score_combined = np.mean(pred_combined == y_test)
    print("模型级一致性约束下的准确率: ", score_combined)
```

上述代码计算了图像和文本特征之间的相似性（余弦相似度），然后使用支持向量机模型分别在图像数据和文本数据上进行训练。接着利用模型级一致性约束，采用两个模型的预测结果的平均作为最终预测结果，并通过交叉验证计算了模型在不同数据子集上的准确率，以验证模态间的一致性。执行代码后会输出：

```
特征级一致性（余弦相似度）:
[[1. 1. 1. ... 1. 1. 1.]
 [1. 1. 1. ... 1. 1. 1.]
 [1. 1. 1. ... 1. 1. 1.]
 ...
 [1. 1. 1. ... 1. 1. 1.]
 [1. 1. 1. ... 1. 1. 1.]
 [1. 1. 1. ... 1. 1. 1.]]
模型级一致性约束下的准确率:  0.05
模型级一致性约束下的准确率:  0.25
模型级一致性约束下的准确率:  0.4
模型级一致性约束下的准确率:  0.2
模型级一致性约束下的准确率:  0.3
```

上面的输出结果展示了特征级一致性的余弦相似度矩阵以及模型级一致性约束下的准确率，这里的准确率仅作演示用，在实际应用中可能需要更复杂的模型和特征工程。另外，上面的特征级一致性（余弦相似度）矩阵显示所有图像和文本特征之间的余弦相似度均为1，这是因为在例子中使用了随机生成的数据，并且简单地计算了特征的均值，因此相似度为1。

7.2.6 模型鲁棒性验证

验证模型在不同环境和条件下的表现，以确保其鲁棒性。

- 噪声和干扰测试：在数据中加入噪声和干扰，评估模型的鲁棒性。例如，在图像数据中加入噪点，或在音频数据中加入背景噪声。
- 缺失模态测试：验证模型在某些模态数据缺失情况下的性能，确保其在部分模态缺失时仍能有效工作。

实例 7-8 演示了在多模态大模型应用中进行模型鲁性验证的过程，包括噪声干扰以及模态缺失测试功能。

实例 7-8　模型的噪声干扰以及模态缺失测试
源码路径：codes/7/que.py

实例文件 que.py 的具体实现代码如下所示。

```python
import numpy as np
from sklearn.model_selection import KFold
from sklearn.svm import SVC
from sklearn.metrics import accuracy_score

# 模拟图像和文本数据
np.random.seed(42)
image_data = np.random.rand(100, 10)   # 100 个图像样本，每个样本有 10 个特征
text_data = np.random.rand(100, 20)    # 100 个文本样本，每个样本有 20 个特征
labels = np.random.randint(2, size=100)    # 随机生成 0 和 1 的标签


# 噪声和干扰测试
def add_noise(data, noise_level=0.1):
    noisy_data = data + noise_level * np.random.randn(*data.shape)
    return noisy_data


# 缺失模态测试
def remove_modality(data, missing_rate=0.5):
    mask = np.random.rand(*data.shape) > missing_rate
    return data * mask


# 加入噪声的图像和文本数据
noisy_image_data = add_noise(image_data)
noisy_text_data = add_noise(text_data)

# 去除部分模态的图像和文本数据
missing_image_data = remove_modality(image_data)
missing_text_data = remove_modality(text_data)


# 模型训练和验证
def train_and_evaluate(X_image, X_text, y):
    kf = KFold(n_splits=5, shuffle=True, random_state=42)
    accuracies = []

    for train_index, test_index in kf.split(X_image):
        X_train_image, X_test_image = X_image[train_index], X_image[test_index]
        X_train_text, X_test_text = X_text[train_index], X_text[test_index]
        y_train, y_test = y[train_index], y[test_index]

        # 在图像数据上训练模型
        svm_image = SVC().fit(X_train_image, y_train)
        # 在文本数据上训练模型
        svm_text = SVC().fit(X_train_text, y_train)
```

```
        # 预测并计算准确率
        pred_image = svm_image.predict(X_test_image)
        pred_text = svm_text.predict(X_test_text)

        # 简单地平均两个模型的预测结果
        pred_combined = (pred_image + pred_text) / 2
        pred_combined = np.round(pred_combined).astype(int)

        accuracy = accuracy_score(y_test, pred_combined)
        accuracies.append(accuracy)

    return np.mean(accuracies)

# 原始数据的准确率
accuracy_original = train_and_evaluate(image_data, text_data, labels)
print("原始数据的准确率:", accuracy_original)

# 噪声数据的准确率
accuracy_noisy = train_and_evaluate(noisy_image_data, noisy_text_data, labels)
print("噪声数据的准确率:", accuracy_noisy)

# 缺失模态数据的准确率
accuracy_missing = train_and_evaluate(missing_image_data, missing_text_data, labels)
print("缺失模态数据的准确率:", accuracy_missing)
```

上述代码通过模拟文本数据和图片数据，展示了添加噪声和缺失部分模态的情况对分类模型性能的影响。首先生成随机文本数据及其二分类标签，然后定义函数，向数据添加噪声和移除模态数据。接着，利用交叉验证方法训练和评估在何种条件下支持向量机分类器的准确率更高，比较原始数据、噪声数据和缺失模态数据的准确率。执行代码后会输出：

```
原始数据的准确率: 0.5700000000000001
噪声数据的准确率: 0.5800000000000001
缺失模态数据的准确率: 0.53
```

7.2.7 验证指标

在多模态大模型应用中，使用多种指标来综合评估模型的验证效果。
- 准确率、精确率、召回率、F1 值：分别用于评估分类任务中的性能。
- ROC 曲线和 AUC：评估二分类或多分类任务中的性能。
- BLEU、ROUGE 等：评估文本生成任务的性能。
- MSE、MAE：用于回归任务中的性能评估。

实例 7-9 演示了使用多种指标综合评估模型性能的过程。首先，生成和处理数据，包括添加噪声和移除部分模态数据。接着，使用交叉验证方法训练支持向量机模型，并预测分类结果。最后，计算并打印多种评估指标（准确率、精确率、召回率、F1 值、ROC 曲线和 AUC），以全面评估模型的验证效果，并进行实验。

实例 7-9　使用多种指标综合评估模型的性能
源码路径：codes/7/zong.py

实例文件 zong.py 的具体实现代码如下所示。

```
import numpy as np
from sklearn.model_selection import KFold
from sklearn.svm import SVC
from sklearn.metrics import accuracy_score, precision_score, recall_score, f1_score, roc_auc_score, roc_curve
```

```python
import matplotlib.pyplot as plt

# 模拟图像和文本数据
np.random.seed(42)
image_data = np.random.rand(100, 10)  # 100 个图像样本，每个样本有 10 个特征
text_data = np.random.rand(100, 20)   # 100 个文本样本，每个样本有 20 个特征
labels = np.random.randint(2, size=100)  # 随机生成 0 和 1 的标签

# 噪声和干扰测试
def add_noise(data, noise_level=0.1):
    noisy_data = data + noise_level * np.random.randn(*data.shape)
    return noisy_data

# 缺失模态测试
def remove_modality(data, missing_rate=0.5):
    mask = np.random.rand(*data.shape) > missing_rate
    return data * mask

# 加入噪声的图像和文本数据
noisy_image_data = add_noise(image_data)
noisy_text_data = add_noise(text_data)

# 去除部分模态的图像和文本数据
missing_image_data = remove_modality(image_data)
missing_text_data = remove_modality(text_data)

# 模型训练和验证
def train_and_evaluate(X_image, X_text, y):
    kf = KFold(n_splits=5, shuffle=True, random_state=42)
    metrics = {
        'accuracy': [],
        'precision': [],
        'recall': [],
        'f1': [],
        'roc_auc': []
    }

    all_y_true = []
    all_y_pred = []
    all_y_scores = []

    for train_index, test_index in kf.split(X_image):
        X_train_image, X_test_image = X_image[train_index], X_image[test_index]
        X_train_text, X_test_text = X_text[train_index], X_text[test_index]
        y_train, y_test = y[train_index], y[test_index]

        # 在图像数据上训练模型
        svm_image = SVC(probability=True).fit(X_train_image, y_train)
        # 在文本数据上训练模型
        svm_text = SVC(probability=True).fit(X_train_text, y_train)

        # 预测并计算准确率
        pred_image = svm_image.predict(X_test_image)
        pred_text = svm_text.predict(X_test_text)

        # 获取概率分数
        scores_image = svm_image.predict_proba(X_test_image)[:, 1]
        scores_text = svm_text.predict_proba(X_test_text)[:, 1]

        # 简单地平均两个模型的预测结果
        pred_combined = (pred_image + pred_text) / 2
        scores_combined = (scores_image + scores_text) / 2
```

```python
            pred_combined = np.round(pred_combined).astype(int)

            all_y_true.extend(y_test)
            all_y_pred.extend(pred_combined)
            all_y_scores.extend(scores_combined)

            # 计算各种指标
            metrics['accuracy'].append(accuracy_score(y_test, pred_combined))
            metrics['precision'].append(precision_score(y_test, pred_combined))
            metrics['recall'].append(recall_score(y_test, pred_combined))
            metrics['f1'].append(f1_score(y_test, pred_combined))
            metrics['roc_auc'].append(roc_auc_score(y_test, scores_combined))

    return metrics, all_y_true, all_y_pred, all_y_scores

# 原始数据的评估指标
metrics_original, y_true_original, y_pred_original, y_scores_original = train_
and_evaluate(image_data, text_data, labels)

# 打印评估结果
print("原始数据的评估结果:")
for metric in metrics_original:
    print(f"{metric}: {np.mean(metrics_original[metric])}")

# 绘制 ROC 曲线
fpr, tpr, _ = roc_curve(y_true_original, y_scores_original)
plt.figure()
plt.plot(fpr, tpr, color='darkorange', lw=2, label='ROC curve (area = %0.2f)' % np.mean
(metrics_original['roc_auc']))
plt.plot([0, 1], [0, 1], color='navy', lw=2, linestyle='--')
plt.xlim([0.0, 1.0])
plt.ylim([0.0, 1.05])
plt.xlabel('False Positive Rate')
plt.ylabel('True Positive Rate')
plt.title('Receiver Operating Characteristic (ROC)')
plt.legend(loc="lower right")
plt.show()
```

上述代码的实现流程如下。

（1）模拟生成了文本数据以及它们的二分类标签，模拟生成了 100 个图像样本和 100 个文本样本，每个样本分别具有 10 个和 20 个特征。同时，为这些样本随机生成了二分类标签（0 或 1）。

（2）构建添加噪声和移除噪声的函数 add_noise 和 remove_modality 用于数据处理，分别对数据添加噪声和移除噪声的模态数据。

（3）实现了模型训练和验证的函数 train_and_evaluate，此函数实现了如下操作：

- 使用 K 折交叉验证（KFold）来评估模型性能，将 K 的值设置为 5。
- 在每一折中，数据被分为训练集和测试集。
- 分别在图像特征（X_train_image）和文本特征（X_train_text）上训练两个支持向量机（SVM）分类器。
- 使用训练好的模型对测试集进行预测，得到分类结果和概率分数。
- 将两个模型的预测结果和概率分数进行简单的平均，以得到综合的预测结果和概率分数。
- 计算并收集各种评估指标，包括准确率、精确率、召回率、F1 值和 ROC 曲线下的面积（AUC）。

（4）对原始数据进行了模型训练和评估，计算了准确率、精确率、召回率、F1 值和 AUC 等评估指标，同时以 MongoDB 展示模型的分类性能并绘制 ROC 曲线。

执行代码后会打印输出准确率、精确率、召回率、F1 值和 AUC 等评估指标，并绘制如图 7-8 所示的 ROC 曲线。

原始数据的评估结果：
```
accuracy: 0.5700000000000001
precision: 0.7333333333333334
recall: 0.19
f1: 0.2756798756798757
roc_auc: 0.43198809523809534
```

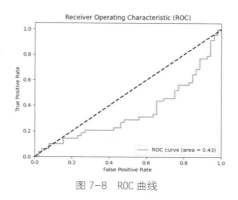

图 7-8 ROC 曲线

7.3 多模态大模型评估基准

多模态大模型评估基准（Multimodal Large Model Evaluation Benchmark，MLMEB）是用于评估多模态大模型性能的标准化测试集或工具集。多模态大模型通常用于处理不同模态的数据，如图像、文本、音频等，因此评估基准通常包含来自多个模态的任务和数据，以衡量模型在不同模态上的表现。通常来说，MLMEB 包括以下方面。

- 任务类型：包括视觉问答、图像字幕生成、文本生成、情感分析、语音识别等。
- 数据集：使用公开的多模态数据集，如 COCO、Flickr30k、VQA 等。
- 评估指标：如准确率、BLEU、CIDEr、F1-score 等，用于量化模型的性能。
- 基准测试：与现有模型进行对比，以评估新模型的相对性能。
- 通用性和鲁棒性：评估模型在不同任务、领域和数据分布下的表现稳定性。

评估基准可以帮助研究人员和开发者了解模型的强项和弱点，从而指导改进模型的设计和训练。本节将详细讲解目前几个常用的多模态大模型评估基准。

7.3.1 MM-Vet

MM-Vet（Multimodal Model Evaluation Tool）是一个专门为评估多模态大模型集成能力而设计的评估基准。MM-Vet 关注模型在不同核心视觉-语言能力集成上的表现，这类能力包含识别、OCR（光学字符识别）、知识、语言生成、空间感知和数学能力等。MM-Vet 已用于测试 GPT-4V 等模型在这些能力上的综合表现。

MM-Vet 的核心目标是衡量模型在集成不同类型数据（如图像、文本和其他模态）方面的表现，其主要特点如下所示。

- 多模态集成评估：MM-Vet 评估模型在整合和处理不同类型的数据（如视觉和语言）方面的能力。这类能力包括识别、OCR、知识、语言生成、空间感知和数学运算等。
- 综合能力测试：通过一系列标准化的测试，MM-Vet 可以评估模型在这些核心视觉-语言能力上的综合表现，确保模型能在各种任务中发挥最佳性能。
- 模型比较：MM-Vet 提供了一种方法，来比较不同多模态模型（如 GPT-4V）在各种能力上的表现，从而帮助研究人员和开发者选择最适合的模型。
- 基准测试：MM-Vet 为多模态模型提供了一个基准测试平台，可以系统地评估模型在实际应用中的表现，包括处理复杂的视觉和语言任务的能力。

MM-Vet 为多模态模型的开发和评估提供了一个全面、标准化的框架，帮助研究人员了解模型在集成能力上的优势和不足。使用 MM-Vet 评估大模型的基本步骤如下所示。

1. 安装和设置

（1）安装依赖：使用 pip 安装所需的 Python 包。确保安装 openai 包：

```
pip install openai>=1
```

（2）获取 GPT-4 或 GPT-3.5 API 的访问权限，如果没有访问权限，可以使用 MM-Vet 的在线评估工具，但可能需要等待较长时间。

2. 下载和准备数据

（1）下载数据：从 MM-Vet 的 Hugging Face 数据集页面下载数据集。

（2）解压数据集：使用如下命令解压数据集。

```
unzip mm-vet.zip
```

（3）准备模型输出：使用 MM-Vet 提供的推理脚本生成模型的输出，使用以下示例脚本实现。

```
python inference/gpt4v.py --mmvet_path /path/to/mm-vet --image_detail high
python inference/gemini_vision.py --mmvet_path /path/to/mm-vet
```

确保将模型输出保存为 JSON 文件，例如 llava_llama2_13b_chat.json。

3. 评估模型

（1）克隆 MM-Vet 仓库：通过如下命令实现。

```
git clone https://github.com/yuweihao/MM-Vet.git
cd MM-Vet
```

（2）运行评估：使用提供的评估脚本运行模型评估。

```
python mm-vet_evaluator.py --mmvet_path /path/to/mm-vet --result_file results/llava_llama2_
13b_chat.json
```

如果无法访问 GPT-4，可以上传模型输出的 JSON 文件到 MM-Vet 的在线评估工具（Hugging Face Space）获取评分结果。

4. 查看和分析结果

（1）查看报告：MM-Vet 将生成一份评估报告，报告包含模型在不同任务和能力上的表现。

（2）分析结果：根据报告中的数据，分析模型在各项能力上的表现，识别模型的优点和不足，以便进一步优化。

5. 调整和重复评估

（1）优化模型：根据评估结果对模型进行调整和优化。

（2）重新评估：使用改进后的模型再次进行评估，验证改进效果。

通过以上步骤，可以有效地使用 MM-Vet 对大模型进行综合评估，确保其在多模态任务中的表现符合要求。

7.3.2 MMEvalPro

MMEvalPro 是由北京大学、中国医学科学院、香港中文大学和阿里巴巴的研究团队于 2024

年提出的多模态大模型评估基准，旨在提供更可信和高效的评估方法，解决现有多模态评估基准中存在的问题。现有基准在评估多模态大模型时存在系统性偏差，即使没有视觉感知能力的大型语言模型，也能在这些基准上取得非凡的性能，这削弱了这些评估的可信度。MMEvalPro 通过增加两个"锚"问题（一个感知问题和一个知识问题）来改进现有的评估方法，形成测试模型多模态理解不同方面的"问题三元组"。

MEvalPro 的主要评估指标是"真实准确性"（Genuine Accuracy），要求模型必须正确回答三元组中的所有问题才能获得分数。这个过程包括多阶段的审核和质量检查，以确保问题清晰、相关、具有挑战性。最终的基准包含 2138 个问题三元组，共 6414 个涵盖不同主题和难度级别的问题。

7.3.3 MMT-Bench

MMT-Bench（Multimodal Tasks Benchmark）基准测试框架用于评估多模态模型在多种任务上的表现。该基准测试框架旨在为研究人员和开发者提供统一的标准，以衡量多模态模型在处理视觉、语言及其交互任务方面的能力。

MMT-Bench 的主要目标是通过一系列精心设计的任务，全面评估多模态模型在不同任务中的综合表现。这些任务包括但不限于视觉问答、图像字幕生成、图像和文本匹配等。具体来说，MMT-Bench 的核心功能如下。

（1）多样化任务集：MMT-Bench 包含多个任务集，每个任务集都用于评估模型的不同能力。这些任务集包括视觉问答、图像字幕生成、图像分类、文本生成等。任务的多样性确保了评估的全面性，能够揭示模型在不同场景下的优势和不足。

（2）跨模态任务评估：评估模型在跨模态任务，如视觉-文本匹配、图像-文本生成等任务中的表现。通过跨模态任务，测试模型在不同模态之间进行信息融合和推理的能力。

（3）统一的评估标准：MMT-Bench 提供了一套统一的评估标准，这套标准包括准确率、召回率、F1 分数、BLEU 分数等。这些标准能帮助用户量化模型的表现，并在不同模型之间进行对比。统一的标准确保了评估结果的可比性和可靠性。

（4）自动化评估工具：MMT-Bench 提供了自动化的评估工具，用户只需简单配置，即可进行模型的批量评估。该自动化评估工具支持多种编程语言和框架，方便用户在不同平台上使用。

（5）详细的评估报告：MMT-Bench 可以生成详细的评估报告，报告内容包含模型在各个任务上的表现、误差分析，以及模型在多模态任务中的综合表现评分等，能帮助用户全面了解模型的能力。

7.4 CLIP 模型的增强训练与评估

本项目是一个基于视觉感知预训练模型（如 LaCLIP/CLIP）和 LLAMA 的图像描述生成系统，包括模型训练、零样本评估以及对生成文本的重写功能，通过整合多种模型和技术，实现对图像内容的理解和生成多样化描述的功能，为图像与文本之间的关联性建模提供一个端到端的解决方案。实例 7-10 是实现了使用 TFT 处理 CSV 文件中的缺失值的例子。

实例 7-10 使用 TFT 处理 CSV 文件中的缺失值
源码路径：codes/7/que.py

7.4.1 项目介绍

本项目实现了对预训练模型 LaCLIP/CLIP 的训练工作，以及 LLAMA 模型的训练和过程。通过这些模型，实现了对图像和文本之间语义关联的建模，可以进行零样本评估，生成多样化和丰富化的图像描述，并提供文本生成的多样化风格选项，拓展多模态大模型在图像与文本处理领域的应用和功能。

本项目的主要功能模块如下。

1. **模型训练模块**
- 包括模型训练的主要逻辑，如模型初始化、损失函数定义、优化器设置、数据加载等。
- 支持分布式训练，使用了分布式数据并行和混合精度训练（AMP）技术。
- 提供了模型训练过程中的日志记录和模型参数保存功能。
2. **零样本评估模块（eval_zeroshot_imagenet.py）**
- 实现了对图像零样本分类的评估功能，使用预训练的模型对图像进行分类，同时支持多种文本模板。
- 包括数据加载、模型预测和准确率计算等功能。
3. **LLAMA 模型加载与应用模块**
- 提供了 LLAMA 模型的加载和应用功能。
- 支持并行加载多个模型，用于多 GPU 或分布式环境下的推理。
- 实现了对给定文本的生成，支持不同的样式选择和文本风格的扩展。
4. **模型并行设置模块**

实现了模型并行的设置，用于在多 GPU 或分布式环境下进行模型的初始化和加载，包括分布式训练的初始化、设备分配等功能。

5. **辅助功能模块**
- 包含了一些辅助函数，如平均计量器（AverageMeter）、进度条显示器（ProgressMeter）等，用于训练过程中的统计和显示。
- 实现了一些工具函数，用于模型的初始化、数据加载等。

上述功能模块共同构成了项目的主要功能，包括模型训练、零样本评估和 LLAMA 模型的加载与应用等。

7.4.2 定义数据集

文件 data.py 定义了两个数据集类 CsvDatasetAugCap 和 CsvDataset，分别用于从 CSV 文件中加载图像及其对应的文本描述。同时提供了函数 get_csv_dataset 和 get_data 用于获取训练和验证数据集。

```
class CsvDatasetAugCap(Dataset):
    def __init__(self, input_filename, transforms, tokenizer=None, root=None,
augmented_caption_filelist=None):
        logging.debug(f'Loading csv data from {input_filename}.')
        self.images = []
        self.captions = []
        self.root = root
        assert input_filename.endswith('.csv')
        assert augmented_caption_filelist is not None, 'augmented_caption_filelist
is None, use csvdataset instead'

        num_augcap = len(augmented_caption_filelist)
        augmented_captions = []
        file_length = []
        for f in augmented_caption_filelist:
            with open(f, 'r') as file:
                cur_captions = file.readlines()
                file_length.append(len(cur_captions))
                augmented_captions.append(cur_captions)
        assert len(augmented_captions) == num_augcap, 'number of augmented captions is not
```

```
equal to num_augcap'

            for i in range(num_augcap):
                assert file_length[i] == file_length[0], 'number of captions in each file is not
                    the same'
            num_samples = file_length[0]

            with open(input_filename, 'r') as csv_file:
                csv_reader = csv.reader(csv_file)
                row_index = 0
                for row in tqdm(csv_reader):
                    image = row[0]
                    prompt = row[1]
                    if image.endswith(('.png', '.jpg', '.jpeg')):
                        image_path = os.path.join(self.root, image)
                        self.images.append(image_path)

                        if row_index < num_samples:
                            self.captions.append([prompt])
                            for augcap_idx in range(num_augcap):
                                self.captions[row_index].append(augmented_captions
                                    [augcap_idx][row_index].replace('\n',''))
                            assert len(self.captions[row_index]) == num_augcap + 1, 'number of
                                captions is not equal to num_augcap + 1'
                        row_index += 1
                assert row_index % num_samples == 0, 'number of samples in csv is not equal to
                    num_samples in new caption'

        self.num_samples = num_samples
        self.transforms = transforms
        logging.debug('Done loading data.')

        self.tokenizer = tokenizer

    def __len__(self):
        return len(self.images)

    def __getitem__(self, idx):
        images = self.transforms(Image.open(str(self.images[idx])))
        caption_list = self.captions[idx%self.num_samples]
        caption = random.choice(caption_list)
        if len(caption.split(' ')) < 2:
            caption = caption_list[0]
        texts = caption
        texts = self.tokenizer(str(texts))
        return images, texts


class CsvDataset(Dataset):
    def __init__(self, input_filename, transforms, tokenizer=None, root=None):
        logging.debug(f'Loading csv data from {input_filename}.')
        self.images = []
        self.captions = []
        self.root = root
        assert input_filename.endswith('.csv')
        with open(input_filename, 'r') as csv_file:
            csv_reader = csv.reader(csv_file)
            for row in tqdm(csv_reader):
                image = row[0]
                prompt = row[1]
                if image.endswith(('.png', '.jpg', '.jpeg')):
                    image_path = os.path.join(self.root, image)
                    self.images.append(image_path)
```

```python
                    self.captions.append(prompt)
        self.transforms = transforms
        logging.debug('Done loading data.')

        self.tokenizer = tokenizer

    def __len__(self):
        return len(self.captions)

    def __getitem__(self, idx):
        images = self.transforms(Image.open(str(self.images[idx])))
        texts = self.tokenizer(str(self.captions[idx]))
        return images, texts


@dataclass
class DataInfo:
    dataloader: DataLoader
    sampler: DistributedSampler = None

    def set_epoch(self, epoch):
        if self.sampler is not None and isinstance(self.sampler, DistributedSampler):
            self.sampler.set_epoch(epoch)


def get_csv_dataset(args, preprocess_fn, is_train, tokenizer=None, aug_text=False):
    input_filename = args.train_data if is_train else args.val_data
    assert input_filename
    if args.aug_text:
        augmented_caption_filelist = args.augmented_caption_filelist
        dataset = CsvDatasetAugCap(
            input_filename,
            preprocess_fn,
            root=args.root,
            tokenizer=tokenizer,
            augmented_caption_filelist=augmented_caption_filelist,
        )

    else:
        dataset = CsvDataset(
            input_filename,
            preprocess_fn,
            root=args.root,
            tokenizer=tokenizer
        )

    num_samples = len(dataset)
    sampler = DistributedSampler(dataset) if args.distributed and is_train else None
    shuffle = is_train and sampler is None

    dataloader = DataLoader(
        dataset,
        batch_size=args.batch_size,
        shuffle=shuffle,
        num_workers=args.workers,
        pin_memory=True,
        sampler=sampler,
        drop_last=is_train,
    )
    dataloader.num_samples = num_samples
    dataloader.num_batches = len(dataloader)

    return DataInfo(dataloader, sampler)
```

```python
def get_data(args, preprocess_fns, tokenizer=None):
    preprocess_train, preprocess_val = preprocess_fns
    data = {"train": get_csv_dataset(args, preprocess_train, is_train=True, tokenizer=tokenizer)}

    return data
```

对上述代码的具体说明如下。

（1）类 CsvDatasetAugCap 的功能是从 CSV 文件中加载图像及其对应的原始文本描述和增强文本描述。它会将原始文本描述和多个增强文本描述合并成一个列表，支持随机选择其中一个文本描述作为输出。

（2）类 CsvDataset 的功能是从 CSV 文件中加载图像及其对应的单个原始文本描述。它用于加载普通的图像数据集，每个图像只有一个文本描述。

（3）类 DataInfo 是一个数据类，用于存储数据加载器和分布式采样器。它提供了设置 epoch 的方法，用于分布式训练时更新采样器的 epoch。

（4）函数 get_csv_dataset 的功能是根据参数返回相应的数据集对象，支持训练和验证数据集的加载，并根据是否启用增强文本描述来选择使用 CsvDatasetAugCap 或 CsvDataset 类。它还设置了数据加载器，并根据是否分布式训练选择是否使用分布式采样器。

（5）函数 get_data 的功能是获取训练数据集，并返回一个字典，包含训练数据集的信息。

7.4.3 创建模型

文件 models.py 创建了"视觉-文本检索（CLIP）"模型的各个组件和不同规模变体的构建方法，包括自注意力机制、残差注意力块、Transformer 架构以及 Vision Transformer，并提供了不同规模的 CLIP 模型构建方法，以便在视觉和文本之间进行编码和检索。文件 models.py 的具体实现流程如下。

（1）实现自定义的类 LayerNorm，此类继承自 PyTorch 中的 nn.LayerNorm 类，用于处理 fp16 数据。在 forward 方法中，将输入张量转换为 float32 类型后再调用父类的 forward 方法进行处理，并最终将结果转换回原始的数据类型。

```python
class LayerNorm(nn.LayerNorm):
    """子类化 torch 的 LayerNorm 以处理 fp16。"""

    def forward(self, x: torch.Tensor):
        orig_type = x.dtype
        ret = super().forward(x.type(torch.float32))
        return ret.type(orig_type)
```

（2）在类 QuickGELU 中定义了一个快速的 GELU（Gaussian Error Linear Unit）激活函数，对输入张量 x 进行 GELU 操作，即 xsigmoid(1.702×x)，用于神经网络的非线性变换。

```python
class QuickGELU(nn.Module):
    def forward(self, x: torch.Tensor):
        return x * torch.sigmoid(1.702 * x)
```

（3）类 ResidualAttentionBlock 定义了一个残差注意力块（Residual Attention Block），其中包括多头注意力机制和前馈神经网络（MLP）。该块首先对输入进行层归一化，然后通过多头注意力机制处理输入，接着再进行层归一化，并通过一个包含快速 GELU 激活函数的前馈神经网络处理。最终将这些处理后的结果与输入进行残差连接并返回，增强了模型的表示能力和训练稳定性。

```python
class ResidualAttentionBlock(nn.Module):
    def __init__(self, d_model: int, n_head: int, attn_mask: torch.Tensor = None):
```

```python
        super().__init__()
        self.attn = nn.MultiheadAttention(d_model, n_head)
        self.ln_1 = LayerNorm(d_model)
        self.mlp = nn.Sequential(OrderedDict([
            ("c_fc", nn.Linear(d_model, d_model * 4)),
            ("gelu", QuickGELU()),
            ("c_proj", nn.Linear(d_model * 4, d_model))
        ]))
        self.ln_2 = LayerNorm(d_model)
        self.attn_mask = attn_mask

    def attention(self, x: torch.Tensor):
        self.attn_mask = self.attn_mask.to(dtype=x.dtype, device=x.device) if self.attn_mask is not None else None
        return self.attn(x, x, x, need_weights=False, attn_mask=self.attn_mask)[0]

    def forward(self, x: torch.Tensor):
        x = x + self.attention(self.ln_1(x))
        x = x + self.mlp(self.ln_2(x))
        return x
```

（4）类 Transformer 定义了一个 Transformer 模块，包括多个残差注意力块。每个残差注意力块包括多头注意力机制和前馈神经网络。该模块初始化时，指定的层数、宽度和头数用于构建这些残差注意力块。在前向传递中，输入将依次通过这些残差注意力块进行处理，最终返回处理后的结果。

```python
class Transformer(nn.Module):
    def __init__(self, width: int, layers: int, heads: int, attn_mask: torch.Tensor = None):
        super().__init__()
        self.width = width
        self.layers = layers
        self.resblocks = nn.Sequential(*[ResidualAttentionBlock(width, heads, attn_mask) for _ in range(layers)])

    def forward(self, x: torch.Tensor):
        return self.resblocks(x)
```

（5）类 CLIP 定义了一个 CLIP 模型，用于将视觉和文本信息编码为嵌入向量。模型包括视觉模型和文本模型，使用了 Transformer 机制来处理文本序列，并通过注意力机制进行处理。模型初始化时设置了各种参数和嵌入层，在前向传递过程中分别编码图像和文本，并返回它们的嵌入向量和一个用于缩放对数概率的参数。

```python
class CLIP(nn.Module):
    def __init__(self,
                 embed_dim: int,
                 # 视觉
                 vision_width: int,
                 vision_model: nn.Module,
                 # 文本
                 context_length: int,
                 vocab_size: int,
                 transformer_width: int,
                 transformer_heads: int,
                 transformer_layers: int,
                 **kwargs,
                 ):
        super().__init__()

        self.context_length = context_length
```

```python
        self.vision_width = vision_width

        self.visual = vision_model

        self.transformer = Transformer(
            width=transformer_width,
            layers=transformer_layers,
            heads=transformer_heads,
            attn_mask=self.build_attention_mask(),
        )

        self.vocab_size = vocab_size
        self.token_embedding = nn.Embedding(vocab_size, transformer_width)
        self.positional_embedding = nn.Parameter(torch.empty(self.context_length, transformer_width))
        self.ln_final = LayerNorm(transformer_width)

        self.image_projection = nn.Parameter(torch.empty(vision_width, embed_dim))
        self.text_projection = nn.Parameter(torch.empty(transformer_width, embed_dim))
        self.logit_scale = nn.Parameter(torch.ones([]) * np.log(1 / 0.07))

        self.initialize_parameters()

    def initialize_parameters(self):
        nn.init.normal_(self.token_embedding.weight, std=0.02)
        nn.init.normal_(self.positional_embedding, std=0.01)

        proj_std = (self.transformer.width ** -0.5) * ((2 * self.transformer.layers) ** -0.5)
        attn_std = self.transformer.width ** -0.5
        fc_std = (2 * self.transformer.width) ** -0.5
        for block in self.transformer.resblocks:
            nn.init.normal_(block.attn.in_proj_weight, std=attn_std)
            nn.init.normal_(block.attn.out_proj.weight, std=proj_std)
            nn.init.normal_(block.mlp.c_fc.weight, std=fc_std)
            nn.init.normal_(block.mlp.c_proj.weight, std=proj_std)

        nn.init.normal_(self.image_projection, std=self.vision_width ** -0.5)
        nn.init.normal_(self.text_projection, std=self.transformer.width ** -0.5)

    def build_attention_mask(self):
        # 延迟创建因果注意掩码，视觉令牌之间的完全注意力
        # pytorch 使用加性注意掩码；填充为 -inf
        mask = torch.empty(self.context_length, self.context_length)
        mask.fill_(float("-inf"))
        mask.triu_(1)  # 清除下对角线
        return mask

    def encode_image(self, image):
        x = self.visual(image)
        x = x @ self.image_projection

        return x

    def encode_text(self, text):
        x = self.token_embedding(text)  # [batch_size, n_ctx, d_model]
        x = x + self.positional_embedding
        x = x.permute(1, 0, 2)  # NLD -> LND
        x = self.transformer(x)
        x = x.permute(1, 0, 2)  # LND -> NLD
        x = self.ln_final(x)

        # x.shape = [batch_size, n_ctx, transformer.width]
```

```
            # 从 eot 嵌入中提取特征（eot_token 是每个序列中最大的数）
            x = x[torch.arange(x.shape[0]), text.argmax(dim=-1)] @ self.text_projection

            return x

        def forward(self, image, text):
            image_embed = self.encode_image(image)
            text_embed = self.encode_text(text)

            return {'image_embed': image_embed,
                    'text_embed': text_embed,
                    'logit_scale': self.logit_scale.exp()}
```

（6）类 Attention 定义了一个注意力机制模块，用于计算输入张量的自注意力。通过查询（query）、键（key）、值（value）向量的线性变换和归一化，模块计算注意力权重并应用于值向量，最终输出经过投影和丢弃层处理的结果。该模块支持多头注意力，并可以选择性地在查询和键上应用 LayerNorm 以提高训练稳定性。

```
class Attention(nn.Module):
    def __init__(self, dim, num_heads=8, qkv_bias=False, qk_norm=False, attn_drop=0.,
                 proj_drop=0.):
        super().__init__()
        self.num_heads = num_heads
        head_dim = dim // num_heads
        self.scale = head_dim ** -0.5

        # todo: 添加 q 和 k 的归一化以提高训练稳定性
        self.q_norm = nn.LayerNorm(head_dim, eps=1e-6) if qk_norm else nn.Identity()
        self.k_norm = nn.LayerNorm(head_dim, eps=1e-6) if qk_norm else nn.Identity()

        self.qkv = nn.Linear(dim, dim * 3, bias=qkv_bias)
        self.attn_drop = nn.Dropout(attn_drop)
        self.proj = nn.Linear(dim, dim)
        self.proj_drop = nn.Dropout(proj_drop)

    def forward(self, x):
        B, N, C = x.shape
        qkv = self.qkv(x).reshape(B, N, 3, self.num_heads, C // self.num_heads).permute(2, 0,
        3, 1, 4)
        q, k, v = qkv.unbind(0)    # 使 torchscript 满意（不能将张量用作元组）

        # todo: 对查询和键应用归一化
        q = self.q_norm(q)
        k = self.k_norm(k)

        attn = (q @ k.transpose(-2, -1)) * self.scale
        attn = attn.softmax(dim=-1)
        attn = self.attn_drop(attn)

        x = (attn @ v).transpose(1, 2).reshape(B, N, C)
        x = self.proj(x)
        x = self.proj_drop(x)
        return x
```

（7）类 Block 定义了一个 Transformer 块，每个块由一个归一化层、一个注意力层、一个 MLP 层（包含隐藏层和激活函数），以及一个可选的随机深度的 drop-path 组成。在前向传播中，输入张量先经过注意力层和 MLP 层的处理，并分别添加 skip connection，以提高网络的稳定性和性能。

```
class Block(nn.Module):
```

```
        def __init__(self, dim, num_heads, mlp_ratio=4., qkv_bias=False, qk_norm=False, drop=0., attn_
drop=0., drop_path=0., act_layer=nn.GELU, norm_layer=nn.LayerNorm):
            super().__init__()
            self.norm1 = norm_layer(dim)
            self.attn = Attention(dim, num_heads=num_heads, qkv_bias=qkv_bias, qk_norm=qk_norm,
                                  attn_drop=attn_drop, proj_drop=drop)
            # 注意:对于随机深度的drop-path,我们将看到这是否比dropout更好
            self.drop_path = DropPath(drop_path) if drop_path > 0. else nn.Identity()
            self.norm2 = norm_layer(dim)
            mlp_hidden_dim = int(dim * mlp_ratio)
            self.mlp = Mlp(in_features=dim, hidden_features=mlp_hidden_dim,
                           act_layer=act_layer, drop=drop)

        def forward(self, x):
            x = x + self.drop_path(self.attn(self.norm1(x)))
            x = x + self.drop_path(self.mlp(self.norm2(x)))
            return x
```

（8）类 VisionTransformer 定义了一个支持全局平均池化的视觉 Transformer。如果启用了 qk_norm（query 和 key 的归一化），它会重新初始化 Transformer 的块（blocks），使用提供的参数（如 embed_dim、num_heads、mlp_ratio 等）创建一系列的 Block 对象，并将它们组织成一个顺序容器。

```
class VisionTransformer(timm.models.vision_transformer.VisionTransformer):
    def __init__(self, qk_norm=False, **kwargs):
        super(VisionTransformer, self).__init__(**kwargs)
        if qk_norm:
            del self.blocks
            embed_dim = kwargs['embed_dim']
            num_heads = kwargs['num_heads']
            mlp_ratio = kwargs['mlp_ratio']
            qkv_bias = kwargs['qkv_bias']
            depth = kwargs['depth']
            drop_rate = 0.
            attn_drop_rate = 0.
            drop_path_rate = 0.
            dpr = [x.item() for x in torch.linspace(0, drop_path_rate, depth)]
            norm_layer = partial(nn.LayerNorm, eps=1e-6)
            act_layer = nn.GELU
            self.blocks = nn.Sequential(*[
                Block(
                    dim=embed_dim, num_heads=num_heads, mlp_ratio=mlp_ratio, qkv_bias=qkv_bias,
                    qk_norm=qk_norm,
                    drop=drop_rate, attn_drop=attn_drop_rate, drop_path=dpr[i], norm_layer=
                    norm_layer, act_layer=act_layer)
                for i in range(depth)])
```

（9）函数 vit_small_patch16_224 定义了一个小型的 Vision Transformer 模型，其特征包括 16×16 的图像块大小、384 的嵌入维度、12 层深度、12 个注意力头和 4 的 MLP 比例。这个模型使用偏置的 qkv（query、key、value）以及带有 epsilon 值为 1e-6 的 LayerNorm 归一化层。该函数通过将这些参数传递给 VisionTransformer 类并返回实例化的模型来创建和返回 Vision Transformer 模型。

```
def vit_small_patch16_224(**kwargs):
    model = VisionTransformer(
        patch_size=16, embed_dim=384, depth=12, num_heads=12, mlp_ratio=4, qkv_bias=True,
        norm_layer=partial(nn.LayerNorm, eps=1e-6), **kwargs)
    return model
```

（10）函数 vit_base_patch16_224 定义了一个基础版的 Vision Transformer 模型，其特征包括 16×16 的图像块大小、768 的嵌入维度、12 层深度、12 个注意力头和 4 的 MLP 比例。这

个模型使用偏置的 qkv（query、key、value）以及带有 epsilon 值为 1e-6 的 LayerNorm 归一化层。该函数通过将这些参数传递给 VisionTransformer 类并返回实例化的模型来创建和返回 Vision Transformer 模型。

```
def vit_base_patch16_224(**kwargs):
    model = VisionTransformer(
        patch_size=16, embed_dim=768, depth=12, num_heads=12, mlp_ratio=4, qkv_bias=True,
        norm_layer=partial(nn.LayerNorm, eps=1e-6), **kwargs)
    return model
```

（11）函数 vit_large_patch16_224 定义了一个大型的 Vision Transformer 模型，其特征包括 16×16 的图像块大小、1024 的嵌入维度、24 层深度、16 个注意力头和 4 的 MLP 比例。这个模型使用偏置的 qkv（query、key、value）以及带有 epsilon 值为 1e-6 的 LayerNorm 归一化层。该函数通过将这些参数传递给 VisionTransforme 类并返回实例化的模型来创建和返回 Vision Transformer 模型。

```
def vit_large_patch16_224(**kwargs):
    model = VisionTransformer(
        patch_size=16, embed_dim=1024, depth=24, num_heads=16, mlp_ratio=4, qkv_bias=True,
        norm_layer=partial(nn.LayerNorm, eps=1e-6), **kwargs)
    return model
```

（12）函数 CLIP_VITS16 定义了一个 CLIP 模型，使用了一个小型的 Vision Transformer（ViT）模型作为视觉编码器。具体来说，函数首先创建一个具有 qk_norm 的 vit_small_patch16_224 模型，设置 num_classes=0。然后，这个 ViT 模型被传递给 CLIP 类进行实例化，并配置了嵌入维度、视觉宽度、上下文长度、词汇表大小、Transformer 的宽度、头数和层数等参数。函数最后返回这个配置好的 CLIP 模型。

```
def CLIP_VITS16(**kwargs):
    vision_model = vit_small_patch16_224(qk_norm=True, num_classes=0)
    model = CLIP(embed_dim=512, vision_width=384, vision_model=vision_model, context_length=77, vocab_size=49408,
        transformer_width=512, transformer_heads=8, transformer_layers=12, **kwargs)

    return model
```

（13）函数 CLIP_VITB16 定义了一个 CLIP 模型，使用了一个基础规模的 Vision Transformer（ViT）模型作为视觉编码器。具体来说，函数 CLIP_VITB16 首先创建一个具有 qk_norm 的 vit_base_patch16_224 模型，设置 num_classes=0。然后，这个 ViT 模型被传递给 CLIP 类进行实例化，并配置了嵌入维度、视觉宽度、上下文长度、词汇表大小、Transformer 的宽度、头数和层数等参数。函数最后返回这个配置好的 CLIP 模型。

```
def CLIP_VITB16(**kwargs):
    vision_model = vit_base_patch16_224(qk_norm=True, num_classes=0)
    model = CLIP(embed_dim=512, vision_width=768, vision_model=vision_model, context_length=77, vocab_size=49408,
        transformer_width=512, transformer_heads=8, transformer_layers=12, **kwargs)

    return model
```

（14）函数 CLIP_VITL16 定义了一个 CLIP 模型，使用了一个大规模的 Vision Transformer（ViT）模型作为视觉编码器。具体来说，函数 CLIP_VITL16 首先创建了一个具有 qk_norm 的 vit_large_patch16_224 模型，设置 num_classes=0。然后，这个 ViT 模型被传递给 CLIP 类进行实例化，并配置了嵌入维度、视觉宽度、上下文长度、词汇表大小、Transformer 的宽度、头数和层数等参数。函数最后返回这个配置好的 CLIP 模型。

```python
def CLIP_VITL16(**kwargs):
    vision_model = vit_large_patch16_224(qk_norm=True, num_classes=0)
    model = CLIP(embed_dim=512, vision_width=1024, vision_model=vision_model, context_length=77,
vocab_size=49408, transformer_width=512, transformer_heads=8, transformer_layers=12, **kwargs)
    return model
```

7.4.4 分词器

文件 tokenizer.py 实现了一个简单的分词器 SimpleTokenizer，用于将文本编码成 BPE（Byte Pair Encoding）标记，并提供解码功能。该分词器可用于为输入文本生成 BPE 标记序列，以便在模型中进行处理，同时支持对生成的标记序列进行解码还原成原始文本。

```python
@lru_cache()
def default_bpe():
    return os.path.join(os.path.dirname(os.path.abspath(__file__)), "bpe_simple_vocab_16e6.txt.gz")


@lru_cache()
def bytes_to_unicode():
    bs = list(range(ord("!"), ord("~")+1))+list(range(ord("¡"), ord("¬")+1))+list(range(ord("®"), ord("ÿ")+1))
    cs = bs[:]
    n = 0
    for b in range(2**8):
        if b not in bs:
            bs.append(b)
            cs.append(2**8+n)
            n += 1
    cs = [chr(n) for n in cs]
    return dict(zip(bs, cs))


def get_pairs(word):
    """Return set of symbol pairs in a word.
    Word is represented as tuple of symbols (symbols being variable-length strings).
    """
    pairs = set()
    prev_char = word[0]
    for char in word[1:]:
        pairs.add((prev_char, char))
        prev_char = char
    return pairs


def basic_clean(text):
    text = ftfy.fix_text(text)
    text = html.unescape(html.unescape(text))
    return text.strip()


def whitespace_clean(text):
    text = re.sub(r'\s+', ' ', text)
    text = text.strip()
    return text


class SimpleTokenizer(object):
    def __init__(self, bpe_path: str = default_bpe()):
```

```python
            self.byte_encoder = bytes_to_unicode()
            self.byte_decoder = {v: k for k, v in self.byte_encoder.items()}
            merges = gzip.open(bpe_path).read().decode("utf-8").split('\n')
            merges = merges[1:49152-256-2+1]
            merges = [tuple(merge.split()) for merge in merges]
            vocab = list(bytes_to_unicode().values())
            vocab = vocab + [v+'</w>' for v in vocab]
            for merge in merges:
                vocab.append(''.join(merge))
            vocab.extend(['<|startoftext|>', '<|endoftext|>'])
            self.encoder = dict(zip(vocab, range(len(vocab))))
            self.decoder = {v: k for k, v in self.encoder.items()}
            self.bpe_ranks = dict(zip(merges, range(len(merges))))
            self.cache = {'<|startoftext|>': '<|startoftext|>', '<|endoftext|>': '<|endoftext|>'}
            self.pat = re.compile(r"""<\|startoftext\|>|<\|endoftext\|>|'s|'t|'re|'ve|'m|'ll|'d|[\p{L}]+|[\p{N}]|[^\s\p{L}\p{N}]+""",
            re.IGNORECASE)

        def bpe(self, token):
            if token in self.cache:
                return self.cache[token]
            word = tuple(token[:-1]) + ( token[-1] + '</w>',)
            pairs = get_pairs(word)

            if not pairs:
                return token+'</w>'

            while True:
                bigram = min(pairs, key = lambda pair: self.bpe_ranks.get(pair, float('inf')))
                if bigram not in self.bpe_ranks:
                    break
                first, second = bigram
                new_word = []
                i = 0
                while i < len(word):
                    try:
                        j = word.index(first, i)
                        new_word.extend(word[i:j])
                        i = j
                    except:
                        new_word.extend(word[i:])
                        break

                    if word[i] == first and i < len(word)-1 and word[i+1] == second:
                        new_word.append(first+second)
                        i += 2
                    else:
                        new_word.append(word[i])
                        i += 1
                new_word = tuple(new_word)
                word = new_word
                if len(word) == 1:
                    break
                else:
                    pairs = get_pairs(word)
            word = ' '.join(word)
            self.cache[token] = word
            return word

        def encode(self, text):
            bpe_tokens = []
```

```
        text = whitespace_clean(basic_clean(text)).lower()
        for token in re.findall(self.pat, text):
            token = ''.join(self.byte_encoder[b] for b in token.encode('utf-8'))
            bpe_tokens.extend(self.encoder[bpe_token] for bpe_token in self.bpe(token).split(' '))
        return bpe_tokens

    def decode(self, tokens):
        text = ''.join([self.decoder[token] for token in tokens])
        text = bytearray([self.byte_decoder[c] for c in text]).decode('utf-8', errors="replace").
        replace('</w>', ' ')
        return text

    def __call__(self, texts, context_length=77):
        if isinstance(texts, str):
            texts = [texts]

        sot_token = self.encoder["<|startoftext|>"]
        eot_token = self.encoder["<|endoftext|>"]
        all_tokens = [[sot_token] + self.encode(text) + [eot_token] for text in
        texts]
        result = torch.zeros(len(all_tokens), context_length, dtype=torch.long)

        for i, tokens in enumerate(all_tokens):
            if len(tokens) > context_length:
                tokens = tokens[:context_length]  # Truncate
                tokens[-1] = eot_token
            result[i, :len(tokens)] = torch.tensor(tokens)

        if len(result) == 1:
            return result[0]
        return result
```

上述代码的主要功能如下。

（1）BPE 编码和解码：通过给定的 BPE 文件对文本进行编码和解码。

（2）文本清洗：提供了基本的文本清洗功能，包括修复文本、去除 HTML 实体、去除多余的空白字符等。

（3）编码和解码方法：encode 方法将文本编码成 BPE 标记的索引列表，decode 方法将索引列表解码成文本。

（4）上下文处理：支持指定上下文长度，超出部分进行截断。

（5）提供了一些辅助函数和常量：如 default_bpe 用于获取默认的 BPE 文件路径，bytes_to_unicode 用于字节到 Unicode 字符的映射，get_pairs 用于获取单词中的符号对等。

7.4.5 损失函数

文件 losses.py 定义了一个用于 CLIP 模型的损失函数，其中包括计算图像和文本之间相似度的损失以及准确率。

```
class CLIPLoss(nn.Module):
    def __init__(self):
        super().__init__()
        # 初始化标签和批大小以进行分布式训练
        self.labels = None
        self.last_local_batch_size = None
```

```python
def forward(self, outputs):
    # 从模型输出中提取图像和文本嵌入以及 logit 缩放比例
    image_embed = outputs['image_embed']   # 图像嵌入
    text_embed = outputs['text_embed']     # 文本嵌入
    logit_scale = outputs['logit_scale']   # logits 缩放比例，用于调整 logits
    local_batch_size = image_embed.size(0) # 当前进程的批大小

    # 如果批大小发生变化，则更新标签
    if local_batch_size != self.last_local_batch_size:
        # 为当前批次生成标签
        self.labels = local_batch_size * utils.get_rank() + torch.arange(
            local_batch_size, device=image_embed.device
        )
        self.last_local_batch_size = local_batch_size

    # 对图像和文本嵌入进行标准化
    image_embed = F.normalize(image_embed, dim=-1, p=2)
    text_embed = F.normalize(text_embed, dim=-1, p=2)

    # 收集来自不同进程的所有嵌入
    image_embed_all = torch.cat(torch.distributed.nn.all_gather(image_embed), dim=0)
    text_embed_all = torch.cat(torch.distributed.nn.all_gather(text_embed), dim=0)

    # 使用余弦相似度计算 logits
    logits_per_image = logit_scale * image_embed @ text_embed_all.t()
    logits_per_text = logit_scale * text_embed @ image_embed_all.t()

    # 计算损失，这里使用交叉熵损失
    loss = (F.cross_entropy(logits_per_image, self.labels) + \
        F.cross_entropy(logits_per_text, self.labels)) / 2

    # 计算准确率
    with torch.no_grad():
        pred = torch.argmax(logits_per_image, dim=-1)
        correct = pred.eq(self.labels).sum()
        acc = 100 * correct / local_batch_size

    return {'loss': loss, 'clip_loss': loss, 'clip_acc': acc}
```

7.4.6 模型训练

文件 train.py 的功能是训练 LaCLIP/CLIP 模型，包括模型初始化、定义损失函数和优化器、数据加载、训练循环和验证过程，旨在训练图像和文本之间的对齐模型，并评估模型在验证集上的性能。文件 train.py 的具体实现流程如下。

（1）定义用于训练 LaCLIP/CLIP 模型的命令行参数解析器，并初始化一个记录最佳准确率的变量。

```python
import argparse

def get_args_parser():
    # 创建参数解析器
    parser = argparse.ArgumentParser(description='LaCLIP/CLIP 训练', add_help=False)
    parser.add_argument(
        "--train-data",
        type=str,
        default=None,
        help="训练数据的路径.",
    )
    parser.add_argument(
        '--root',
        type=str,
```

```python
        default='./data/',
        help='图像的根目录.'
    )
    # 增强标注文件的文件名列表
    parser.add_argument('--augmented_caption_filelist', nargs='+', help='增强标注文件的文件名列表，以空格分隔')
    parser.add_argument('--aug-text', action='store_true', help='设置为 True 以进行 LaCLIP 训练')

    parser.add_argument('--imagenet-root', default='data/imagenet', type=str, help='imagenet 数据集的路径')
    parser.add_argument('--output-dir', default='./output', type=str, help='输出目录')

    parser.add_argument('--model', default='CLIP_VITB16', type=str)
    parser.add_argument('--resume', default='', type=str, help='恢复训练的模型路径')

    parser.add_argument('--epochs', default=25, type=int)
    parser.add_argument('--warmup-epochs', default=1, type=int)
    parser.add_argument('--start-epoch', default=0, type=int)
    parser.add_argument('--batch-size', default=64, type=int, help='每个 GPU 的样本数量')
    parser.add_argument('--lr', default=3e-3, type=float)
    parser.add_argument('--lr-start', default=1e-6, type=float, help='初始学习率用于热身阶段')
    parser.add_argument('--lr-end', default=1e-5, type=float, help='最终学习率的最小值')
    parser.add_argument('--update-freq', default=1, type=int, help='优化器更新频率（梯度累积步数）')
    parser.add_argument('--wd', default=0.1, type=float)
    parser.add_argument('--betas', default=(0.9, 0.98), nargs=2, type=float)
    parser.add_argument('--eps', default=1e-8, type=float)
    parser.add_argument('--disable-amp', action='store_true', help='禁用混合精度训练（需要更多内存和计算资源）')

    parser.add_argument('--print-freq', default=10, type=int, help='打印频率')
    parser.add_argument('-j', '--workers', default=10, type=int, metavar='N',
                        help='每个进程的数据加载工作线程数')
    parser.add_argument('--world-size', default=1, type=int,
                        help='用于分布式训练的节点数量')
    parser.add_argument('--rank', default=0, type=int,
                        help='用于分布式训练的节点排名')
    parser.add_argument("--local_rank", type=int, default=0)
    parser.add_argument('--dist-url', default='env://', type=str,
                        help='用于设置分布式训练的 URL')
    parser.add_argument('--dist-backend', default='nccl', type=str)
    parser.add_argument('--seed', default=0, type=int)
    parser.add_argument('--gpu', default=None, type=int, help='要使用的 GPU ID.')
    return parser


best_acc1 = 0
```

（2）函数 main 实现了模型训练的主要功能，它会根据参数设置初始化模型和优化器，加载数据集，进行训练和验证，并保存训练过程中的日志和最佳模型检查点。

```python
def main(args):
    # 初始化分布式模式
    utils.init_distributed_mode(args)

    global best_acc1

    # 为了可重现性固定种子
    seed = args.seed + utils.get_rank()
    torch.manual_seed(seed)
    np.random.seed(seed)

    # 创建模型
```

```python
print("=> 创建模型: {}".format(args.model))
model = getattr(models, args.model)()
model.cuda(args.gpu)

if args.distributed:
    model = torch.nn.parallel.DistributedDataParallel(model, device_ids= [args.gpu],
        bucket_cap_mb=200)

# 定义损失函数 (criterion) 和优化器
criterion = losses.CLIPLoss().cuda(args.gpu)

p_wd, p_non_wd = [], []
for n, p in model.named_parameters():
    if not p.requires_grad:
        continue   # 冻结权重
    if p.ndim < 2 or 'bias' in n or 'ln' in n or 'bn' in n:
        p_non_wd.append(p)
    else:
        p_wd.append(p)

optim_params = [{"params": p_wd, "weight_decay": args.wd},
                {"params": p_non_wd, "weight_decay": 0}]

optimizer = torch.optim.AdamW(optim_params, lr=args.lr, betas=args.betas,
                              eps=args.eps, weight_decay=args.wd)
scaler = amp.GradScaler(enabled=not args.disable_amp)

# 可选地从检查点恢复 (优先于自动恢复)
if args.resume:
    if os.path.isfile(args.resume):
        print("=> 加载恢复的检查点 '{}'".format(args.resume))
        checkpoint = torch.load(args.resume, map_location='cpu')
        epoch = checkpoint['epoch'] if 'epoch' in checkpoint else 0
        args.start_epoch = epoch
        result = model.load_state_dict(checkpoint['state_dict'], strict=False)
        print(result)
        optimizer.load_state_dict(checkpoint['optimizer']) if 'optimizer' \
            in checkpoint else ()
        scaler.load_state_dict(checkpoint['scaler']) if 'scaler' in checkpoint \
            else ()
        best_acc1 = checkpoint['best_acc1']
        print("=> 加载完毕的恢复检查点 '{}' (epoch {})"
                .format(args.resume, epoch))
    else:
        print("=> 在 '{}' 找不到检查点".format(args.resume))
else:
    # 从输出目录中的最新检查点自动恢复
    latest = os.path.join(args.output_dir, 'checkpoint.pt')
    if os.path.isfile(latest):
        print("=> 加载最新的检查点 '{}'".format(latest))
        latest_checkpoint = torch.load(latest, map_location='cpu')
        args.start_epoch = latest_checkpoint['epoch']
        model.load_state_dict(latest_checkpoint['state_dict'])
        optimizer.load_state_dict(latest_checkpoint['optimizer'])
        scaler.load_state_dict(latest_checkpoint['scaler'])
        best_acc1 = latest_checkpoint['best_acc1']
        print("=> 加载完毕的最新检查点 '{}' (epoch {})"
                .format(latest, latest_checkpoint['epoch']))

cudnn.benchmark = True

# 数据加载代码
print("=> 创建数据集")
```

```python
tokenizer = SimpleTokenizer()
normalize = transforms.Normalize(mean=[0.485, 0.456, 0.406],
                                 std=[0.229, 0.224, 0.225])
train_transform = transforms.Compose([
        transforms.RandomResizedCrop(224, scale=(0.5, 1.0)),
        transforms.ToTensor(),
        normalize
    ])
val_transform = transforms.Compose([
        transforms.Resize(224),
        transforms.CenterCrop(224),
        transforms.ToTensor(),
        normalize
    ])

val_dataset = datasets.ImageFolder(os.path.join(args.imagenet_root, 'val'),
transform=val_transform)

if args.distributed:
    val_sampler = torch.utils.data.distributed.DistributedSampler(val_dataset)
else:
    val_sampler = None

val_loader = torch.utils.data.DataLoader(
    val_dataset, batch_size=args.batch_size, shuffle=(val_sampler is None),
    num_workers=args.workers, pin_memory=True, sampler=val_sampler, drop_last=False)

data = get_data(args, (train_transform, val_transform), tokenizer=tokenizer)
print('数据集大小: %d' % data['train'].dataloader.num_samples)
train_loader = data['train'].dataloader

loader_len = train_loader.num_batches

lr_schedule = utils.cosine_scheduler(args.lr, args.lr_end, args.epochs,
                                    loader_len // args.update_freq,
                                    warmup_epochs=args.warmup_epochs,
                                    start_warmup_value=args.lr_start)

if utils.is_main_process() and args.output_dir is not None:
    args.log_dir = os.path.join(args.output_dir, 'tb_logs')
    os.makedirs(args.log_dir, exist_ok=True)
    log_writer = SummaryWriter(log_dir=args.log_dir)
else:
    log_writer = None

print(args)

print("=> 开始训练")
for epoch in range(args.start_epoch, args.epochs):
    if args.distributed:
        data['train'].set_epoch(epoch)
    train_loader = data['train'].dataloader

    # 训练一个周期
    train_stats = train(train_loader, log_writer, model, criterion, optimizer,
    scaler, epoch, lr_schedule, args)

    val_stats = validate_zeroshot(val_loader, model, tokenizer, args)
    acc1 = val_stats['acc1']

    is_best = acc1 > best_acc1
    best_acc1 = max(acc1, best_acc1)
    print("=> 保存检查点")
```

```python
            utils.save_on_master({
                'epoch': epoch + 1,
                'state_dict': model.state_dict(),
                'optimizer' : optimizer.state_dict(),
                'scaler': scaler.state_dict(),
                'best_acc1': best_acc1,
                'args': args,
            }, is_best, args.output_dir)

        log_stats = {**{f'train_{k}': v for k, v in train_stats.items()},
                     **{f'test_{k}': v for k, v in val_stats.items()},
                     'epoch': epoch}

        # 将测试统计信息记录到 log_writer（tensorboard）
        if log_writer is not None:
            for k, v in log_stats.items():
                if k.startswith('test'):
                    log_writer.add_scalar(k, v, epoch)

        if utils.is_main_process():
            with open(os.path.join(args.output_dir, 'log.txt'), 'a') as f:
                f.write(json.dumps(log_stats) + '\n')
```

（3）函数 train 实现了模型的训练过程，包括数据加载、前向传播、反向传播、优化器更新等步骤。在训练过程中，会记录训练损失和指标，并将其写入 Tensorboard 日志中。

```python
def train(train_loader, log_writer, model, criterion, optimizer, scaler, epoch, lr_schedule,
          args):
    # 记录平均时间和内存使用
    batch_time = AverageMeter('时间', ':6.2f')
    data_time = AverageMeter('数据', ':6.2f')
    mem = AverageMeter('内存 (GB)', ':6.1f')
    metric_names = ['loss', 'clip_loss', 'clip_acc']  # 指标名称

    loader_len = train_loader.num_batches
    iters_per_epoch = loader_len // args.update_freq
    # 记录指标的平均值
    metrics = OrderedDict([(name, AverageMeter(name, ':.2e')) for name in metric_names])
    progress = ProgressMeter(
        iters_per_epoch,
        [batch_time, data_time, mem, *metrics.values()],
        prefix="Epoch: [{}]".format(epoch))  # 显示进度条

    # 切换到训练模式
    model.train()

    end = time.time()
    for data_iter, inputs in enumerate(train_loader):
        optim_iter = data_iter // args.update_freq

        # 记录数据加载时间
        data_time.update(time.time() - end)

        # 根据调度更新权重衰减和学习率
        it = iters_per_epoch * epoch + optim_iter  # 全局训练迭代次数
        for k, param_group in enumerate(optimizer.param_groups):
            param_group['lr'] = lr_schedule[it]

        inputs = [tensor.cuda(args.gpu, non_blocking=True) for tensor in inputs]

        # 计算输出
        with amp.autocast(enabled=not args.disable_amp):
```

```python
            outputs = model(*inputs)
            loss_dict = criterion(outputs)
            loss = loss_dict['loss']
            loss /= args.update_freq

        if not math.isfinite(loss.item()):
            print("Loss is {}, stopping training".format(loss.item()))
            sys.exit(1)

        scaler.scale(loss).backward()

        if (data_iter + 1) % args.update_freq != 0:
            continue

        # 计算梯度并进行 SGD 步骤
        scaler.step(optimizer)
        scaler.update()
        model.zero_grad(set_to_none=True)

        # 限制 logit 的范围为 [0, 100]
        utils.get_model(model).logit_scale.data.clamp_(0, 4.6052)
        logit_scale = utils.get_model(model).logit_scale.exp().item()

        for k in loss_dict:
            metrics[k].update(loss_dict[k].item(), args.batch_size)

        # 记录经过的时间
        batch_time.update(time.time() - end)
        end = time.time()

        mem.update(torch.cuda.max_memory_allocated() // 1e9)

        # 记录到 log_writer (tensorboard)
        if log_writer is not None:
            for k, v in loss_dict.items():
                log_writer.add_scalar(k, v.item(), it)
            log_writer.add_scalar('scaler', scaler.get_scale(), it)
            log_writer.add_scalar('logit', logit_scale, it)
            log_writer.add_scalar('lr', optimizer.param_groups[0]['lr'], it)

        if optim_iter % args.print_freq == 0:
            progress.display(optim_iter)

    progress.synchronize()
    return {**{k: v.avg for k, v in metrics.items()},
            'lr': optimizer.param_groups[0]['lr'],
            'logit_scale': logit_scale}
```

（4）函数 validate_zeroshot 实现了零样本图像分类的验证过程，使用模型对标题进行编码，并计算图像与标题之间的余弦相似度，将其作为预测结果，然后评估 top1 和 top5 准确率。

```python
def validate_zeroshot(val_loader, model, tokenizer, args):
    # 记录平均时间和 top1、top5 准确率
    batch_time = AverageMeter('时间', ':6.3f')
    top1 = AverageMeter('准确率@1', ':6.2f')
    top5 = AverageMeter('准确率@5', ':6.2f')
    progress = ProgressMeter(
        len(val_loader),
        [batch_time, top1, top5],
        prefix='测试: ')

    # 切换到评估模式
```

```python
model.eval()

print('=> 编码标题')
cwd = os.path.dirname(os.path.realpath(__file__))
templates = [
    "一个{}的照片。",
    "{}的糟糕照片。",
    "一个{}的折纸。",
    "{}的大照片。",
    "视频游戏中的{}。",
    "{}的艺术作品。",
    "一个{}的小照片。"
]

with open(os.path.join(cwd, 'imagenet_labels.json')) as f:
    labels = json.load(f)

with torch.no_grad():
    text_features = []
    for l in labels:
        texts = [t.format(l) for t in templates]
        texts = tokenizer(texts).cuda(args.gpu, non_blocking=True)
        class_embeddings = utils.get_model(model).encode_text(texts)
        class_embeddings = class_embeddings / class_embeddings.norm(dim=-1, keepdim=True)
        class_embeddings = class_embeddings.mean(dim=0)
        class_embeddings = class_embeddings / class_embeddings.norm(dim=-1, keepdim=True)
        text_features.append(class_embeddings)
    text_features = torch.stack(text_features, dim=0)

    end = time.time()
    for i, (images, target) in enumerate(val_loader):
        images = images.cuda(args.gpu, non_blocking=True)
        target = target.cuda(args.gpu, non_blocking=True)

        # 编码图像
        image_features = utils.get_model(model).encode_image(images)
        image_features = image_features / image_features.norm(dim=-1, keepdim=True)

        # 计算余弦相似度作为 logits
        logits_per_image = image_features @ text_features.t()

        # 计算准确率并记录准确率
        acc1, acc5 = accuracy(logits_per_image, target, topk=(1, 5))
        acc1, acc5 = utils.scaled_all_reduce([acc1, acc5])
        top1.update(acc1.item(), images.size(0))
        top5.update(acc5.item(), images.size(0))

        # 记录经过的时间
        batch_time.update(time.time() - end)
        end = time.time()

        if i % args.print_freq == 0:
            progress.display(i)

progress.synchronize()
print('0-shot * 准确率@1 {top1.avg:.3f} 准确率@5 {top5.avg:.3f}'
      .format(top1=top1, top5=top5))
return {'acc1': top1.avg, 'acc5': top5.avg}
```

（5）类 AverageMeter 用于计算和存储数值的平均值和当前值，可以用于跟踪训练或验证过程中的指标。方法 reset 用于重置计数器，方法 update 用于更新值和计数，方法 synchronize 用于在分布式环境中同步计数器的值，方法 __str__ 用于返回格式化的字符串表示。

```python
class AverageMeter(object):
    """计算和存储平均值和当前值"""
    def __init__(self, name, fmt=':f'):
        self.name = name
        self.fmt = fmt
        self.reset()

    def reset(self):
        self.val = 0
        self.avg = 0
        self.sum = 0
        self.count = 0

    def update(self, val, n=1):
        self.val = val
        self.sum += val * n
        self.count += n
        self.avg = self.sum / self.count

    def synchronize(self):
        # 如果未初始化分布式环境，直接返回
        if not utils.is_dist_avail_and_initialized():
            return
        t = torch.tensor([self.sum, self.count], dtype=torch.float64, device='cuda')
        dist.barrier()
        dist.all_reduce(t)
        t = t.tolist()
        self.sum = int(t[0])
        self.count = t[1]
        self.avg = self.sum / self.count

    def __str__(self):
        fmtstr = '{name} {val' + self.fmt + '} ({avg' + self.fmt + '})'
        return fmtstr.format(**self.__dict__)
```

（6）类 ProgressMeter 用于显示训练或验证进度，包括当前批次数和各种指标的值。__init__ 方法用于初始化进度条，display 方法用于显示进度条，synchronize 方法用于同步进度条中的各个计量器，_get_batch_fmtstr 方法用于生成批次数的格式化字符串。

```python
class ProgressMeter(object):
    def __init__(self, num_batches, meters, prefix=""):
        self.batch_fmtstr = self._get_batch_fmtstr(num_batches)
        self.meters = meters
        self.prefix = prefix

    def display(self, batch):
        entries = [self.prefix + self.batch_fmtstr.format(batch)]
        entries += [str(meter) for meter in self.meters]
        print('\t'.join(entries))

    def synchronize(self):
        for meter in self.meters:
            meter.synchronize()

    def _get_batch_fmtstr(self, num_batches):
        num_digits = len(str(num_batches // 1))
        fmt = '{:' + str(num_digits) + 'd}'
        return '[' + fmt + '/' + fmt.format(num_batches) + ']'
```

（7）函数 accuracy 用于计算给定 topk 值下的预测准确率。

```python
def accuracy(output, target, topk=(1,)):
    """计算指定 topk 值的预测准确率"""
    with torch.no_grad():
        maxk = max(topk)
        batch_size = target.size(0)

        _, pred = output.topk(maxk, 1, True, True)
        pred = pred.t()
        correct = pred.eq(target.reshape(1, -1).expand_as(pred))

        res = []
        for k in topk:
            correct_k = correct[:k].reshape(-1).float().sum(0, keepdim=True)
            res.append(correct_k.mul_(100.0 / batch_size))
        return res
```

（8）下面代码是文件的主程序，首先解析命令行参数，然后创建输出目录（如果不存在），最后调用 main 函数开始 LaCLIP/CLIP 训练。

```python
if __name__ == '__main__':
    parser = argparse.ArgumentParser('LaCLIP/CLIP training', parents=[get_args_parser()])
    args = parser.parse_args()
    os.makedirs(args.output_dir, exist_ok=True)
    main(args)
```

7.4.7 模型评估

（1）文件 eval_zeroshot_imagenet.py 用于在 ImageNet 数据集上进行零样本评估操作。代码首先加载预训练的模型权重，然后加载 ImageNet 数据集进行评估，最后输出零样本准确率。

```python
def get_model(model):
    if isinstance(model, torch.nn.DataParallel) \
      or isinstance(model, torch.nn.parallel.DistributedDataParallel):
        return model.module
    else:
        return model

def main(args):
    ckpt_path = args.ckpt_path
    ckpt = torch.load(ckpt_path, map_location='cpu')

    state_dict = OrderedDict()
    for k, v in ckpt['state_dict'].items():
        state_dict[k.replace('module.', '')] = v

    print("creating model: {}".format(args.model))
    print(f"loading checkpoint '{args.ckpt_path}")

    model = getattr(models, args.model)(rand_embed=False)

    model.cuda()
    model.load_state_dict(state_dict, strict=True)

    cudnn.benchmark = True

    with open('imagenet_labels.json') as f:
        labels = json.load(f)

    # 数据加载
```

```python
    print("... creating dataset")
    tokenizer = SimpleTokenizer()
    val_transform = transforms.Compose([
            transforms.Resize(224),
            transforms.CenterCrop(224),
            lambda x: x.convert('RGB'),
            transforms.ToTensor(),
            transforms.Normalize(mean=[0.485, 0.456, 0.406],
                                 std=[0.229, 0.224, 0.225])
    ])

    val_dataset = datasets.ImageFolder(os.path.join(args.imagenet_root, 'val'), transform=val_transform)
    val_loader = torch.utils.data.DataLoader(
        val_dataset, batch_size=args.batch_size, shuffle=False,
        num_workers=args.workers, pin_memory=True, drop_last=False)

    templates = [
        "itap of a {}.",
        "a bad photo of the {}.",
        "a origami {}.",
        "a photo of the large {}.",
        "a {} in a video game.",
        "art of the {}.",
        "a photo of the small {}."
    ]

    acc = validate_zeroshot(val_loader, templates, labels, model, tokenizer)
    print(f'ImageNet zero-shot accuracy: {acc}')


def validate_zeroshot(val_loader, templates, labels, model, tokenizer):
    # 切换到评估模式
    model.eval()
    total_top1 = 0
    total_images = 0

    print('... getting classifier')
    with torch.no_grad():
        text_features = []
        for label in labels:
            if isinstance(label, list):
                texts = [t.format(l) for t in templates for l in label]
            else:
                texts = [t.format(label) for t in templates]
            texts = tokenizer(texts).cuda(non_blocking=True)
            texts = texts.view(-1, 77).contiguous()
            class_embeddings = get_model(model).encode_text(texts)
            class_embeddings = class_embeddings / class_embeddings.norm(dim=-1, keepdim=True)
            class_embeddings = class_embeddings.mean(dim=0)
            class_embeddings = class_embeddings / class_embeddings.norm(dim=-1, keepdim=True)
            text_features.append(class_embeddings)
        text_features = torch.stack(text_features, dim=0)

        for images, target in val_loader:
            images = images.cuda(non_blocking=True)
            target = target.cuda(non_blocking=True)

            #对图像进行编码
            image_features = get_model(model).encode_image(images)
            image_features = image_features / image_features.norm(dim=-1,
            keepdim=True)
```

```python
            #将余弦相似度作为对数概率
            logits_per_image = image_features @ text_features.t()

            #测量准确率并记录损失值
            pred = logits_per_image.argmax(dim=1)
            correct = pred.eq(target).sum()
            total_top1 += correct.item()
            total_images += images.size(0)
    return 100 * total_top1 / total_images


if __name__ == '__main__':
    parser = argparse.ArgumentParser(description='ImageNet zero-shot evaluations', add_help=False)
    parser.add_argument('--imagenet-root', default='data/imagenet', type=str, help='path to imagenet dataset')
    parser.add_argument('--ckpt-path', default='checkpoints/cc12m_laclip.ckpt', type=str, help='model to test')
    parser.add_argument('--batch-size', default=256, type=int, help='batch_size')
    parser.add_argument('--model', default='CLIP_VITB16', type=str, help='model architecture')
    parser.add_argument('-j', '--workers', default=10, type=int)
    args = parser.parse_args()
    main(args)
```

（2）文件 eval_zeroshot_imagenet_laion.py 用于在 ImageNet 数据集上进行零样本评估，与之前的评估脚本相比，它使用了 OpenCLIP 模型进行评估。首先创建了 OpenCLIP 模型和数据预处理方法，然后加载预训练模型的权重。接着加载 ImageNet 数据集进行评估，最后输出零样本准确率。

```python
from tokenizer import SimpleTokenizer
from open_clip import create_model_and_transforms
from eval_zeroshot_imagenet import validate_zeroshot

def main(args):
    model, preprocess_train, preprocess_val = create_model_and_transforms(
        args.model,
        '',
        precision='amp',
        device='cuda',
        jit=False,
        force_quick_gelu=args.quickgelu,
        force_custom_text=False,
        force_patch_dropout=None,
        force_image_size=224,
        pretrained_image=False,
        image_mean=None,
        image_std=None,
        aug_cfg={},
        output_dict=True,
    )
    with open(args.ckpt_path, 'rb') as f:
        checkpoint = torch.load(f, map_location='cpu')
    sd = checkpoint["state_dict"]
    model.load_state_dict(sd)
    device = "cuda:0" if torch.cuda.is_available() else "cpu"
    model = model.to(device)
    model.eval()

    cudnn.benchmark = True

    with open('imagenet_labels.json') as f:
```

```
        labels = json.load(f)

    # 加载数据
    print("... creating dataset")
    tokenizer = SimpleTokenizer()


    val_dataset = datasets.ImageFolder(os.path.join(args.imagenet_root, 'val'), transform=
    preprocess_val)
    val_loader = torch.utils.data.DataLoader(
        val_dataset, batch_size=args.batch_size, shuffle=False,
        num_workers=args.workers, pin_memory=True, drop_last=False)

    templates = json.load(open('imagenet_templates.json'))
    acc = validate_zeroshot(val_loader, templates, labels, model, tokenizer)
    print(f'ImageNet zero-shot accuracy: {acc}')

if __name__ == '__main__':
    parser = argparse.ArgumentParser(description='ImageNet zero-shot evaluations', add_help=
    False)
    parser.add_argument('--imagenet-root', default='data/imagenet', type=str, help='path to
    imagenet dataset')
    parser.add_argument('--ckpt-path', default='checkpoints/cc12m_laclip.ckpt', type=str,
    help='model to test')
    parser.add_argument('--batch-size', default=256, type=int, help='batch_size')
    parser.add_argument('--model', default='ViT-B-32', type=str, help='model
    architecture')
    parser.add_argument('--quickgelu', action='store_true', help='Use quickgelu')
    parser.add_argument('-j', '--workers', default=10, type=int)
    args = parser.parse_args()
    main(args)
```

7.4.8 文本重写

文件 llama_rewrite.py 对生成的文本进行重写，利用预训练的 LLAMA 模型生成更多样化的图像描述。

```
def setup_model_parallel() -> Tuple[int, int]:
    local_rank = int(os.environ.get("LOCAL_RANK", -1))
    world_size = int(os.environ.get("WORLD_SIZE", -1))

    torch.distributed.init_process_group("nccl")
    initialize_model_parallel(world_size)
    torch.cuda.set_device(local_rank)

    #在所有进程中必须使用相同的种子
    torch.manual_seed(1)
    return local_rank, world_size


def load(
    ckpt_dir: str,
    tokenizer_path: str,
    local_rank: int,
    world_size: int,
    max_seq_len: int,
    max_batch_size: int,
) -> LLaMA:
    start_time = time.time()
    checkpoints = sorted(Path(ckpt_dir).glob("*.pth"))
    assert world_size == len(
        checkpoints
    ), f"Loading a checkpoint for MP={len(checkpoints)} but world size is {world_size}"
    ckpt_path = checkpoints[local_rank]
```

```python
    print("Loading")
    checkpoint = torch.load(ckpt_path, map_location="cpu")
    with open(Path(ckpt_dir) / "params.json", "r") as f:
        params = json.loads(f.read())

    model_args: ModelArgs = ModelArgs(
        max_seq_len=max_seq_len, max_batch_size=max_batch_size, **params
    )
    tokenizer = Tokenizer(model_path=tokenizer_path)
    model_args.vocab_size = tokenizer.n_words
    torch.set_default_tensor_type(torch.cuda.HalfTensor)
    model = Transformer(model_args)
    torch.set_default_tensor_type(torch.FloatTensor)
    model.load_state_dict(checkpoint, strict=False)

    generator = LLaMA(model, tokenizer)
    print(f"Loaded in {time.time() - start_time:.2f} seconds")
    return generator


def main(
    ckpt_dir: str,
    tokenizer_path: str,
    temperature: float = 0.9,
    top_p: float = 0.95,
    max_seq_len: int = 512,
    max_batch_size: int = 32,
    prompt_filename: str = 'text/source.txt',
    output_filename: str = 'text/target.txt',
    sample_mode: str = 'chatgpt',
):
    print('current sample mode is: ', sample_mode)
    local_rank, world_size = setup_model_parallel()
    if local_rank > 0:
        sys.stdout = open(os.devnull, "w")

    generator = load(
        ckpt_dir, tokenizer_path, local_rank, world_size, max_seq_len, max_batch_size
    )

    #逐行从文件中读取文本到列表
    new_prompt_filename = output_filename
    # 更改父目录为输出目录
    with open(prompt_filename, 'r') as f:
        original_prompts = f.readlines()

    new_prompts = []
    num_batches = math.ceil(len(original_prompts) / max_batch_size)

    for batch_idx in tqdm(range(num_batches)):
        prompts = []
        current_batch = original_prompts[batch_idx * max_batch_size: (batch_idx + 1)
        * max_batch_size]

        for prompt_idx, original_prompt in enumerate(current_batch):
            chosen_source_caption_list = []
            chosen_target_caption_list = []
            if sample_mode == 'chatgpt':
                num_caps = len(chatgpt_source_caption_list)
                chosen_idx = random.sample(range(num_caps), 3)
                for idx in chosen_idx:
                    chosen_source_caption_list.append(chatgpt_source_caption_list[idx])
                    chosen_target_caption_list.append(chatgpt_target_caption_list[idx])
```

```
        elif sample_mode == 'bard':
            num_caps = len(bard_source_caption_list)
            chosen_idx = random.sample(range(num_caps), 3)
            for idx in chosen_idx:
                chosen_source_caption_list.append(bard_source_caption_list[idx])
                chosen_target_caption_list.append(bard_target_caption_list[idx])
        elif sample_mode == 'coco':
            num_caps = len(coco_caption_list)
            chosen_idx = random.sample(range(num_caps), 3)
            for idx in chosen_idx:
                coco_chosen_idx = random.sample(range(len(coco_caption_list
                    [idx])), 2)
                chosen_source_caption_list.append(coco_caption_list[idx]
                    [coco_chosen_idx[0]])
                chosen_target_caption_list.append(coco_caption_list[idx]
                    [coco_chosen_idx[1]])
        elif sample_mode == 'human':
            num_caps = len(human_source_caption_list)
            chosen_idx = random.sample(range(num_caps), 3)
            for idx in chosen_idx:
                chosen_source_caption_list.append(human_source_caption_list[idx])
                chosen_target_caption_list.append(human_target_caption_list[idx])
        else:
            raise ValueError('sample mode not supported')

        current_prompt = """write image captions differently,

{} => {}

{} => {}

{} => {}

{} =>""".format(
            chosen_source_caption_list[0], chosen_target_caption_list[0],
            chosen_source_caption_list[1], chosen_target_caption_list[1],
            chosen_source_caption_list[2], chosen_target_caption_list[2],
            original_prompt.replace('\n', ''))
        prompt_tokens = generator.tokenizer.encode(current_prompt, bos=True,
        eos=False)
        if len(prompt_tokens) <= max_seq_len-5:
            prompts.append(current_prompt)
        else:
            cut_len = max_seq_len - 10
            prompt_tokens = prompt_tokens[:cut_len]
            current_prompt = generator.tokenizer.decode(prompt_tokens) + ' =>'
            prompts.append(current_prompt)

results = generator.generate(
    prompts, max_gen_len=77, temperature=temperature, top_p=top_p
)

for result in results:
    prompt_line = result.split('\n')[8].strip()
    new_prompt = prompt_line.split('=>')[1].strip()
    new_prompts.append(new_prompt)

if local_rank == 0:
    with open(new_prompt_filename, 'w') as f:
        f.writelines([p.strip().replace('\n', ' ') + '\n' for p in
        new_prompts])
```

```
if __name__ == "__main__":
    fire.Fire(main)
```

在本项目中，这个文件的作用是使用预训练的生成模型来生成文本，具体说明如下。

（1）文本生成：该文件中的 main 函数负责加载预训练的生成模型，并根据给定的输入文本生成新的文本。生成的文本可以用于各种应用，如对话系统、文本摘要等。

（2）支持多种采样模式：根据用户指定的采样模式（chatgpt、bard、coco 和 human），生成不同风格或来源的文本，使生成文本更加多样化。

（3）分布式训练支持：代码中通过设置模型的并行化，支持在多个 GPU 上进行分布式训练，提高了生成速度和效率。

（4）参数设置和命令行接口：使用 Fire 库进行命令行参数解析，使用户可以方便地通过命令行指定参数来运行生成文本的过程，提高了代码的易用性和灵活性。

综上所述，这个文件在项目中用于生成文本，为文本生成任务提供了便利的工具和接口。到此为止，整个项目的主要源码介绍完毕，执行效果如图 7-9 所示。

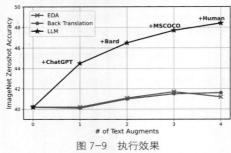

图 7-9　执行效果

第 8 章 基于多模态大模型的翻译系统

本章将详细讲解一个多模态翻译系统的实现过程。这个多模态翻译系统实现了基于图像和文本的多模态翻译任务，包括图像描述生成、文本重构、图像到文本翻译以及三元组训练等阶段。本项目使用 CLIP 模型进行图像和文本的编码，并结合 MBart 模型进行翻译任务；同时，通过分布式训练和混合精度训练等技术，实现高效训练和推理，并提供灵活的参数配置选项，使模型可以适应不同的数据集和任务要求。

8.1 背景介绍

随着人工智能和自然语言处理技术的快速发展，多模态翻译作为一种将图像和文本结合起来的新兴任务，受到了越来越多的关注。传统的文本翻译系统主要依赖于文本数据，而多模态翻译系统则可以同时利用图像和文本信息，提供更加丰富和准确的翻译结果。

多模态翻译系统的应用场景非常广泛，比如将图像中的物体描述翻译成不同语言的文本、实现跨语言的图像标注和翻译、辅助视觉障碍者理解图像内容等。为了构建高效的多模态翻译系统，需要整合图像处理、自然语言处理和深度学习等领域的技术，并针对不同的任务和数据集进行模型设计和优化。

本项目旨在开发一个多模态翻译系统，利用最新的 CLIP 模型结合 MBart 模型，实现图像到文本的翻译任务，并提供了预训练阶段、文本重构、图像描述生成和三元组训练等不同阶段的功能。通过分布式训练和模型优化，实现了高效的模型训练和推理，为多模态翻译任务的研究和应用提供了一个强大的工具。

8.2 系统分析

系统分析是指对一个系统进行全面深入的研究和评估,旨在全面理解系统的设计和运行原理，为系统的开发、优化和应用提供指导和支持。

8.2.1 系统需求分析

系统需求分析明确了系统的功能和性能要求，本项目的系统需求分析如下。
- 多语言支持：系统需要支持多种源语言和目标语言的翻译，能够处理不同语言之间的文本和图像翻译任务。
- 多模态数据处理：能够处理文本和图像的多模态数据，实现图像到文本的翻译以及文本到图像的生成任务。
- 分布式训练：能够支持分布式训练，利用多个 GPU 进行模型训练，提高训练效率。

- 模型预训练：提供预训练阶段，对模型进行预训练以适应特定的任务和数据。
- 文本重构和图像描述：实现文本重构和图像描述生成功能，从给定的文本生成对应的图像描述或者从图像生成对应的文本。
- 高性能和高效率：系统需要具备高性能和高效率，能够处理大规模的数据集并快速生成翻译结果。

8.2.2 技术架构分析

技术架构分析说明了系统各组件的实现方式和技术选择，本项目的技术架构分析如下。
- 数据处理层：使用 PyTorch 数据加载和处理工具，处理多模态数据，包括文本和图像数据。
- 模型层：基于 CLIP 和 MBart 模型构建多模态翻译模型，使用 Transformer 架构实现文本和图像之间的翻译。
- 分布式训练层：使用 PyTorch 的分布式训练工具进行分布式训练，利用多个 GPU 加速模型的训练过程。
- 预训练和微调：提供预训练阶段，可以使用预训练模型进行微调，适应特定的翻译任务和数据集。
- 评估和优化：利用 BLEU 分数等指标对模型进行评估，并根据评估结果进行模型优化和调整。
- 工具库和框架：使用 PyTorch、Hugging Face Transformers 等开源工具库和框架进行模型开发和训练。
- 分布式环境管理：使用 Torch Distributed 库进行分布式训练环境的管理和协调。
- 性能优化：使用 GPU 进行加速，利用分布式训练和混合精度训练等技术优化模型的训练速度和性能。

8.2.3 项目介绍

本项目是一个多模态翻译系统，旨在实现文本和图像之间的翻译任务，包括图像到文本的描述生成和文本到图像的生成功能。通过结合 CLIP 和 MBart 模型，实现了跨语言和跨模态的翻译，使系统能够处理多种源语言和目标语言之间的翻译任务。

1. 功能特点
- 多语言支持：支持多种源语言和目标语言之间的翻译，用户可以灵活选择所需的语言组合进行翻译。
- 多模态处理：能够处理文本和图像的多模态数据，实现图像描述生成和文本到图像的翻译功能。
- 分布式训练：使用分布式训练技术，充分利用多个 GPU 进行模型训练，提高训练速度和效率。
- 预训练和微调：提供预训练模型，用户可以根据自己的需求对模型进行微调，适应特定的翻译任务和数据集。
- 文本重构和图像描述：实现文本重构和图像描述生成功能，用户可以输入文本生成对应的图像描述，或者输入图像生成对应的文本描述。
- 高性能和高效率：系统具备高性能和高效率，能够处理大规模的数据集并快速生成翻译结果。

2. 技术实现
- 使用 PyTorch 深度学习框架构建模型，结合 CLIP 和 MBart 模型进行多模态翻译。

- 利用 Hugging Face Transformers 库加载预训练模型，并进行微调以适应特定任务。
- 使用分布式训练技术，通过 Torch Distributed 库实现多 GPU 训练，提高训练速度。
- 使用 PyTorch DataLoader 进行数据加载和处理，支持多模态数据输入。
- 通过 BLEU 分数等指标对模型进行评估，并根据评估结果进行模型优化和调整。

本项目可用于构建多语言翻译系统、图像描述生成系统等应用，具有广泛的应用前景，例如跨语言信息检索、图像标注和翻译服务等领域。

8.3 准备数据集

本项目使用了 Multi30k 数据集，这是一个多语言、多模态数据集，广泛用于图像描述生成和多模态机器翻译（multimodal machine translation）等任务。

8.3.1 Multi30k 数据集介绍

Multi30k 数据集由多种语言的图像描述对齐构成，包括英文、德文、法文等。下面是对 Multi30k 数据集的详细介绍。

1. 数据集组成

- 图像数据：数据集包含来自 Flickr30k 和 MSCOCO 的图像，这些图像经过人工标注，配有多语言的描述。
- 文本数据：每张图像都有相应的描述，这些描述被翻译成多种语言，主要包括英文、德文、法文、西班牙文和罗马尼亚文。

2. 数据集结构

Multi30k 数据集组织在 data/multi30k 文件夹中，具体结构如下。

- flickr30k-images.tar.gz：包含训练集、验证集和 2016 测试集的图像。
- test_2017-flickr-images.gz：包含 2017 测试集的 Flickr 图像。
- images_mscoco.task1.tar.gz：包含 MSCOCO 图像。

3. 数据集的用途

Multi30k 数据集的主要用途如下。

- 图像描述生成：根据图像生成相应的自然语言描述。
- 多模态机器翻译：利用图像和文本信息进行多语言翻译。
- 跨模态检索：通过图像检索文本描述或通过文本描述检索图像。

4. 数据集的重要性

Multi30k 数据集在多模态学习和多语言处理研究中起到了重要作用。它提供了丰富的资源，可以用于训练和评估多模态模型，推动了图像描述生成和多模态机器翻译等领域的进展。

总之，通过使用 Multi30k 数据集，研究人员可以探索如何有效地利用图像和文本的互补信息，提升模型在多语言生成任务中的表现。这对于开发能够理解和生成多语言描述的智能系统具有重要意义。

8.3.2 下载 Multi30k 数据集

文件 setup_multi30k.sh 的功能是下载并组织 Multi30k 数据集文件，具体步骤包括克隆数据集的 Git 仓库、解压缩图像文件、组织文件结构以及重命名和移动文件，以便为后续的多模态训练和测试任务做好准备。

```
git clone --recursive https://github.com/multi30k/dataset.git multi30k-dataset
```

```
mv multi30k-dataset/* .
rm -rf multi30k-dataset
mkdir images text
mv data text
mv scripts text
cd text/data/task1/raw
gunzip *.gz
for file in val.*; do mv "$file" "test_2016_val.${file##*.}"; done
cd ../image_splits
mv val.txt test_2016_val.txt
cd ../../../..
mv flickr30k-images.tar.gz images
mv images_mscoco.task1.tar.gz images
mv test_2017-flickr-images.gz images
cd images
tar -xvzf images_mscoco.task1.tar.gz
mv translated_images test_2017_mscoco
tar -xvzf test_2017-flickr-images.tar.gz
mv task1 test_2017_flickr
tar -xvzf flickr30k-images.tar.gz
mv flickr30k-images train
mkdir test_2016_flickr test_2016_val
mv ../create_test_val_flickr.py .
python create_test_val_flickr.py ../text/data/task1/image_splits/test_2016_val.txt train test_2016_val
python create_test_val_flickr.py ../text/data/task1/image_splits/test_2016_flickr.txt train test_2016_flickr
```

通过运行如下命令下载 Multi30k 数据集的文本数据并组织文件结构：

```
bash setup_multi30k.sh
```

在文件 setup_multi30k.sh 中调用了程序文件 create_test_val_flickr.py，文件 create_test_val_flickr.py 的功能是根据一个包含文件名列表的文本文件，将指定目录中的文件移动到目标目录中。它首先读取文件名列表，检查每个文件是否存在，然后将这些文件从源目录移动到目标目录。如果文件不存在，会打印相应的提示信息。这个脚本主要用于组织 Multi30k 数据集中的测试和验证图像文件。文件 create_test_val_flickr.py 的具体代码如下所示。

```
import os
import sys

def main(list_file_path, source_folder, destination_folder):
    with open(list_file_path, 'r') as file_list:
        for line in file_list:
            # 去除行尾的换行符和任何前后空格
            file_name = line.strip()

            # 在移动文件前检查文件是否存在
            if os.path.isfile(file_name):
                # 使用 os.path.basename 获取不带路径的文件名
                file_name_only = os.path.basename(file_name)

                # 将文件移动到目标文件夹
                source_path = os.path.join(source_folder, file_name_only)
                destination_path = os.path.join(destination_folder, file_name_only)
                os.rename(source_path, destination_path)
            else:
                print(f"文件未找到：{file_name}")

if __name__ == '__main__':
    file_list_path = sys.argv[1]
```

```
source_folder = sys.argv[2]
destination_folder = sys.argv[3]
main(file_list_path, source_folder, destination_folder)
```

8.3.3 下载 WIT 数据集中的图像数据

在本项目中，下载 WIT 数据集中的图像数据是为了在项目中使用这些图像进行多模态机器翻译的训练和评估。WIT 数据集是一个用于多语言机器翻译的数据集，其中包含了丰富的多语言文本数据和相应的图像数据。在多模态机器翻译任务中，图像与文本是相互对应的，图像可以作为文本的视觉补充，帮助模型更好地理解文本含义。

文件 download_images.py 的功能是下载 WIT 数据集中的图像数据，通过 Python 的 requests 库从指定的 URL 下载图像，并将其保存到指定的文件夹中。根据提供的参数，可以选择下载不同语言对和不同数据集划分的图像数据。文件 download_images.py 的具体代码如下所示。

```
import requests
from tqdm import tqdm
import sys
import os
import hashlib
import time

# 设置请求 API 时的用户名和邮箱
NAME = ''
EMAIL = ''

# 确保用户名和邮箱已被设置
assert len(NAME) > 0 and len(EMAIL) > 0, 'Please set your name and email in the code. 
This will be used when you call the API for images and will help the hosting service to
manage server load.'

# 读取包含图片 URL 的文件
with open(f'mmt/{sys.argv[1]}/{sys.argv[2]}_url.txt') as f:
    urls = f.read().splitlines()

# 创建保存图片的目录
save_at = f'mmt/{sys.argv[1]}/images'
os.makedirs(save_at, exist_ok=True)

names = []
ignore_indices = []

# 下载图片
for i, url in enumerate(tqdm(urls)):
    while True:
        try:
            # 使用用户代理请求图片
            img_bytes = requests.get(url, headers={'User-Agent': f'{NAME}/0.0({EMAIL})'})
        except requests.exceptions.ConnectionError as e:
            code = 403  # 达到最大重试次数
            break
        code = img_bytes.status_code
        if code in [429, 502]:  # 超过限额，服务器错误
            time.sleep(1)
        else:
            break

    # 生成图片名称
    names.append('\n' + str(hashlib.md5(url.encode("utf-8")).hexdigest()) + '.jpg')
    if code != 200:
        print(f'{url} | {code}')
```

```
            ignore_indices.append(str(i))

        # 将图片保存到指定目录
        with open(f'{save_at}/{names[-1][1:]}', 'wb') as f:
            f.write(img_bytes.content)

# 记录下载失败的图片索引和成功下载的图片名称
with open(f'mmt/{sys.argv[1]}/{sys.argv[2]}_image.txt', 'w') as f:
    f.writelines([','.join(ignore_indices)] + names)
```

在上述代码中，首先读取包含图片 URL 的文件，然后逐个 URL 进行请求，处理可能的连接错误和服务器响应代码。如果请求成功，将图片保存到本地目录，并生成一个基于 URL 的 MD5 哈希值作为图片的文件名。同时，记录下载失败的图片索引和成功下载的图片文件名，最后将这些信息写入一个文本文件中。

8.4 数据集处理

本项目的"数据集处理"模块包括自定义的 PyTorch 数据集类（DocDataset、ConcatDataset 和 MultiModalDataset）、数据处理和后处理函数、数据集填充函数以及针对不同数据集（Multi30k、WIT 和 WMT）的数据获取和预处理函数，实现了对文本和图像数据的分词、嵌入表示生成以及数据集的统一处理和准备工作。

本项目涉及多种数据集，使用这么多种数据集的主要原因是要训练多模态大模型，这些模型可以同时处理文本和图像等不同类型的数据。具体原因如下。

（1）数据多样性：不同的数据集提供了不同类型和不同领域的数据，可以增加模型对于多样化场景的适应能力。

（2）多模态学习：多模态数据集可以用于训练能够同时处理文本和图像等多种数据类型的模型，促进文本和图像之间的语义理解和关联学习。

（3）泛化能力：通过在多个数据集上进行训练，模型可以学习到更广泛、更通用的特征，提升其在不同任务和数据上的泛化能力。

（4）应用场景需求：实际应用中可能涉及多种类型的数据，如跨语言文本理解、图像描述生成等任务，需要模型具备处理多模态数据的能力。

因此，使用多种数据集训练多模态大模型有助于提高模型的性能和适用范围，使其能够在更广泛的任务和场景中发挥作用。

8.4.1 PyTorch 数据集类

文件 dataset.py 定义了 3 个自定义的 PyTorch 数据集类，分别是 DocDataset、ConcatDataset 和 MultiModalDataset。文件 dataset.py 的具体实现代码如下所示。

```
import torch
from torch.utils.data import Dataset
import math

class DocDataset(Dataset):
    def __init__(self, docs):
        super().__init__()
        self.docs = docs

    def __len__(self):
        return len(self.docs)
```

```python
    def __getitem__(self, idx):
        return self.docs[idx]

class ConcatDataset(Dataset):
    def __init__(self, dataset1, dataset2):
        super().__init__()
        len1 = len(dataset1)
        len2 = len(dataset2)
        if len1 > len2: # self.dataset1 is the longer dataset
            self.dataset1 = dataset1
            self.dataset2 = dataset2
            self.length = len2 + math.ceil((len1 - len2)/2)
        elif len1 < len2:
            self.dataset1 = dataset2
            self.dataset2 = dataset1
            self.length = len1 + math.ceil((len2 - len1)/2)
        else:
            self.dataset1 = dataset2
            self.dataset2 = dataset1
            self.length = len1
        self.len1 = len(self.dataset1)
        self.len2 = len(self.dataset2)

    def __len__(self):
        return self.length

    def __getitem__(self, idx):
        if idx >= self.len2:
            idx2 = idx - self.len2 + 1
            item1 = self.dataset1[idx]
            if idx == self.len1 - idx2: # 避免一批退回一件物品两次
                return (*item1,)
            else:
                item2 = self.dataset1[-idx2]
                batch = item1 + item2
                return (*batch,)
        else:
            batch = self.dataset1[idx] + self.dataset2[idx]
            return (*batch,)

class MultiModalDataset(Dataset):
    def __init__(self, params, clip_embs, tok_data, raw_data, clip_tok_data, mask_token
    = None, mask_inputs = False, is_pretraining = False):
        super().__init__()
        self.is_pretraining = is_pretraining
        self.mask_prob = params.mask_prob
        self.mask_inputs = mask_inputs
        self.mask_token = mask_token
        self.use_clip_tok = params.unfreeze_clip
        if self.is_pretraining:
            self.input_tok  = tok_data
            self.output_raw = raw_data
            self.langs = list(self.input_tok.keys())
            self.clip_embs  = clip_embs
        else:
            self.input_tok = tok_data[params.src_lang]
            self.output_tok = tok_data[params.tgt_lang]
            self.output_raw = raw_data[params.tgt_lang]
            self.clip_embs  = clip_embs[params.src_lang] if isinstance(clip_embs, dict)
            else clip_embs
            self.clip_tok = clip_tok_data[params.src_lang]
        self.length = len(self.clip_embs[params.src_lang]) if isinstance(self.clip_embs,
        dict) else len(self.clip_embs)
```

```python
        self.use_all_langs = params.use_all_langs
        if self.use_all_langs:
            self.single_length = self.length
            self.length *= len(tok_data)
            self.langs = list(tok_data.keys())
        else:
            langs = [params.src_lang, params.tgt_lang]
            self.caption_lang = 'en' if 'en' in langs else 'es'

    def __len__(self):
        return self.length

    def mask_inputs_(self, inputs):
        if self.mask_inputs:
            mask_inputs = torch.rand(inputs['input_ids'].shape) > self.mask_prob
            mask_inputs[0][0] = mask_inputs[0][-1] = True # Keep LID and EOS
            return {'input_ids': inputs['input_ids'][mask_inputs], 'attention_mask':
                inputs['attention_mask'][mask_inputs]}, mask_inputs
        return inputs, None

    def __getitem__(self, idx):
        batch = []
        if self.is_pretraining:
            if hasattr(self, 'caption_lang'):
                lang = self.caption_lang
            else:
                lang_idx = -1
                while idx >= self.single_length:
                    idx -= self.single_length
                    lang_idx -= 1
                lang = self.langs[lang_idx]

            if isinstance(self.clip_embs, dict):
                batch.append(self.clip_embs[lang][idx])
            else:
                batch.append(self.clip_embs[idx])
            inputs, mask = self.mask_inputs_(self.input_tok[lang][idx])
            batch.append(inputs)
            batch.append(mask)
            batch.append(self.input_tok[lang][idx])
            batch.append(self.output_raw[lang][idx])
        else:
            if self.use_clip_tok:
                batch.append(self.clip_tok[idx])
            else:
                batch.append(self.clip_embs[idx])
            inputs, _ = self.mask_inputs_(self.input_tok[idx])
            batch.append(inputs)
            outputs = self.output_tok[idx]
            mask = torch.ones_like(outputs['input_ids']).bool() # Mask doesnt do anything
            while finetuning. Only here to keep the collate_fn and
            forward same.
            batch.append(mask)
            batch.append(outputs)
            batch.append(self.output_raw[idx])
        return (*batch,)
```

对上述代码的具体说明如下。

（1）类 DocDataset 的功能是将文档数据集包装成 PyTorch Dataset 对象，使其可以被 PyTorch 的 DataLoader 使用，实现了__len__和__getitem__方法。

（2）类 ConcatDataset 的功能是将两个数据集按照一定的规则拼接在一起，确保合并后的数据集长度能够覆盖两个数据集中的所有样本，同时尽量保持平衡。它也实现了__len__和__getitem__方法。

（3）类 MultiModalDataset 用于处理多模态数据集，能够根据参数的不同在预训练和微调阶段做出不同的数据处理，并提供了__len__和__getitem__方法。其功能如下。

- 加载预训练或微调所需的数据。
- 在预训练阶段，支持数据的掩码处理（用于预测任务）。
- 在微调阶段，提供输入数据和输出数据，包括掩码用于保持接口一致性。
- 支持不同语言对的数据加载和处理。

8.4.2 数据处理和后处理

文件 data_utils.py 包含了一些用于数据处理和后处理的函数，包括从图像文件中提取图像的嵌入表示、读取图像文件并进行预处理、对文本进行分词并保存结果、使用编码器对数据集进行编码并保存结果，以及对训练和测试数据进行后处理以保持数据的一致性。文件 data_utils.py 的具体代码如下所示。

```
import torch
from tqdm import tqdm
import pickle as pkl
from PIL import Image
import os
from model_utils import get_lang_code, send_to_cuda
from collate_fns import *
from dataset import DocDataset
from torch.utils.data import DataLoader

def get_image_embs(clip, folder, image_splits, image_embs_f, desc, preprocessor, ignore_indices=[]):
    """
    从图像文件中提取图像的嵌入表示。

    Args:
        clip (CLIP): CLIP 模型实例。
        folder (str): 包含图像文件的文件夹路径。
        image_splits (list): 图像文件列表。
        image_embs_f (str): 嵌入表示结果的文件路径。
        desc (str): 进度条的描述信息。
        preprocessor: 图像预处理函数。
        ignore_indices (list): 需要忽略的图像索引列表。

    Returns:
        torch.Tensor: 图像的嵌入表示。
    """
    try:
        img_embs = torch.load(image_embs_f)
    except:
        img_paths = [os.path.join(folder, f) for f in image_splits]
        imgs = []
        for i, path in enumerate(tqdm(img_paths, desc='Reading images')):
            if i in ignore_indices:
                continue
            if '#' in path:
                path = path[:path.index('#')]  # 对于 multi30k mscoco 测试数据集的处理
            try:
                img = read_image(path, preprocessor)
                imgs.append(img)
            except:
```

```python
                    ignore_indices.append(i)
        img_ds = DocDataset(imgs)
        img_dl = DataLoader(img_ds, batch_size=128, shuffle=False, num_workers=4, pin_memory
        =True, collate_fn=collate_images)
        img_embs = create_embeddings(img_dl, clip, image_embs_f, desc)
    return img_embs

def read_image(img_path, preprocessor):
    """
    读取图像文件并进行预处理。

    Args:
        img_path (str): 图像文件路径。
        preprocessor: 图像预处理函数。

    Returns:
        torch.Tensor: 处理后的图像数据。
    """
    img = Image.open(img_path)
    try:
        img = torch.from_numpy(preprocessor(images=img)['pixel_values'][0])
        # 与 Huggingface 兼容
    except:
        img = preprocessor(img)    # 与 OpenAI CLIP 兼容

    return img

def tokenize(texts, tokenizer, lang, outfile, desc=''):
    """
    对文本进行分词并保存结果。

    Args:
        texts (list): 待处理的文本列表。
        tokenizer: 分词器。
        lang (str): 文本所属语言。
        outfile (str): 结果保存路径。
        desc (str): 进度条的描述信息。

    Returns:
        list: 分词后的文本数据。
    """
    if not os.path.exists(outfile):
        data = []
        tokenizer.src_lang = get_lang_code(lang)
        for text in tqdm(texts, desc=desc):
            data.append(tokenizer(text, return_tensors='pt', truncation=True))

        with open(outfile, 'wb') as f:
            pkl.dump(data, f)
    else:
        with open(outfile, 'rb') as f:
            data = pkl.load(f)
    return data

def create_embeddings(dl, encoder, outfile, desc=''):
    """
    使用编码器对数据集进行编码并保存结果。

    Args:
        dl: 数据加载器。
        encoder: 编码器模型。
        outfile (str): 结果保存路径。
        desc (str): 进度条的描述信息。
```

```python
    Returns:
        torch.Tensor: 编码后的结果。
    """
    embs = torch.tensor([]).cuda()
    with torch.no_grad():
        with torch.autocast(device_type='cuda'):
            for data in tqdm(dl, desc=desc):
                embs = torch.cat((embs, encoder(send_to_cuda(data))), dim=0)
    torch.save(embs.cpu(), outfile)
    return embs

def postprocess_pairs(train_texts, test_texts, train_tok_mbart, test_tok_mbart,
train_img_embs, test_img_embs, train_text_embs, test_text_embs, params, train_ignore_indices,
test_ignore_indices, train_tok_mclip, test_tok_mclip, force_pretraining):
    """
    对训练和测试数据进行后处理，用于保持数据的一致性。

    Args:
        train_texts (dict): 训练文本数据。
        test_texts (dict): 测试文本数据。
        train_tok_mbart (dict): 训练 MBART 分词结果。
        test_tok_mbart (dict): 测试 MBART 分词结果。
        train_img_embs (torch.Tensor): 训练图像嵌入表示。
        test_img_embs (torch.Tensor): 测试图像嵌入表示。
        train_text_embs (dict): 训练文本嵌入表示。
        test_text_embs (dict): 测试文本嵌入表示。
        params: 参数配置。
        train_ignore_indices (list): 训练集忽略的索引列表。
        test_ignore_indices (list): 测试集忽略的索引列表。
        train_tok_mclip (dict): 训练 M-CLIP 分词结果。
        test_tok_mclip (dict): 测试 M-CLIP 分词结果。
        force_pretraining (bool): 是否强制进行预训练。

    Returns:
        tuple: 后处理后的数据。
    """
    if params.stage in ['caption', 'text_recon'] or force_pretraining:
        # 需要配对图像的阶段，删除忽略的测试索引
        for index in sorted(train_ignore_indices[::-1], reverse=True):
            for lang in train_texts.keys():
                train_texts[lang].pop(index)
                train_tok_mbart[lang].pop(index)

        train_mask = torch.isin(torch.arange(train_text_embs[params.src_lang].shape[0]),
        torch.tensor(train_ignore_indices),
        invert=True)
        for lang in train_texts.keys():
            train_text_embs[lang] = train_text_embs[lang][train_mask]

        for index in sorted(test_ignore_indices[::-1], reverse=True):
            for lang in test_texts.keys():
                test_texts[lang].pop(index)
                test_tok_mbart[lang].pop(index)

        test_mask = torch.isin(torch.arange(test_text_embs[params.src_lang].shape[0]),
        torch.tensor(test_ignore_indices), invert=True)
        for lang in test_texts.keys():
            test_text_embs[lang] = test_text_embs[lang][test_mask]

        for lang in train_texts.keys():
            assert len(train_texts[lang]) == len(train_tok_mbart[lang]) == len
            (train_text_embs[lang]), 'Misalignment in train text pairs'
```

```
            if params.stage == 1:
                assert len(train_texts[lang]) == len(train_tok_mbart[lang]) == len(train_
text_embs[lang]) == len(train_img_embs), 'Misalignment in train text pairs with images'
        for lang in test_texts.keys():
            assert len(test_texts[lang]) == len(test_tok_mbart[lang]) == len(test_text_embs
            [lang]), 'Misalignment in train text pairs'
            if params.stage == 1:
                assert len(test_texts[lang]) == len(test_tok_mbart[lang]) == len (test_text_
                embs[lang]) == len(test_img_embs), 'Misalignment in test text pairs with images'

    return train_texts, test_texts, train_tok_mbart, test_tok_mbart, train_img_embs,
    test_img_embs, train_text_embs, test_text_embs,
    train_tok_mclip, test_t
```

在上述代码中定义了一些用于数据处理和后处理的函数。

（1）get_image_embs：从图像文件中提取图像的嵌入表示。

（2）read_image：读取图像文件并进行预处理。

（3）tokenize：对文本进行分词并保存结果。

（4）create_embeddings：使用编码器对数据集进行编码并保存结果。

（5）postprocess_pairs：对训练和测试数据进行后处理，用于保持数据的一致性。

8.4.3 数据集填充

文件 collate_fns.py 定义了用于实现数据集填充处理的函数，其中函数 collate_texts 用于对文本数据进行填充处理，函数 collate_images 用于对图像数据进行堆叠，函数 collate_multi 用于多模态数据的填充处理，确保多模态数据的各部分格式的一致性。文件 collate_fns.py 的具体代码如下所示。

```
import torch
from torch.nn.utils.rnn import pad_sequence
import transformers

def collate_texts(batch):    # 用于对文本数据进行填充处理，用于创建嵌入表示
    # 定义一个辅助函数，用于对指定键的数据进行填充
    collate_util = lambda key, b, pad_val: pad_sequence([x[key].squeeze() for x in b],
     batch_first=True, padding_value=pad_val)
    input_ids = collate_util('input_ids', batch, 1)
    att_mask = collate_util('attention_mask', batch, 0)
    return {'input_ids': input_ids, 'attention_mask': att_mask}

def collate_images(batch):    # 用于对图像数据进行堆叠
    return torch.stack(batch)

def collate_multi(batch):    # 用于多模态数据的填充处理，调用上述两个函数，因为格式相同
    out, labels, masks = {'clip': [], 'mbart': [], 'raw': []}, [], []
    for b in batch:
        out['clip'].append(b[0])
        out['mbart'].append(b[1])
        masks.append(b[2])
        labels.append(b[3])
        out['raw'].append(b[4])
        if len(b) > 5:
            out['clip'].append(b[5])
            out['mbart'].append(b[6])
            masks.append(b[7])
            labels.append(b[8])
            out['raw'].append(b[9])
```

```
# 判断 clip 和 mbart 数据的类型，分别调用对应的填充函数
out['clip'] = collate_images(out['clip']) if not isinstance(out['clip'][0],
transformers.tokenization_utils_base.BatchEncoding)
else collate_texts(out['clip'])
out['mbart'] = collate_texts(out['mbart'])
# 处理 mbart 的 labels 和 mask_decoder_input_ids
out['mbart']['labels'] = collate_texts(labels)['input_ids']
out['mbart']['mask_decoder_input_ids'] = pad_sequence([x.squeeze() for x in masks],
 batch_first=True, padding_value=True)
return out
```

8.4.4 获取 Multi30k 数据集的数据

文件 multi30k.py 用于从 Multi30k 数据集中获取文本数据和图像数据，并对它们进行预处理，包括分词、生成嵌入表示等操作。它能够读取训练集和测试集的文本数据，对其进行分词和嵌入表示的处理，并获取对应的图像数据，使用 CLIP 模型生成图像的嵌入表示，最终返回处理后的训练集和测试集的文本和图像数据。文件 multi30k.py 的具体代码如下所示。

```
import sys
sys.path.append('..')
import torch
from tqdm import tqdm
import pickle as pkl
import os
from data_utils import *
from collate_fns import *
from dataset import DocDataset
from torch.utils.data import DataLoader
from ddp import *

def get_Multi30k(params, model, test=('2017', 'mscoco'), force_pretraining=False):
    # 根据参数获取 Multi30k 数据集的文本和图像数据，包括训练集和测试集，生成对应的文本和图像的嵌入表示

    if force_pretraining:
        langs = ['en']   # 仅在英文图像上进行预训练
        test = ('2017', 'mscoco')   # 不会被使用，设置一个默认值
    else:
        langs = [params.src_lang, params.tgt_lang]

    datapath = os.path.join(params.data_dir, 'multi30k')
    os.makedirs(os.path.join(datapath, f'text/data/task1/mbart'), exist_ok=True)
    os.makedirs(os.path.join(datapath, f'text/data/task1/{params.image_encoder}'), exist_ok=
    True)

    # 读取训练文件
    train_texts = {lang: open(os.path.join(datapath, f'text/data/task1/raw/train.{lang}')).
    read().splitlines() for lang in langs}
    try:
        train_tok_mbart = {lang: pkl.load(open(os.path.join(datapath, f'text/data/task1/
        mbart/train.{lang}.pkl'), 'rb')) for lang in langs}
    except:
        print('Did not find mbart train tokenized data. Creating...')
        train_tok_mbart = {lang: tokenize(train_texts[lang], model.tokenizer, lang, os.
        path.join(datapath, f'text/data/task1/
        mbart/train.{lang}.pkl'), f'Tokenizing train {lang} with mbart') for lang in langs}

    try:
        train_tok_mclip = {lang: pkl.load(open(os.path.join(datapath, f'text/data/task1/
        {params.image_encoder}/train.{lang}.pkl'), 'rb')) for lang in langs}
    except:
        print('Did not find mclip train tokenized data. Creating...')
```

```python
    train_tok_mclip = {lang: tokenize(train_texts[lang], model.clip.text_preprocessor,
lang, os.path.join(datapath, f'text/data/task1/{params.image_encoder}/train.{lang}.pkl'),
f'Tokenizing train {lang} with {params.image_encoder}') for lang in langs}

    # 读取测试文件
    test_texts = {lang: open(os.path.join(datapath, f'text/data/task1/raw/test_
    {test[0]}_{test[1]}.{lang}')).read().splitlines() for lang in langs}
    try:
        test_tok_mbart = {lang: pkl.load(open(os.path.join(datapath, f'text/data/task1/
            mbart/test_{test[0]}_{test[1]}.{lang}.pkl'),
            'rb')) for lang in langs}
    except:
        print('Did not find mbart test tokenized data. Creating...')
        test_tok_mbart = {lang: tokenize(test_texts[lang], model.tokenizer, lang, os.
path.join(datapath, f'text/data/task1/mbart/test_{test[0]}_{test[1]}.{lang}.pkl'),
f'Tokenizing test {lang} with mbart') for lang in langs}

    try:
        test_tok_mclip = {lang: pkl.load(open(os.path.join(datapath, f'text/data/task1/{
params.image_encoder}/test_{test[0]}_{test[1]}.{lang}.pkl'), 'rb')) for lang in langs}
    except:
        print('Did not find mclip test tokenized data. Creating...')
        test_tok_mclip = {lang: tokenize(test_texts[lang], model.clip.text_preprocessor,
 lang, os.path.join(datapath, f'text/data/task1/{params.image_encoder}/test_{test[0]}_
{test[1]}.{lang}.pkl'), f'Tokenizing test {lang} with
{params.image_encoder}') for lang in langs}

    train_image_splits = open(os.path.join(datapath, f'text/data/task1/image_
    splits/train.txt')).read().splitlines()
    test_image_splits = open(os.path.join(datapath, f'text/data/task1/image_splits/
    test_{test[0]}_{test[1]}.txt')).read().splitlines()

    # 获取图像嵌入表示
    print('Loaded all text files. Getting images...')
    train_img_embs = get_image_embs(model.clip, os.path.join(datapath, 'images/train'),
train_image_splits, os.path.join(datapath, f'text/data/task1/{params.image_encoder}/
train.pth'), 'Embedding train images', model.clip.image_preprocessor)
    test_img_embs = get_image_embs(model.clip, os.path.join(datapath, f'images/test_
{test[0]}_{test[1]}'), test_image_splits, os.path.join(datapath, f'text/data/task1/{params.
image_encoder}/test_{test[0]}_{test[1]}.pth'), f'Embedding test_{test[0]}_{test[1]} images',
model.clip.image_preprocessor)

    train_text_embs, test_text_embs = {}, {}
    for lang in langs:
        embs_f = os.path.join(datapath, f'text/data/task1/{params.image_encoder}/
        train.{lang}.pth')
        try:
            train_text_embs[lang] = torch.load(embs_f)
        except:
            text_ds = DocDataset(train_tok_mclip[lang])
            text_dl = DataLoader(text_ds, batch_size=256, shuffle=False,
            num_workers=0, pin_memory=True, collate_fn=collate_texts)
            train_text_embs[lang] = create_embeddings(text_dl, model.clip, embs_f,
            f'Embedding train.{lang} mclip')

        embs_f = os.path.join(datapath, f'text/data/task1/{params.image_encoder}/test_{test[0]}
_{test[1]}.{lang}.pth')
        try:
            test_text_embs[lang] = torch.load(embs_f)
        except:
            text_ds = DocDataset(test_tok_mclip[lang])
            text_dl = DataLoader(text_ds, batch_size=256, shuffle=False, num_workers=4,
            pin_memory=True, collate_fn=collate_texts)
```

```
            test_text_embs[lang] = create_embeddings(text_dl, model.clip, embs_f, f'Embedding
            test_{test[0]}_{test[1]}.{lang} mclip')

    return train_texts, test_texts, train_tok_mbart, test_tok_mbart, train_img_embs,
        test_img_embs, train_text_embs, test_text_embs, train_tok_mclip, test_tok_mclip
```

8.4.5　获取 WIT 数据集的数据

文件 wit.py 用于从 WIT 数据集中获取文本和图像数据，并对它们进行预处理，包括分词、生成嵌入表示等操作。它能够读取训练集和测试集的文本数据，对其进行分词和嵌入表示的处理，同时获取对应的图像数据，并使用 CLIP 模型生成图像的嵌入表示。文件 wit.py 的具体代码如下所示。

```
import sys
sys.path.append('..')

import torch
from tqdm import tqdm
import pickle as pkl
import os
from data_utils import *
from collate_fns import *
from dataset import DocDataset
from torch.utils.data import DataLoader
import warnings
from ddp import *

def get_WIT(params, model, test_ds=['val'], force_pretraining=False):
    # 设置测试数据集
    test_ds = test_ds[0]
    # 获取源语言和目标语言
    langs = [params.src_lang, params.tgt_lang]
    # 数据路径
    datapath = os.path.join(params.data_dir, f'wit/mmt/{params.src_lang}_{params.tgt_lang}')
    # 创建目录
    os.makedirs(os.path.join(datapath, 'mbart'), exist_ok=True)
    os.makedirs(os.path.join(datapath, params.image_encoder), exist_ok=True)

    # 读取训练文件
    train_texts = {lang: open(os.path.join(datapath, f'train.{lang}')).read().
        splitlines() for lang in langs}
    try:
        train_tok_mbart = {lang: pkl.load(open(os.path.join(datapath, f'mbart/
            train.{lang}.pkl'), 'rb')) for lang in langs}
    except:
        print('Did not find mbart train tokenized data. Creating...')
        train_tok_mbart = {lang: tokenize(train_texts[lang], model.tokenizer,
            lang, os.path.join(datapath, f'mbart/train.
            {lang}.pkl'), f'Tokenizing train {lang} with mbart') for lang in langs}

    try:
        train_tok_mclip = {lang: pkl.load(open(os.path.join(datapath, f'{params.image_encoder}/
            train.{lang}.pkl'), 'rb')) for
            lang in langs}
    except:
        print('Did not find mclip train tokenized data. Creating...')
        train_tok_mclip = {lang: tokenize(train_texts[lang], model.clip.text_preprocessor,
            lang, os.path.join(datapath, f'{params.image_encoder}/train.{lang}.pkl'), f'Tokenizing
    train {lang} with {params.image_encoder}') for lang in langs}

    # 读取测试文件
```

```python
    test_texts = {lang: open(os.path.join(datapath, f'{test_ds}.{lang}')).read().
    splitlines() for lang in langs}
    try:
        test_tok_mbart = {lang: pkl.load(open(os.path.join(datapath, f'mbart/{test_ds}.
{lang}.pkl'), 'rb')) for lang in langs}
    except:
        print('Did not find mbart test tokenized data. Creating...')
        test_tok_mbart = {lang: tokenize(test_texts[lang], model.tokenizer, lang, os.path.
join(datapath, f'mbart/{test_ds}.{lang}.pkl'), f'Tokenizing test {lang} with mbart') for
lang in langs}

    try:
        test_tok_mclip = {lang: pkl.load(open(os.path.join(datapath, f'{params.
        image_encoder}/{test_ds}.{lang}.pkl'), 'rb')) for lang in langs}
    except:
        print('Did not find mclip test tokenized data. Creating...')
        test_tok_mclip = {lang: tokenize(test_texts[lang], model.clip.text_preprocessor,
lang, os.path.join(datapath, f'{params.image_encoder}/{test_ds}.{lang}.pkl'), f'Tokenizing
 test {lang} with {params.image_encoder}') for lang in langs}

    # 读取图片文件列表
    train_image_splits = open(os.path.join(datapath, f'train_image.txt')).read().splitlines()
    test_image_splits = open(os.path.join(datapath, f'{test_ds}_image.txt')).read().
    splitlines()

    # 获取图片嵌入表示
    print('Loaded all text files. Getting images...')
    train_ignore_indices = [int(x) for x in train_image_splits[0].split(',')]
    train_img_embs = get_image_embs(model.clip, os.path.join(datapath, 'images'),
    train_image_splits[1:], os.path.join(datapath, f'{params.image_encoder}/
    train.pth'), 'Embedding train images', model.clip.image_preprocessor, train_
    ignore_indices)
    train_ignore_indices_f = os.path.join(datapath, 'images/train_ignore_indices.
    pkl')
    if not os.path.exists(train_ignore_indices_f):
        with open(train_ignore_indices_f, 'wb') as f:
            pkl.dump(train_ignore_indices, f)
        warnings.warn('train_ignore_indices.pkl was created. If the images were not read
 and embedded prior to this, this is an error. You would need
        to preprocess the image data again, ' +
        'since train_ignore_indices.pkl is not correctly aligned since it is the same as
 what is in train_image.txt, but does not include indices which
            failed to get read by PIL.')
    else:
        with open(train_ignore_indices_f, 'rb') as f:
            train_ignore_indices = pkl.load(f)

    test_ignore_indices = [int(x) for x in test_image_splits[0].split(',')]
    test_img_embs = get_image_embs(model.clip, os.path.join(datapath, 'images'),
    test_image_splits[1:], os.path.join(datapath,
    f'{params.image_encoder}/{test_ds}.pth'), f'Embedding {test_ds} images', model.
    clip.image_preprocessor, test_ignore_indices)
    test_ignore_indices_f = os.path.join(datapath, f'images/{test_ds}_ignore_indices.pkl')
    if not os.path.exists(test_ignore_indices_f):
        with open(test_ignore_indices_f, 'wb') as f:
            pkl.dump(test_ignore_indices, f)
        warnings.warn(f'{test_ds}_ignore_indices.pkl was created. If the images were not
 read and embedded prior to this, this is an error. You would need to preprocess the
 image data again, ' +
                    f'since {test_ds}_ignore_indices.pkl is not correctly aligned since
 it is the same as what is in {test_ds}_image.txt, but does not include indices which
 failed to get read by PIL.')
    else:
```

```python
        with open(test_ignore_indices_f, 'rb') as f:
            test_ignore_indices = pkl.load(f)

    # 获取文本嵌入表示
    train_text_embs, test_text_embs = {}, {}
    for lang in langs:
        # 训练集文本嵌入表示
        embs_f = os.path.join(datapath, f'{params.image_encoder}/train.{lang}.pth')
        try:
            train_text_embs[lang] = torch.load(embs_f)
        except:
            text_ds = DocDataset(train_tok_mclip[lang])
            text_dl = DataLoader(text_ds, batch_size=128, shuffle=False, num_workers=0,
                pin_memory=True, collate_fn=collate_texts)
            train_text_embs[lang] = create_embeddings(text_dl, model.clip, embs_f,
                f'Embedding train.{lang} mclip')

        # 测试集文本嵌入表示
        embs_f = os.path.join(datapath, f'{params.image_encoder}/{test_ds}.{lang}.pth')
        try:
            test_text_embs[lang] = torch.load(embs_f)
        except:
            text_ds = DocDataset(test_tok_mclip[lang])
            text_dl = DataLoader(text_ds, batch_size=128, shuffle=False, num_workers=4,
                pin_memory=True, collate_fn=collate_texts)
            test_text_embs[lang] = create_embeddings(text_dl, model.clip, embs_f,
                f'Embedding {test_ds}.{lang} mclip')

    # 返回处理后的数据
    return postprocess_pairs(train_texts, test_texts, train_tok_mbart, test_tok_mbart,
        train_img_embs, test_img_embs, train_text_embs, test_text_embs, params, train_ignore_
        indices, test_ignore_indices, train_tok_mclip, test_tok_mclip, force_pretraining)
```

8.4.6 获取 WMT 数据集的文本数据

文件 wmt.py 用于从 WMT 数据集中获取文本数据，并对其进行预处理，包括分词、生成嵌入表示等操作；能够读取训练集和测试集的文本数据，对其进行分词和嵌入表示的处理。文件 wmt.py 的具体代码如下所示。

```python
import sys
sys.path.append('..')

import torch
from tqdm import tqdm
import pickle as pkl
import os
from data_utils import *
from collate_fns import *
from dataset import DocDataset
from torch.utils.data import DataLoader
from ddp import *

def get_WMT(params, model, test_ds=['val'], force_pretraining=False):
    # 设置测试数据集
    test_ds = test_ds[0]
    # 获取源语言和目标语言
    langs = [params.src_lang, params.tgt_lang]
    # 数据路径
    datapath = os.path.join(params.data_dir, f'wmt/wmt_{params.src_lang}_{params.
        tgt_lang}')
    # 创建目录
    os.makedirs(os.path.join(datapath, f'mbart'), exist_ok=True)
```

```python
os.makedirs(os.path.join(datapath, f'{params.image_encoder}'), exist_ok=True)

# 读取训练文件
train_texts = {lang: open(os.path.join(datapath, f'train.{lang}')).read().
splitlines() for lang in tqdm(langs, desc='Loading raw
text files', disable=not is_main_process())}
try:
    train_tok_mbart = {lang: pkl.load(open(os.path.join(datapath, f'mbart/
    train.{lang}.pkl'), 'rb')) for lang in tqdm(langs,
    desc='Loading mbart tokenized train files', disable=not is_main_process())}
except:
    print('Did not find mbart train tokenized data. Creating...')
    train_tok_mbart = {lang: tokenize(train_texts[lang], model.tokenizer,
    lang, os.path.join(datapath, f'mbart/train.{lang}.pkl'), f'Tokenizing
    train {lang} with mbart') for lang in langs}

try:
    train_tok_mclip = {lang: pkl.load(open(os.path.join(datapath, f'{params.
    image_encoder}/train.{lang}.pkl'), 'rb')) for
    lang in tqdm(langs, desc='Loading mclip tokenized train files', disable=
    not is_main_process())}
except:
    print('Did not find mclip train tokenized data. Creating...')
    train_tok_mclip = {lang: tokenize(train_texts[lang], model.clip.text_
    preprocessor, lang, os.path.join(datapath, f'{params.image_encoder}/
    train.{lang}.pkl'), f'Tokenizing train {lang} with mclip') for lang in
    langs}

# 读取测试文件
test_texts = {lang: open(os.path.join(datapath, f'{test_ds}.{lang}')).read().
splitlines() for lang in tqdm(langs, desc='Loading test raw files', disable=
not is_main_process())}
try:
    test_tok_mbart = {lang: pkl.load(open(os.path.join(datapath, f'mbart/
    {test_ds}.{lang}.pkl'), 'rb')) for lang in tqdm(langs, desc='Loading
    mbart tokenized test files', disable=not is_main_process())}
except:
    print('Did not find mbart test tokenized data. Creating...')
    test_tok_mbart = {lang: tokenize(test_texts[lang], model.tokenizer,
    lang, os.path.join(datapath, f'mbart/{test_ds}.{lang}.pkl'), f'Tokenizing
    test {lang} with mbart') for lang in langs}

try:
    test_tok_mclip = {lang: pkl.load(open(os.path.join(datapath, f'{params.
    image_encoder}/{test_ds}.{lang}.pkl'), 'rb')) for lang in tqdm(langs,
    desc='Loading mclip tokenized test files', disable=not is_main_process())}
except:
    print('Did not find mclip test tokenized data. Creating...')
    test_tok_mclip = {lang: tokenize(test_texts[lang], model.clip.text_
    preprocessor, lang, os.path.join(datapath, f'{params.image_encoder}/
    {test_ds}.{lang}.pkl'), f'Tokenizing test {lang} with mclip') for lang
    in langs}

# 初始化训练集和测试集的文本嵌入表示字典
train_text_embs, test_text_embs = {}, {}

# 获取训练集和测试集的文本嵌入表示
for lang in langs:
    # 训练集
    embs_f = os.path.join(datapath, f'{params.image_encoder}/train.{lang}.pth')
    try:
        train_text_embs[lang] = torch.load(embs_f)
    except:
```

```
        text_ds = DocDataset(train_tok_mclip[lang])
        text_dl = DataLoader(text_ds, batch_size=512, shuffle=False, num_
        workers=4, pin_memory=True, collate_fn=collate_texts)
        train_text_embs[lang] = create_embeddings(text_dl, model.clip, embs_f, f'Embedding
        train.{lang} mclip')

    # 测试集
    embs_f = os.path.join(datapath, f'{params.image_encoder}/{test_ds}.{lang}.
    pth')
    try:
        test_text_embs[lang] = torch.load(embs_f)
    except:
        text_ds = DocDataset(test_tok_mclip[lang])
        text_dl = DataLoader(text_ds, batch_size=512, shuffle=False,
        num_workers=4, pin_memory=True, collate_fn=collate_texts)
        test_text_embs[lang] = create_embeddings(text_dl, model.clip, embs_f,
        f'Embedding {test_ds}.{lang} mclip')

# 返回处理后的数据
return train_texts, test_texts, train_tok_mbart, test_tok_mbart, None, None, train_text_embs,
test_text_embs, train_tok_mclip, test_tok_mclip
```

8.5 多模态大模型

接下来将介绍多模态模型的实现方法，包括实用功能函数、适配器模型、多模态模型类、文本生成模型、分布式训练支持以及模型训练和测试框架。通过集成文本和图像处理能力，实现模型的高效训练和推断，同时支持跨模态适配和分布式计算，以增强模型性能和扩展性。主程序作为入口点，负责整体流程的协调和执行。

8.5.1 功能函数

文件 model_utils.py 定义了用于实现模型训练和推断功能的几个实用功能函数，其中函数 freeze_params 用于将模型参数设置为不需要梯度更新，函数 get_lang_code 的功能是根据输入的语言代码返回对应的语言标识码，函数 send_to_cuda 的功能是将输入的数据 batch 移动到 GPU 上（如果可用），以加速计算。文件 model_utils.py 的具体代码如下所示。

```
import torch

def freeze_params(module):
    for param in module.parameters():
        param.requires_grad = False

def get_lang_code(lang):
    LANG_MAP = {
                'ar': 'ar_AR', 'cs': 'cs_CZ', 'de': 'de_DE', 'en': 'en_XX', 'es':
                'es_XX', 'et': 'et_EE',
                'fi': 'fi_FI', 'fr': 'fr_XX', 'gu': 'gu_IN', 'hi': 'hi_IN',
                'it': 'it_IT', 'ja': 'ja_XX',
                'kk': 'kk_KZ', 'ko': 'ko_KR', 'lt': 'lt_LT', 'lv': 'lv_LV', 'my': 'my_MM',
'ne': 'ne_NP', 'nl': 'nl_XX', 'ro': 'ro_RO', 'ru': 'ru_RU', 'si': 'si_LK', 'tr':
                'tr_TR', 'vi': 'vi_VN', 'zh': 'zh_CN', 'af': 'af_ZA'
                }
    return LANG_MAP[lang]

def send_to_cuda(batch):
    if torch.is_tensor(batch):
        batch = batch.cuda()
```

```
        else:
            for key in batch:
                if torch.is_tensor(batch[key]):
                    batch[key] = batch[key].cuda()
        return batch
```

8.5.2 适配器模型

文件 adapters.py 定义了 3 种不同类型的适配器模型：MLPAdapter、HiddenMLPAdapter 和 TransformerAdapter。这些适配器模型用于将 CLIP 模型的输出适配到与 MBART 模型输入相匹配的维度和形状，以便在多模态任务中使用。

（1）MLPAdapter：使用简单的多层感知机（MLP）作为适配器，将 CLIP 输出线性映射到与 MBART 模型输入相匹配的维度和形状。

（2）HiddenMLPAdapter：在 MLPAdapter 的基础上添加了额外的隐藏层，以增加模型的复杂度和表达能力。

（3）TransformerAdapter：使用了一个基于 Transformer 的适配器模型，它将 CLIP 输出进行线性映射后输入 Transformer 编码器，以捕获更复杂的语义信息。

```python
import torch
import torch.nn as nn

class MLPAdapter(nn.Module):
    def __init__(self, **kwargs):
        super(MLPAdapter, self).__init__()
        self.hidden = nn.Sequential(
            nn.Linear(kwargs['clip_dim'], kwargs['prefix_length'] * kwargs
              ['mbart_dim']),
            nn.PReLU(),
        )
        self.prefix_length, self.mbart_dim = kwargs['prefix_length'], kwargs
          ['mbart_dim']

    def forward(self, x):
        x = x.float()
        return self.hidden(x).view(-1, self.prefix_length, self.mbart_dim)

    def reset(self):
        nn.init.xavier_uniform_(self.hidden[0].weight)
        nn.init.xavier_uniform_(self.projector.weight)

class HiddenMLPAdapter(nn.Module):
    def __init__(self, **kwargs):
        super(HiddenMLPAdapter, self).__init__()
        self.hidden = nn.Sequential(
            nn.Linear(kwargs['clip_dim'], kwargs['mbart_dim']),
            nn.ReLU(),
            nn.Linear(kwargs['mbart_dim'], kwargs['prefix_length'] * kwargs ['mbart_dim']),
            nn.PReLU(),
        )
        self.prefix_length, self.mbart_dim = kwargs['prefix_length'], kwargs
          ['mbart_dim']

    def forward(self, x):
        x = x.float()
        return self.hidden(x).view(-1, self.prefix_length, self.mbart_dim)

    def reset(self):
        nn.init.xavier_uniform_(self.hidden[0].weight)
```

```
            nn.init.xavier_uniform_(self.projector.weight)
class TransformerAdapter(nn.Module):
    def __init__(self, **kwargs):
        super(TransformerAdapter, self).__init__()
        self.prefix_length, self.mbart_dim = kwargs['prefix_length'], kwargs ['mbart_dim']
        self.projector = nn.Linear(kwargs['clip_dim'], self.prefix_length * kwargs
['mbart_dim']) # MLPAdapter(clip_dim = kwargs['clip_dim'], mbart_dim = kwargs['mbart_dim'],
prefix_length = kwargs['prefix_length'])
        encoder_layer = nn.TransformerEncoderLayer(
            d_model = kwargs['mbart_dim'],
            nhead = 2,
            dim_feedforward = kwargs['mbart_dim']//3,
            batch_first = True
        )
        self.transformer = nn.TransformerEncoder(
            encoder_layer = encoder_layer,
            num_layers = kwargs['num_encoder_layers']
        )

    def forward(self, x):
        x = self.projector(x).view(-1, self.prefix_length, self.mbart_dim)
        return self.transformer(x)
```

上述适配器模型可以将 CLIP 模型与 MBART 模型进行有效地集成，用于各种多模态任务中，如图文匹配、图像标注等。

8.5.3 获取文本输入的嵌入表示

文件 mclip.py 定义了一个名为 MClip 的类，它是基于 MultilingualCLIP 的自定义模型。该模型用于获取文本输入的嵌入表示。

```
from multilingual_clip.pt_multilingual_clip import MultilingualCLIP
import transformers
import torch

class MClip(MultilingualCLIP):
    def __init__(self, config, *args, **kwargs):
        super().__init__(config, *args, **kwargs)

    def forward(self, txt_tok): #移除即时分词处理
        embs = self.transformer(**txt_tok)[0]
        att = txt_tok['attention_mask']
        embs = (embs * att.unsqueeze(2)).sum(dim=1) / att.sum(dim=1)[:, None]
        return self.LinearTransformation(embs)
```

对上述代码的具体说明如下。

（1）类 MClip 继承自 MultilingualCLIP，是一个多语言版本的 CLIP 模型。

（2）方法 forward 用于接收文本的 tokenized 输入 txt_tok，并返回文本的嵌入表示。在 forward 方法中，通过模型的 transformer 将 tokenized 的文本输入转换为嵌入表示，并根据注意力掩码 attention_mask 对嵌入进行加权求和以得到文本的整体表示。

（3）最终使用 LinearTransformation 对得到的文本表示进行线性变换处理，以获得最终的输出表示。

8.5.4 多模态模型类

文件 clip_comb.py 定义了多个用于多模态学习的模型类，其中 M_CLIP 模型类基于

Multilingual CLIP，包含文本和图像编码器；CLIP 模型类使用 OpenAI 的 CLIP 模型作为文本和图像编码器；CLIP_RES 模型类使用 CLIP-Res 预训练模型进行文本和图像编码。这些模型可以根据输入类型（文本或图像）进行编码处理，并可选择是否冻结 CLIP 或 ResNet-50 参数。文件 clip_comb.py 的具体代码如下所示。

```python
import torch
import torch.nn as nn
import transformers
import clip
from mclip import MClip
from model_utils import *
from transformers import AutoProcessor, AutoTokenizer, CLIPModel, BlipModel

class M_CLIP(nn.Module):
    def __init__(self, params):
        super(M_CLIP, self).__init__()
        self.text_encoder, self.text_preprocessor = MClip.from_pretrained('M-CLIP/
        XLM-Roberta-Large-Vit-B-32'),
        AutoTokenizer.from_pretrained('M-CLIP/XLM-Roberta-Large-Vit-B-32')
        self.image_encoder, self.image_preprocessor = CLIPModel.from_pretrained ("openai
        /clip-vit-base-patch32"),
        AutoProcessor.from_pretrained("openai/clip-vit-base-patch32")
        if not params.unfreeze_clip:
            freeze_params(self.text_encoder)
        freeze_params(self.image_encoder)

    def forward(self, x):
        if isinstance(x, dict):
            return self.text_encoder(x)
        else:
            return self.image_encoder.get_image_features(x)

class CLIP(nn.Module):
    def __init__(self, params):
        super(CLIP, self).__init__()
        self.text_encoder, self.text_preprocessor = CLIPModel.from_pretrained
        ("openai/clip-vit-base-patch32"),
        AutoTokenizer.from_pretrained("openai/clip-vit-base-patch32")
        self.image_encoder, self.image_preprocessor = self.text_encoder,
        AutoProcessor.from_pretrained
        ("openai/clip-vit-base-patch32")
        if not params.unfreeze_clip:
            freeze_params(self.text_encoder)
        freeze_params(self.image_encoder)

    def forward(self, x):
        if isinstance(x, dict):
            return self.text_encoder.get_text_features(**x)
        else:
            return self.image_encoder.get_image_features(**x)

class CLIP_RES(nn.Module):
    def __init__(self, params):
        super(CLIP_RES, self).__init__()
        self.image_encoder, self.image_preprocessor = clip.load('RN50x64')
        self.text_encoder, self.text_preprocessor = self.image_encoder, AutoTokenizer.
        from_pretrained("openai/clip-vit-base-patch32")
        if not params.unfreeze_clip:
            freeze_params(self.text_encoder)
        freeze_params(self.image_encoder)
```

```
    def forward(self, x):
        if isinstance(x, dict):
            return self.text_encoder.encode_image(x)
        else:
            return self.image_encoder.encode_text(x)
```

8.5.5 多模态文本生成任务模型

文件 model.py 定义了用于多模态文本生成任务的模型类 CLIPTrans 和 CLIPTrans_CLIP，CLIPTrans 模型类结合了 CLIP 图像编码器和 MBart 文本生成模型，可以在训练或测试时生成文本。CLIPTrans_CLIP 模型类在 CLIPTrans 的基础上添加了 CLIP 图像编码器，用于训练过程中的文本生成任务。文件 model.py 的具体代码如下所示。

```
import sys
sys.path.append('src/utils')
import torch
import torch.nn as nn
import transformers
from transformers import MBart50Tokenizer, MBartForConditionalGeneration
from clip_comb import *
from adapters import *
from model_utils import *
import warnings

class CLIPTrans(nn.Module):
    def __init__(self, params):
        super(CLIPTrans, self).__init__()
        if params.image_encoder == 'mclip':
            print('Using MCLIP encoder')
            self.clip = M_CLIP(params)
        elif params.image_encoder == 'clip':
            print('Using CLIP encoder')
            self.clip = CLIP(params)
        elif params.image_encoder == 'clip_res':
            print('Using CLIP-Resnet encoder')
            self.clip = CLIP_RES(params)
        self.prefix_length = params.prefix_length
        if self.prefix_length > 0:
            adapter_inputs = {'clip_dim': 512, 'mbart_dim': 1024, 'prefix_length':
            self.prefix_length}
            if params.mapping_network == 'mlp':
                self.adapter = MLPAdapter(**adapter_inputs)
            elif params.mapping_network == 'transformer':
                self.adapter = TransformerAdapter(num_encoder_layers = 1,
                **adapter_inputs)
            else:
                self.adapter = HiddenMLPAdapter(**adapter_inputs)
        if self.prefix_length not in [0, 10]:# and self.fusion == 'context':
            warnings.warn("prefix_length != 10 or != 0. Change the combined
            attention_mask line in huggingface accordingly")
        self.tokenizer = MBart50Tokenizer.from_pretrained('facebook/mbart-large-50')

        self.mbart = MBartForConditionalGeneration.from_pretrained('facebook/
        mbart-large-50')
        self.target_lang = get_lang_code(params.tgt_lang)
        if not params.test:
            if params.stage in ['caption', 'text_recon']:
                freeze_params(self.mbart.model.encoder)

    def forward(self, batch, mode = 'train'): #[1]重构以包含虚拟（占位）标记和适配器
        if self.prefix_length > 0:
```

```python
                visual_context = batch.pop('clip')
                prefix_embeds = self.adapter(visual_context)
        mask = batch['mbart'].pop('mask_decoder_input_ids')
        #[2]需要强制移除这一点，  因为输入嵌入正在被传递
        if mode == 'train':
            if self.prefix_length > 0:
# https://github.com/huggingface/transformers/blob/9e40bba6ba177cbc3b72b5fc7c8939174ad77899/src/transformers/models/mbart/modeling_mbart.py#L1027
                decoder_inputs_embeds = self.mbart.model.decoder.embed_tokens
(decoder_input_ids) * self.mbart.model.decoder.embed_scale
                decoder_inputs_embeds = torch.cat((decoder_inputs_embeds[:, 0].unsqueeze(1),
prefix_embeds, decoder_inputs_embeds[:, 1:]), dim = 1)
                batch['mbart']['decoder_inputs_embeds'] = decoder_inputs_embeds
                decoder_attention_mask = torch.ones_like(decoder_input_ids).cuda()
                decoder_attention_mask[decoder_input_ids == 1] = 0
                decoder_attention_mask = torch.cat((decoder_attention_mask[:, 0].unsqueeze(1), torch.ones(prefix_embeds.shape[0], self.prefix_length).cuda(),
decoder_attention_mask[:, 1:]), dim = 1)
                batch['mbart']['decoder_attention_mask'] = decoder_attention_mask
                dummy_tokens = (torch.ones((batch['mbart']['labels'].shape[0], self.
prefix_length)) * 1).cuda() # Ones because padding token in mbart is 1.
        #确认 GPT2 的填充标记是 0
                batch['mbart']['labels'] = torch.cat((batch['mbart']['labels'][:, 0].
unsqueeze(1), dummy_tokens, batch['mbart']['labels'][:, 1:]), dim = 1).long()

            batch['mbart']['labels'][batch['mbart']['labels'] == 1] = -100 #[3]忽略填充标记上的损失
            outputs = self.mbart(**batch['mbart'])
        elif mode == 'test':
            batch['mbart'].pop('labels')
            if self.prefix_length > 0:
                decoder_input_ids = (torch.ones(batch['mbart']['input_ids'].shape[0], 1) *
                2).long().cuda()
                decoder_inputs_embeds = self.mbart.model.decoder.embed_tokens
                (decoder_input_ids) * self.mbart.model.decoder.embed_scale
                decoder_inputs_embeds = torch.cat((prefix_embeds, decoder_inputs_embeds),
dim = 1) # [4]将前缀嵌入（prefix_embeds）放置在结束标记（eos）之前因为在解码器中，结束标记现在是起始标记 ID。

                batch['mbart']['decoder_inputs_embeds'] = decoder_inputs_embeds
                batch['mbart']['decoder_attention_mask'] = torch.ones
((decoder_inputs_embeds.shape[0], decoder_inputs_embeds.shape[1])).cuda()

            outputs = self.mbart.generate(**batch['mbart'], forced_bos_token_id = self.to
            kenizer.lang_code_to_id[self.target_lang], max_new_tokens = 60)
        return outputs


class CLIPTrans_CLIP(CLIPTrans):
    def __init__(self, params):
        super().__init__(params)

    def forward(self, batch, mode = 'train'):
        batch['clip'] = self.clip(batch['clip'])
        return super().forward(batch, mode)
```

8.5.6　分布式训练

文件 ddp.py 提供了用于分布式训练的功能，包括初始化分布式后端、检查分布式环境是否可用和已初始化、在主进程上保存模型、获取当前进程的排名以及检查当前进程是否为主进程等功能，使模型训练可以在多个节点和 GPU 上进行并实现同步和通信。文件 ddp.py 的具体代码如下所示。

```
import torch
import torch.distributed as dist
```

```python
import os

def init_distributed():
    # 初始化分布式后端，负责同步节点/GPU
    dist_url = "env://"    # 默认使用环境变量
    # 从环境变量中获取当前进程的排名、总进程数和本地进程排名
    rank = int(os.environ["RANK"])
    world_size = int(os.environ['WORLD_SIZE'])
    local_rank = int(os.environ['LOCAL_RANK'])
    # 初始化进程组，使用 NCCL 后端
    dist.init_process_group(
        backend="nccl",
        init_method=dist_url,
        world_size=world_size,
        rank=rank
    )

    # 设置当前进程的 GPU 设备
    torch.cuda.set_device(local_rank)
    torch.cuda.empty_cache()

    # 等待所有线程同步到这一点
    dist.barrier()

def is_dist_avail_and_initialized():
    # 检查分布式环境是否可用和已初始化
    if not dist.is_available():
        return False

    if not dist.is_initialized():
        return False

    return True

def save_on_master(*args, **kwargs):
    # 在主进程上保存模型
    if is_main_process():
        torch.save(*args, **kwargs)

def get_rank():
    # 返回当前进程的排名
    if not is_dist_avail_and_initialized():
        return 0
    return dist.get_rank()

def is_main_process():
    # 检查当前进程是否为主进程
    return get_rank() == 0
```

对上述代码的具体说明如下。

（1）函数 init_distributed 用于初始化分布式后端，负责同步节点/GPU。
（2）函数 is_dist_avail_and_initialized 用于检查分布式环境是否可用和已初始化。
（3）函数 save_on_master 用于在主进程上保存模型。
（4）函数 get_rank 用于返回当前进程的排名。
（5）函数 is_main_process 用于检查当前进程是否为主进程。

8.5.7 模型训练和测试

文件 runner.py 定义了类 Runner，用于实现多模态模型的训练和测试工作。其中包含了保存和

加载模型、训练一个 epoch、测试模型等功能。在训练过程中，可以进行梯度更新、学习率调度以及在每个 epoch 结束后进行测试，同时支持分布式训练和测试。文件 runner.py 的具体代码如下所示。

```python
class Runner:
    def __init__(self, train_dl, test_dl, params):
        self.train_dl = train_dl
        self.test_dl = test_dl
        self.update_count = params.update_count
        self.test_after = params.test_after * self.update_count
        self.is_pretraining = params.stage in ['caption', 'text_recon']
        self.meteor = evaluate.load('meteor') #[5]将此放入 __init__ 方法的唯一原因是为了从每个时代
        #（epoch）中移除 NLIK（自然语言工具包）的下载信息。

    def save_model(self, model, name, epoch):
        checkpoint = {
            'model': model.state_dict(),
            'optimizer': self.optimizer.state_dict(),
            'scaler': self.scaler.state_dict(),
            'epoch': epoch,
            'best_bleu_test': self.best_bleu_test
        }
        torch.save(checkpoint, name)

    def load_model(self, params, model, name, load_opt):
        if not os.path.exists(name) and os.path.exists(name.replace('.pth', '_1.pth')):
            model_names = sorted(glob.glob(name.replace('.pth', '*')))
            checkpoint = {'model': torch.load(model_names[0], map_location =
            torch.device('cpu')), 'epoch': 0, 'best_bleu_test': 0}
            for model_name in model_names[1:]:
                checkpoint['model'].update(torch.load(model_name, map_location =
                torch.device('cpu')))
        else:
            checkpoint = torch.load(name, map_location = torch.device('cpu'))
        if params.num_gpus == 1:
            from collections import OrderedDict
            new_state_dict = OrderedDict()
            for k, v in checkpoint['model'].items():
                name = k[7:] if 'module' in k[:7] else k #[6]移除 DataParallel/
                DistributedDataParallel 中的 "module." 前缀
                new_state_dict[name] = v
            checkpoint['model'] = new_state_dict

        model.load_state_dict(checkpoint['model'], strict = False)

        if load_opt:
            try:
                self.optimizer.load_state_dict(checkpoint['optimizer'], strict = False)
                self.scaler.load_state_dict(checkpoint['scaler'], strict = False)
            except:
                warnings.warn('Could not load optimizer due to extra parameters')
        else:
            warnings.warn('Not loading optimizer - if you intended to continue
            training(and not just load weights), hard code
            load_opt = True and rerun')
        model.cuda()

        return checkpoint['epoch'], checkpoint['best_bleu_test']

    def fit_one_epoch(self, model, tokenizer, params, epoch):
        model.train()
        train_loss = 0.0
        prob = torch.rand(1).item()
        self.optimizer.zero_grad()
```

```python
        for step, batch in enumerate(tqdm(self.train_dl, desc = f'Epoch {epoch}',
        disable = not is_main_process())):
            batch['mbart'], batch['clip'] = send_to_cuda(batch['mbart']), send_
            to_cuda(batch['clip'])
            with torch.autocast(device_type='cuda'):
                output = model(batch)
                loss = output[0]
            self.scaler.scale(loss).backward()
            if (step + 1) % self.update_count == 0:
                self.scaler.step(self.optimizer)
                self.scaler.update()
                self.cycle_scheduler.step()
                self.optimizer.zero_grad()
            if params.num_gpus > 1:
                loss_collated = [torch.zeros_like(loss).cuda() for _ in range
                (params.num_gpus)]
                dist.all_gather(loss_collated, loss)
                train_loss = sum(loss_collated).item()
            else:
                train_loss = loss.item()
            del batch
            if self.test_after > 0 and (step + 1) % self.test_after == 0:
                self.test(model, tokenizer, params, epoch)

    if is_main_process():
        print(f'Epoch {epoch}: Train Loss: {self.update_count * train_loss/
        (params.num_gpus * len(self.train_dl))}\n')
        if self.is_pretraining:
            self.save_model(model, f'{params.model_name}/model_pretrained.
            pth', epoch)

def test(self, model, tokenizer, params, epoch):
    model.eval()
    test_loss = 0.0
    translated_sentences, target_sentences = [], []
    tokenizer.tgt_lang = get_lang_code(params.tgt_lang)
    with torch.no_grad():
        for i, batch in enumerate(tqdm(self.test_dl, desc = f'Epoch {epoch}',
        disable = not is_main_process())):
            batch['clip'] = send_to_cuda(batch['clip'])
            batch['mbart'] = send_to_cuda(batch['mbart'])
            raw_target_text = batch.pop('raw')
            with torch.autocast(device_type='cuda'):
                output = model(batch, mode = 'test')
            output = tokenizer.batch_decode(output, skip_special_tokens = True)
            if params.num_gpus > 1:
                output_collated = [None for _ in range(params.num_gpus)]
                dist.all_gather_object(output_collated, output)

                targets_collated = [None for _ in range(params.num_gpus)]
                dist.all_gather_object(targets_collated, raw_target_text)
            if is_main_process():
                if params.num_gpus > 1:
                    for gpu_list in output_collated:
                        translated_sentences.extend(gpu_list)
                    for gpu_list in targets_collated:
                        target_sentences.extend(gpu_list)
                else:
                    translated_sentences.extend(output)
                    target_sentences.extend(raw_target_text)

    if is_main_process():
        bleu_score = sacrebleu.corpus_bleu(translated_sentences, [target_
        sentences]).score
```

```python
            meteor_score = self.meteor.compute(predictions = translated_sentences,
            references = target_sentences)['meteor']
            print(f'Epoch {epoch}; Test BLEU: {bleu_score}; Test METEOR: {100 *
            meteor_score}')
            print('------------------------------------------')
            for i, (tra, tgt) in enumerate(zip(translated_sentences[0:5], target_
            sentences[0:5])):
                print(f'Target Sentence {i}: {tgt}')
                print(f'Translated Sentence {i}: {tra}')
                print('------------------------------------------')

            if bleu_score > self.best_bleu_test and not self.is_pretraining and not
            params.test:
                self.best_bleu_test = bleu_score
                self.save_model(model, f'{params.model_name}/model_best_test.pth', epoch)

    def train(self, model, tokenizer, params):
        self.best_bleu_test = -float('Inf')
        self.optimizer = optim.AdamW(model.parameters(), lr = params.lr, betas =
        (0.9, 0.98), eps = 1e-6, weight_decay = 0.0)
        self.scaler = GradScaler()
        steps_per_epoch = math.ceil(len(self.train_dl)/self.update_count)
        last_epoch, last_batch = 0, -1
        self.cycle_scheduler = PolynomialLRDecay(self.optimizer, max_decay_
        steps = 40000, end_learning_rate = params.lr, power = 2.0)
        if params.load_model:
            last_epoch, self.best_bleu_test = self.load_model(params, model, f'
            {params.model_name}/{params.load_model}' if '/' not in params.load_
            model else f'{params.model_dir}/{params.load_model}', load_opt = True)
            if params.continue_training:
                last_batch = (last_epoch - 1) * steps_per_epoch
                for step in range(last_batch):
                    self.cycle_scheduler.step()
            elif not params.test: #[7]加载一个模型重新开始训练
                last_epoch, last_batch = 0, -1
                self.best_bleu_test = -float('Inf')
        if params.test:
            self.test(model, tokenizer, params, last_epoch)
            return
        for epoch in range(last_epoch, params.epochs):
            if params.num_gpus > 1:
                self.train_dl.sampler.set_epoch(epoch)
                self.test_dl.sampler.set_epoch(epoch)
            self.fit_one_epoch(model, tokenizer, params, epoch+1)
            self.test(model, tokenizer, params, epoch+1)
```

对上述代码的具体说明如下。

（1）类 Runner 负责模型的训练和测试，包括保存和加载模型、训练一个 epoch、测试模型等功能。

（2）方法 save_model 用于保存模型及其相关信息，包括模型状态字典、优化器状态字典、当前 epoch 等。

（3）方法 load_model 用于加载模型及其相关信息，包括模型状态字典、优化器状态字典、上次训练的 epoch 等。

（4）方法 fit_one_epoch 用于训练一个 epoch 的数据，包括将数据送入模型进行前向传播、计算损失、梯度反向传播和更新参数等操作。

（5）方法 test 用于对模型进行测试，包括计算测试集上的损失以及生成翻译结果，并计算 BLEU 和 METEOR 分数。

（6）方法 train 用于执行整个训练过程，包括设置优化器、学习率调度器，然后按照设定的

epoch 数进行训练和测试。

8.5.8 主程序

文件 main.py 是本项目的主程序文件,负责设置参数、加载数据、初始化模型,并执行训练或推理。文件 main.py 的具体实现流程如下。

(1)通过解析命令行参数,设置模型名称、数据集路径、超参数等信息。
(2)调用 init_seed 函数初始化随机种子,确保实验的可复现性。
(3)加载模型类 CLIPTrans 或 CLIPTrans_CLIP,并根据是否解冻 CLIP 进行初始化。
(4)根据参数设置当前使用的 GPU 设备。
(5)加载数据集,并根据不同阶段(预训练、翻译等)设置不同的数据处理方式。
(6)根据是否多 GPU 运行,设置分布式数据并行训练。
(7)创建 Runner 实例,进行模型的训练或推理。

文件 main.py 的具体实现代码如下所示。

```
get_ds = {'multi30k': get_Multi30k, 'wit': get_WIT, 'wmt': get_WMT, 'coco': get_coco}
def init_seed(seed):
    random.seed(seed)
    os.environ['PYTHONHASHSEED'] = str(seed)
    np.random.seed(seed)
    torch.manual_seed(seed)
    torch.cuda.manual_seed(seed)
    torch.backends.cudnn.deterministic = True

def main(params):
    init_seed(params.seed)
    os.environ["TOKENIZERS_PARALLELISM"] = "false"
    params.model_name = os.path.join(params.model_dir, f'{params.model_name}-{params.
    src_lang}-{params.tgt_lang}')
    os.makedirs(params.model_name, exist_ok = True)
    MODEL = CLIPTrans if not params.unfreeze_clip else CLIPTrans_CLIP
    model = MODEL(params)
    if params.num_gpus > 1:
        init_distributed()
    else:
        torch.cuda.set_device(params.local_rank)
    model.cuda()
    tokenizer = model.tokenizer
    train_texts, test_texts, train_tok, test_tok, train_image_embs, test_image_embs, 
    train_text_embs, test_text_embs, 
    train_tok_mclip, test_tok_mclip = get_ds[params.ds](params, model, params.test_ds)
    if params.preprocess_only:
        exit()
    train_dataset_inputs = {'params': params, 'tok_data': train_tok, 'raw_data': train_texts, 
'clip_tok_data': train_tok_mclip}
    test_dataset_inputs = {'params': params, 'tok_data': test_tok, 'raw_data': test_texts, 
'clip_tok_data': test_tok_mclip}
    stage_message = {
                    'caption': 'Pretraining on Image Captions',
                    'text_recon': 'Pretraining on Text Reconstruction',
                    'translate': 'Pairwise Translation',
                    'triplet': 'Triplet Training'
                }
    if is_main_process():
        print(stage_message[params.stage])

    train_dataset_inputs['is_pretraining'] = params.stage in ['caption', 'text_recon']
```

```python
        test_dataset_inputs['is_pretraining'] = params.stage in ['caption', 'text_recon']

        train_dataset_inputs['mask_inputs'] = train_dataset_inputs['is_pretraining'] or params.
        noise_train
        test_dataset_inputs['mask_inputs'] = test_dataset_inputs['is_pretraining'] or params.
        noise_test

        if params.stage in ['text_recon', 'translate']:
            train_dataset_inputs['clip_embs'] = train_text_embs
            test_dataset_inputs['clip_embs'] = test_text_embs if not params.noise_
            test else test_image_embs

        elif params.stage in ['caption']:
            train_dataset_inputs['clip_embs'] = train_image_embs
            test_dataset_inputs['clip_embs'] = test_image_embs

        elif params.stage in ['triplet']:
            train_dataset_inputs['clip_embs'] = train_image_embs
            test_dataset_inputs['clip_embs'] = test_text_embs

        train_dataset = MultiModalDataset(**train_dataset_inputs)
        test_dataset = MultiModalDataset(**test_dataset_inputs)


        if not params.unfreeze_clip:
            del model.clip #[8]如果 CLIP 模型始终是冻结的，我们可以将其从内存中移除，因为所有数据都已预处理
        elif is_main_process():
            print('Also finetuning CLIP.')

        if params.num_gpus > 1:
            local_rank = int(os.environ['LOCAL_RANK'])
            model = nn.parallel.DistributedDataParallel(model, device_ids = [local_rank],
            output_device = [local_rank],
            find_unused_parameters=True)
        if params.num_gpus > 1:
            train_sampler = DistributedSampler(dataset=train_dataset, shuffle=True)
            test_sampler = DistributedSampler(dataset=test_dataset, shuffle=False)
            train_dl = DataLoader(train_dataset, batch_size = params.batch_size,
                        num_workers = 0, pin_memory = True, collate_fn = collate_multi,
                        sampler=train_sampler)
            test_dl = DataLoader(test_dataset, batch_size = 2*params.batch_size,
                        num_workers = 6, pin_memory = True, collate_fn = collate_multi,
                        sampler=test_sampler)
        else:
            train_dl = DataLoader(train_dataset, batch_size = params.batch_size, shuffle = True,
                        num_workers = 0, pin_memory = True, collate_fn = collate_multi)
            test_dl = DataLoader(test_dataset, batch_size = params.batch_size, shuffle = False,
                        num_workers = 0, pin_memory = True, collate_fn = collate_multi)
        if is_main_process():
            if is_dist_avail_and_initialized():
                print(model.module.adapter) if hasattr(model.module, 'adapter') elseprint('No
                adapter')
            else:
                print(model.adapter) if hasattr(model, 'adapter') else print('No adapter')
            print('%' * 80)
            print(params)
            print('%' * 80)
        runner = Runner(train_dl, test_dl, params)
        runner.train(model, tokenizer, params)

if __name__ == '__main__':
    parser = argparse.ArgumentParser()
    parser.add_argument('--mn', dest = 'model_name', type = str, default = '', help =
    'Name of the job')
```

```
parser.add_argument('--lm', dest = 'load_model', type = str, default = '', help =
'Name of model to be loaded')
parser.add_argument('--test', action = 'store_true', help = 'to run inference on a saved
model')
parser.add_argument('--ct', dest = 'continue_training', action = 'store_true',help =
'flag to continue training')
parser.add_argument('--bs', dest = 'batch_size', type = int, default = 128)
parser.add_argument('--lr', type = float, default = 1e-5)
parser.add_argument('--epochs', type = int, default = 15)
parser.add_argument('--model_dir', type = str, default = 'models')
parser.add_argument('--data_dir', type = str, default = 'data')
parser.add_argument('--src_lang', type = str, default = 'en')
parser.add_argument('--tgt_lang', type = str, default = 'de')
parser.add_argument('--update_count', type = int, default = 4, help = 'number of steps to
accumulate gradient before
backpropagating')
parser.add_argument('--local_rank', type = int, default = 0, help = "Don't modify,
will be used automatically for DDP")
parser.add_argument('--num_gpus', type = int, default = 1)
parser.add_argument('--test_after', type = int, default = -1)
parser.add_argument('--seed', type = int, default = 29)
parser.add_argument('--ds', type = str, choices = ['multi30k', 'wit', 'wmt','coco'],
default = 'multi30k')
parser.add_argument('--unfreeze_clip', action = 'store_true', help = 'used to also
finetune the CLIP text encoder in stage 2')
parser.add_argument('--test_ds', nargs = '+', type = str, default = ['2016','val'])
parser.add_argument('--mapping_network', type = str, default = 'mlp', choices = ['mlp',
'transformer', 'hidden_mlp'], help =
'Choice of mapping network, refer paper for details')
parser.add_argument('--image_encoder', type = str, default = 'mclip', choices =
['clip_res', 'mclip', 'clip'])
parser.add_argument('--prefix_length', type = int, default = 10)
parser.add_argument('--hidden_dims', type = int, default = 300)
parser.add_argument('--mask_prob', type = float, default = 1)
parser.add_argument('--stage', type = str, required = True, choices = ['caption',
'text_recon', 'translate', 'triplet'])
parser.add_argument('--use_all_langs', action = 'store_true', help = 'Multilingual
captioning in stage 1')
parser.add_argument('--noise_train', action = 'store_true', help = 'Remove mask_prob
% of the tokens while training')
parser.add_argument('--noise_test', action = 'store_true', help = 'Remove mask_prob%
of the tokens while testing')
parser.add_argument('--preprocess_only', action = 'store_true')
params = parser.parse_args()
assert not (params.stage in ['caption', 'text_recon'] and params.ds == 'wmt'),
'While using text-only NMT, you cannot train stage 1. Make sure you load a stage 1
pretrained model'
main(params)
```

对上述代码的具体说明如下。

（1）函数 init_seed（seed）用于初始化随机种子，以确保实验的可复现性。设置 Python、NumPy 和 PyTorch 的随机种子，并配置其他相关环境变量。

（2）主函数 main（params）负责整个训练或推理流程的控制。根据参数设置模型、数据集、优化器等，然后执行训练或推理过程。

第 9 章 基于多模态大模型的音视频广义零样本学习系统

本章将详细讲解实现一个基于多模态大模型的音视频广义零样本学习系统的过程。本项目基于多模态大模型（如 CLIP 和 WavCaps），旨在实现音视频数据的广义零样本学习（GZSL）系统。通过提取音频和视频特征，并结合预训练模型进行训练和评估，系统能够在不同的数据集（如 UCF、ActivityNet、VGGSound）上进行动作分类，即使面对从未见过的样本，也能准确识别其类别。该系统的功能包括数据准备、特征提取、模型训练、模型评估和结果展示，适用于广泛的音视频分析应用。

9.1 背景介绍

在当今的多媒体内容处理和理解中，音频和视频数据的应用变得越来越普遍，然而，在面对广泛的多媒体内容时，传统的机器学习系统往往需要大量标记数据进行训练，这在某些情况下可能不切实际或成本过高。为了解决这个问题，零样本学习（ZSL）被引入，允许模型在没有直接观察过的类别上进行学习和推理。

针对音频-视频广义零样本学习的需求，本项目提出了一种基于多模态大型预训练模型的解决方案。该系统利用了最新的多模态预训练模型（如 OpenAI 的 CLIP）以及音频-视觉嵌入模型（如 WavCaps），实现了在音频和视频数据上进行广义零样本学习的能力。

通过该系统，可以实现对各种多媒体内容进行理解和分类，即使在面对没有直接标记的类别时也能够做出准确的推理。这种技术在视频内容分析、音频内容理解以及跨模态场景下的智能应用等领域具有重要的应用前景。

9.2 系统分析

系统分析是指对一个系统进行全面深入的研究和评估，旨在全面理解系统的设计和运行原理，为系统的开发、优化和应用提供指导和支持。

9.2.1 系统需求分析

本项目旨在开发一个基于多模态大模型的音视频广义零样本学习系统，通过结合音频和视频信息，实现对多种动作的识别和分类，这些动作包括但不限于 FrisbeeCatch、HandstandWalking、PlayingDhol 等。

1. 用户需求
- 数据科学家/研究人员：需要一个功能强大、高效准确的系统，用于进行音视频数据的广义零样本学习研究。
- 开发人员：需要一个清晰的系统架构和模块化的代码，以便于维护、拓展和定制化。

- 系统管理员：需要一个易于部署和管理的系统，能够在不同环境下运行，并具备良好的性能和稳定性。

2. **功能性需求**
- 数据处理：能够处理音频和视频数据，提取特征并进行预处理。
- 模型训练：能够训练多模态模型，结合 CLIP 和 WavCaps 等模型，实现对动作的识别和分类。
- 模型评估：提供对模型性能的评估指标，包括广义零样本学习性能和零样本学习性能。
- 结果展示：能够展示模型训练和评估的结果，以便用户进行分析和比较。

3. **非功能性需求**
- 性能要求：系统需要具备较高的计算性能，能够处理大规模的音视频数据集。
- 可靠性：系统需要具备良好的稳定性和可靠性，能够在长时间运行中保持正常工作。
- 易用性：界面友好，操作简单，用户能够方便地使用系统进行模型训练和评估。
- 可扩展性：系统需要具备良好的扩展性，能够灵活添加新的模型、算法或数据集。

4. **安全性需求**
- 数据安全：对用户数据和模型训练过程中的敏感信息进行保护，防止未授权访问和数据泄露。
- 系统安全：防止系统受到恶意攻击和非法访问，确保系统的安全稳定运行。

5. **部署需求**
- 环境要求：能够在不同的操作系统和硬件环境下运行，包括 Linux、Windows 等。
- 依赖组件：需要安装和配置相关的软件和库，如 PyTorch、CLIP、WavCaps 等。

通过系统需求分析，可以清晰地了解项目的目标、功能、性能、安全性等方面的要求，为系统的设计和开发提供指导。

9.2.2 功能分析

根据前面的系统需求分析，得出本项目的功能如下。

1. **数据准备**
- 提供了用于准备数据的脚本，包括提取音频和视频特征、准备数据集等。
- 支持针对不同数据集（UCF、ActivityNet、VGGSound）进行数据准备。

2. **特征提取**
- 提供了特征提取的代码，包括从音频和视频中提取 CLIP 和 WavCaps 特征。
- 特征提取是为了准备数据以输入到多模态模型中进行训练和评估。

3. **模型训练**
- 实现了多模态模型的训练过程，结合了 CLIP 和 WavCaps 等预训练模型。
- 支持在不同数据集上进行模型训练，如 UCF、ActivityNet、VGGSound。

4. **模型评估**
- 提供了模型评估的代码，包括在训练后自动进行评估或手动评估。
- 支持评估不同阶段（stage-1、stage-2）的模型性能。

5. **结果展示**
- 可以展示模型对新数据的分类结果。
- 提供了评估指标，如 GZSL 和 ZSL 性能指标，以便用户进行分析和比较。

通过以上功能分析，可以实现对音视频数据进行动作分类的系统，支持在不同数据集上进行广义零样本学习任务的训练和评估。

9.3 系统配置

本项目的"系统配置"模块包括 args.py、dataset.py 和 utils.py 3 个文件，提供了项目中的配置、数据集处理和实用功能函数。其中文件 args.py 用于管理实验参数和配置，文件 dataset.py 实现了和数据集相关的处理和加载函数，文件 utils.py 则提供了各种实用功能函数，包括实验设置、模型加载和评估等功能。

9.3.1 命令行接口

文件 args.py 定义并解析了一个命令行接口，用于配置和运行一个解释性音视频低样本学习（Explainable Audio Visual Low Shot Learning）系统。它使用 configargparse 库来处理命令行参数和配置文件，支持各种配置选项，如运行模式、使用的模态（音频、视频或两者）、数据并行、特征提取方法、数据集路径和日志目录等。代码还包括一些布尔值参数的处理函数，并将解析后的参数分组，以便在后续模型训练和评估过程中使用。文件 args.py 的具体代码如下所示。

```
import configargparse
import pathlib

def args_main(*args, **kwargs):
    parser = configargparse.ArgParser(
        description="Explainable Audio Visual Low Shot Learning",
        default_config_files=["config/default.yaml"],
        config_file_parser_class=configargparse.YAMLConfigFileParser
    )
    parser.add_argument('-c', '--cfg', required=False, is_config_file=True, help=
    'config file path')
    parser.add_argument('--run', default='all', type=str, choices=['all', 'stage-1',
    'stage-2', 'eval'])
    parser.add_argument(
        "--modality",
        help="Wether to use embeddings  audio features, video features, or both combined",
        choices=["video", "audio", "both"],
        default='both',
        type=str,
        # required=True
    )
    parser.add_argument(
        "--data_parallel",
        help="Whether to use DataParallel or not",
        type=str_to_bool, nargs='?', const=True
    )
    parser.add_argument(
        "--use_wavcaps_embeddings",
        help="Whether to use text embeddings of wavcaps or not (are concatenated to
        the clip embeddings)",
        default=True,
        type=str_to_bool, nargs='?', const=True
    )
    parser.add_argument(
        "--word_embeddings",
        help="Whether to use text embeddings of wavcaps, clip, or both",
        choices=["clip", "wavcaps", "both"],
        default='both',
        type=str
```

```
)
parser.add_argument(
    "--best_model_criterion",
    help="Choose best model based on best loss or hm score",
    choices=['loss', 'score'],
    type=str
)
###文件系统 ###
parser.add_argument(
    "--root_dir",
    help="Path to dataset directory. Expected subfolder structure: '{root_dir}/
    features/{feature_extraction_method}/{audio,video,text}'",
    required=True,
    type=pathlib.Path
)

parser.add_argument(
    "--log_dir",
    help="Path where to create experiment log dirs",
    type=pathlib.Path
)

parser.add_argument(
    "--exp_name",
    help="Flag to set the name of the experiment",
    type=str
)

parser.add_argument(
    "--feature_extraction_method",
    help="Name of folder containing respective extracted features. Has to match
    {feature_extraction_method} in --root_dir argument.",
    required=False,
    type=pathlib.Path
)

parser.add_argument(
    "--dataset_name",
    help="Name of the dataset to use",
    choices=["AudioSetZSL", "VGGSound", "UCF", "ActivityNet"],
    type=str
)

parser.add_argument(
    "--selavi",
    help="Wether to use selavi features or cls features",
    type=str_to_bool, nargs='?', const=True
)

parser.add_argument(
    "--zero_shot_split",
    help="Name of zero shot split to use.",
    choices=["", "cls_split", "main_split"],
)

parser.add_argument(
    "--manual_text_word2vec",
    help="Flag to use the manual word2vec text embeddings. CARE: Need to create
    cache files again!",
    type=str_to_bool, nargs='?', const=True
)

parser.add_argument(
```

```
        "--val_all_loss",
        help="Validate loss with seen + unseen",
        type=str_to_bool, nargs='?', const=True
    )

    parser.add_argument(
        "--additional_triplets_loss",
        help="Flag for using more triplets loss",
        type=str_to_bool, nargs='?', const=True
    )

    parser.add_argument(
        "--reg_loss",
        help="Flag for setting the regularization loss",
        type=str_to_bool, nargs='?', const=True

    )

    parser.add_argument(
        "--cycle_loss",
        help="Flag for using cycle loss",
        type=str_to_bool, nargs='?', const=True
    )
......
    eval_group.add_argument(
        "--eval_save_performances",
        help="Save class performances to disk",
        type=str_to_bool, nargs='?', const=True
    )
    args = parser.parse_args(*args, **kwargs)
    arg_groups={}
    for group in parser._action_groups:
        group_dict={a.dest:getattr(args,a.dest,None) for a in group._group_actions}
        arg_groups[group.title]=configargparse.Namespace(**group_dict)

    model_args = arg_groups['model']
    eval_args = arg_groups['eval']
    shared_args_list = ['root_dir', 'dataset_name', 'device', 'batch_seqlen_test',
    'batch_seqlen_test_maxlen', 'batch_seqlen_test_trim']
    shared_args_dict = {a:getattr(args,a,None) for a in shared_args_list}
    eval_main_args = configargparse.Namespace(**shared_args_dict, **vars(eval_args),
    **vars(model_args))
    return args, eval_main_args

def str_to_bool(value):
    if isinstance(value, bool):
        return value
    if value.lower() in {'false', 'f', '0', 'no', 'n'}:
        return False
    elif value.lower() in {'true', 't', '1', 'yes', 'y'}:
        return True
    raise ValueError(f'{value} is not a valid boolean value')
```

9.3.2 数据集处理

文件 dataset.py 实现了多个数据集类和辅助功能类，包括 VGGSoundDataset、AudioSetZSLDataset、ContrastiveDataset、DefaultCollator、UCFDataset 和 ActivityNetDataset。这些类用于加载、预处理和管理音频、视频和文本数据集，支持零样本学习模式以及对比学习任务。其中 UCFDataset 和 ActivityNetDataset 类处理 UCF 数据集和 ActivityNet 数据集，提供了动态获取数据、预处理和分割数据的功能；AudioSetZSLDataset 类用于处理 AudioSetZSL 数据集，支持训练、验证和测试数据的零样本学习模式；ContrastiveDataset 类用于生成对比学习任务所需的正样本对和负样本对；DefaultCollator 类用于处理批量数据，确保数据具有一致的长度并准备用于对比学习任务。这些类的设计使对多模态数据集进行处理和准备变得简单而灵活。

文件 dataset.py 的具体实现流程如下所示。

（1）定义一个名为 VGGSoundDataset 的数据集类，用于处理 VGGSound 数据集中的音频、视频和文本数据。通过各种属性和方法，VGGSoundDataset 类可以加载和预处理数据集，将数据按不同模态和零样本学习模式进行分割，并提供训练、验证和测试所需的数据。该类还支持从磁盘读取处理后的特征文件，并根据需要创建这些文件，以加快训练过程。

```python
class VGGSoundDataset(data.Dataset):
    @property
    def training_file(self):
        return self.features_processed_folder / self.feature_extraction_method / f"training{self.zero_shot_split}.pkl"

    @property
    def val_file(self):
        return self.features_processed_folder / self.feature_extraction_method / f"val{self.zero_shot_split}.pkl"

    @property
    def train_val_file(self):
        return self.features_processed_folder / self.feature_extraction_method / f"train_val{self.zero_shot_split}.pkl"

    @property
    def test_file(self):
        return self.features_processed_folder / self.feature_extraction_method / f"test{self.zero_shot_split}.pkl"

    @property
    def targets(self):
        classes_mask = np.where(np.isin(self.data["audio"]["target"], self.classes))[0]
        return self.data["audio"]["target"][classes_mask]

    @property
    def all_data(self):
        classes_mask = np.where(np.isin(self.data["audio"]["target"], self.classes))[0]
        return {
            "audio": np.array(self.data["audio"]["data"])[classes_mask],
            "video": np.array(self.data["video"]["data"])[classes_mask],
            "text": np.array(self.data["text"]["data"])[sorted(self.classes.astype(int))],
            "target": np.array(self.data["audio"]["target"])[classes_mask],
            "url": np.array(self.data["audio"]["url"])[classes_mask],
            "fps": np.array(self.data['audio']['fps'])[classes_mask]
        }

    @property
    def map_embeddings_target(self):
        # w2v_embedding = torch.Tensor(self.data["text"]["data"])[sorted(self.classes)].cuda() # orig
        w2v_embedding = torch.Tensor(self.data["text"]["data"])[sorted(self.classes)].to(self.args.device)
        sorted_classes = sorted(self.classes)
        mapping_dict = {}
        for i in range(len(sorted_classes)):
            mapping_dict[int(sorted_classes[i])] = i
        return w2v_embedding, mapping_dict
```

```python
    @property
    def features_processed_folder(self):
        # return Path().cwd() / "avgzsl_benchmark_non_averaged_datasets/VGGSound/
        _features_processed"
        return self.root / "_features_processed"

    @property
    def all_class_ids(self):
        return np.asarray([self.class_to_idx[name] for name in self.all_class_names])

    @property
    def train_train_ids(self):
        return np.asarray([self.class_to_idx[name] for name in self.train_train_class_names])

    @property
    def val_seen_ids(self):
        return np.asarray([self.class_to_idx[name] for name in self.val_seen_class_names])

    @property
    def val_unseen_ids(self):
        return np.asarray([self.class_to_idx[name] for name in self.val_unseen_class_names])

    @property
    def test_train_ids(self):
        return np.asarray([self.class_to_idx[name] for name in self.test_train_class_names])

    @property
    def test_seen_ids(self):
        return np.asarray([self.class_to_idx[name] for name in self.test_seen_class_names])

    @property
    def test_unseen_ids(self):
        return np.asarray([self.class_to_idx[name] for name in self.
        test_unseen_class_names])

    @property
    def text_label_mapping(self):
        df = pd.read_csv(self.root / "class-split/vggsound_w2v_class_names.csv")
        return {val: df.original[idx] for idx, val in enumerate(df.manual)}

    @property
    def classes(self):
        if self.zero_shot_split:
            return np.sort(np.concatenate((self.seen_class_ids, self.unseen_class_ids)))

        else:
            if self.zero_shot_mode == "all":
                return self.all_class_ids
            elif self.zero_shot_mode == "seen":
                return self.seen_class_ids
            elif self.zero_shot_mode == "unseen":
                return self.unseen_class_ids
            else:
                raise AttributeError(f"Zero shot mode has to be either all, seen or
                unseen. Is {self.zero_shot_mode}")

    @property
    def class_to_idx(self):
        return {_class: i for i, _class in enumerate(sorted(self.all_class_names))}

    @property
    def all_class_names(self):
        return get_class_names(self.root / "class-split/all_class.txt")
```

```python
    @property
    def seen_class_names(self):
        if self.dataset_split == "train":
            return self.train_train_class_names
        elif self.dataset_split == "val":
            return self.val_seen_class_names
        elif self.dataset_split == "train_val":
            return np.concatenate((self.train_train_class_names, self.
            val_unseen_class_names))
        elif self.dataset_split == "test":
            return self.test_seen_class_names
        else:
            raise AttributeError("Dataset split has to be in {train,val,train_val,test}")

    @property
    def unseen_class_names(self):
        if self.dataset_split == "train":
            return np.array([])
        elif self.dataset_split == "val":
            return self.val_unseen_class_names
        elif self.dataset_split == "train_val":
            return np.array([])
        elif self.dataset_split == "test":
            return self.test_unseen_class_names
        else:
            raise AttributeError("Dataset split has to be in {train,val,train_val,test}")

    @property
    # @lru_cache(maxsize=128)
    def seen_class_ids(self):
        return np.asarray([self.class_to_idx[name] for name in self.seen_class_names])

    @property
    # @lru_cache(maxsize=128)
    def unseen_class_ids(self):
        return np.asarray([self.class_to_idx[name] for name in self.unseen_class_names])

    @property
    def train_train_class_names(self):
        return get_class_names(self.root / f"class-split/{self.zero_shot_split}/
        stage_1_train.txt")

    @property
    def val_seen_class_names(self):
        return get_class_names(self.root / f"class-split/{self.zero_shot_split}/
        stage_1_val_seen.txt")

    @property
    def val_unseen_class_names(self):
        return get_class_names(self.root / f"class-split/{self.zero_shot_split}/
        stage_1_val_unseen.txt")

    @property
    def test_train_class_names(self):
        return get_class_names(self.root / f"class-split/{self.zero_shot_split}/
        stage_2_train.txt")

    @property
    def test_seen_class_names(self):
        return get_class_names(self.root / f"class-split/{self.zero_shot_split}/
        stage_2_test_seen.txt")
```

```python
    @property
    def test_unseen_class_names(self):
        return get_class_names(self.root / f"class-split/{self.zero_shot_split}/"
        stage_2_test_unseen.txt")

    def __init__(self, args, dataset_split, zero_shot_mode, download=False, transform=None):
        super(VGGSoundDataset, self).__init__()
        self.logger = logging.getLogger()
        self.logger.info(
            f"Initializing Dataset {self.__class__.__name__}\t"
            f"Dataset split: {dataset_split}\t"
            f"Zero shot mode: {zero_shot_mode}")
        self.args = args
        self.root = args.root_dir
        self.dataset_name = args.dataset_name
        self.feature_extraction_method = args.feature_extraction_method
        self.dataset_split = dataset_split
        self.zero_shot_mode = zero_shot_mode
        self.zero_shot_split = args.zero_shot_split

        self.transform = transform

        self.preprocess()
        self.data = self.get_data()

    def __getitem__(self, item):
        raise NotImplementedError()

    def __len__(self):
        return len(self.data)

    def _check_exists(self):
        return self.training_file.exists() and self.val_file.exists() and self.test_file
        .exists() and self.train_val_file.exists()

    def preprocess(self):
        if self._check_exists():
            return

        (self.features_processed_folder / self.feature_extraction_method).mkdir(parents=
        True, exist_ok=True)

        self.logger.info('Processing extracted features for faster training (only done
        once)...')
        self.logger.info(
            f"Processed files will be stored locally in {(self.features_processed_folder
            / self.feature_extraction_method).resolve()}"
        )

        training_set = self.read_dataset(dataset_type="train")
        val_set = self.read_dataset(dataset_type="val")
        train_val_set = self.read_dataset(dataset_type="train_val")
        test_set = self.read_dataset(dataset_type="test")

        with self.training_file.open('wb') as f:
            self.logger.info(f"Dumping to {self.training_file}")
            pickle.dump(training_set, f, pickle.HIGHEST_PROTOCOL)
        with self.val_file.open('wb') as f:
            self.logger.info(f"Dumping to {self.val_file}")
            pickle.dump(val_set, f, pickle.HIGHEST_PROTOCOL)
        with self.train_val_file.open('wb') as f:
            self.logger.info(f"Dumping to {self.train_val_file}")
            pickle.dump(train_val_set, f, pickle.HIGHEST_PROTOCOL)
```

```python
        with self.test_file.open('wb') as f:
            self.logger.info(f"Dumping to {self.test_file}")
            pickle.dump(test_set, f, pickle.HIGHEST_PROTOCOL)

    if not self._check_exists():
        raise RuntimeError("Dataset not found after preprocessing!")
    self.logger.info("Successfully finished preprocessing.")

def get_data(self):
    if self.dataset_split == "train":
        data_file = self.training_file
    elif self.dataset_split == "val":
        data_file = self.val_file
    elif self.dataset_split == "train_val":
        data_file = self.train_val_file
    elif self.dataset_split == "test":
        data_file = self.test_file
    else:
        raise AttributeError("Dataset_split has to be either train, val or test.")

    load_path = (self.features_processed_folder / data_file).resolve()
    self.logger.info(f"Loading processed data from disk from {load_path}")
    with load_path.open('rb') as f:
        return pickle.load(f)

def read_dataset(self, dataset_type):
    result_audio = self.get_data_by_modality(modality="audio", dataset_type=dataset_type)
    result_video = self.get_data_by_modality(modality="video", dataset_type=dataset_type)
    assert torch.equal(result_audio["target"], result_video["target"])
    assert np.array_equal(result_audio["url"], result_video["url"])
    result_text = self.get_data_by_modality(modality="text", dataset_type=dataset_type)
    return {"audio": result_audio, "video": result_video, "text": result_text}

def get_data_by_modality(self, modality, dataset_type="train"):
    result = {"data": [], "target": [], "url": [], "fps": []}
    if modality == "text":
        data_raw = np.load(
            (
                    self.root / "features" / self.feature_extraction_method / "text/word_embeddings_vggsound_normed.npy").resolve(),allow_pickle=True).item()
        data_raw_sorted = dict(sorted(data_raw.items()))

        if self.args.use_wavcaps_embeddings == True:
            data_raw_wavcaps = np.load(
                (
                        self.root / "features" / self.feature_extraction_method / "text/wavcaps_word_embeddings_vggsound_normed.npy").resolve(),allow_pickle=True).item()
            data_raw_wavcaps_sorted = dict(sorted(data_raw_wavcaps.items()))
            data_raw_list = list(data_raw_sorted.values())
            data_raw_wavcaps_list = list(data_raw_wavcaps_sorted.values())

            data_list = [np.concatenate((data_raw_list[i], data_raw_wavcaps_list[i])
            , axis=0) for i in range(len(data_raw_list))]

            result['data'] = data_list

        else:
            result["data"] = list(data_raw_sorted.values())
        result["target"] = [self.class_to_idx[self.text_label_mapping[key].lower()]
for key in list(data_raw_sorted.keys())]
```

```python
        elif modality == "audio" or modality == "video":
            split_names = []
            if dataset_type == "train":
                split_names.append("stage_1_train")
            elif dataset_type == "val":
                split_names.append("stage_1_val_seen")
                split_names.append("stage_1_val_unseen")
            elif dataset_type == "train_val":
                split_names.append("stage_1_train")
                split_names.append("stage_1_val_seen")
                split_names.append("stage_1_val_unseen")
            elif dataset_type == "test":
                split_names.append("stage_2_test_seen")
                split_names.append("stage_2_test_unseen")
            else:
                raise AttributeError("Dataset type incompatible. Has to be either train, 
                val or test.")
            
            for split_name in split_names:
                modality_path = (
                        self.root / "features" / self.feature_extraction_method /
                        f"{modality}/{split_name}").resolve()
                files = modality_path.iterdir()
                for file in tqdm(files, total=len(list(modality_path.glob('*'))),
                                 desc=f"{dataset_type}:{modality}:{split_name}"):
                    data, url, fps = read_features(file)
                    #assert len(data[
                    #0]) == self.args.input_size, f"Feature size {len(data[0])} is not 
                    #compatible with specified --input_size {self.args.input_size}"
                    for i, d in enumerate(data):
                        result["data"].append(d)
                        result["target"].append(self.class_to_idx[file.stem])
                        result["url"].append(url[i])
                        result["fps"].append(fps[i])
    else:
        raise AttributeError("Modality has to be either audio, video or text")
    result["data"] = result["data"]
    result["target"] = torch.LongTensor(result["target"])
    result["url"] = np.array(result["url"])
    return result
```

（2）定义一个名为 AudioSetZSLDataset 的数据集类，用于处理 AudioSetZSL 数据集。类 AudioSetZSLDataset 继承自 data.Dataset 类，并提供属性和方法来加载、预处理和访问音频、视频和文本模态的数据。该类通过读取预处理后的特征文件，加快训练过程，并支持对训练、验证和测试数据进行零样本学习模式的划分。该类还实现了数据下载、预处理和从磁盘加载处理后数据的功能。

（3）定义一个名为 ContrastiveDataset 的数据集类，用于对比学习任务。类 ContrastiveDataset 基于 AudioSetZSLDataset 数据集生成对比样本，包含正样本和负样本对。在训练和验证阶段，正样本对由同一类别的数据构成，而负样本对则包含不同类别的数据。对于训练集，负样本是随机生成的，而验证和测试集则使用预先计算的样本对。这种设计有助于模型在训练过程中学习到不同类别之间的差异，从而提高分类性能。

（4）定义一个名为 DefaultCollator 的类，用于处理批量数据的整理和处理，特别是为对比学习任务准备数据。在类 DefaultCollator 中，音频和视频数据会根据指定的模式（最大长度或固定长度）进行填充或截断，以确保所有样本在同一个批次中具有相同的长度。类 DefaultCollator 的主要功能如下。

- 初始化：__init__ 方法初始化了 DefaultCollator 类的实例，设置了模式（mode）、最大长度（max_len）、截断方式（trim）、视频和音频的速率。
- 获取最大序列长度：get_max_seq_len 方法遍历一批数据，找到正样本和负样本中音频和视频数据的最大长度，用于后续的填充或截断。
- 处理批量数据：__call__ 方法接收一个批次的数据，并将数据整理成模型所需的格式，根据模式决定使用固定长度还是最大长度。初始化填充掩码和时间步长数组。遍历每个样本，按照需要进行填充或截断。最终将处理后的数据转换成张量，并整理成 data_final 和 target_final 返回。
- 填充和截断逻辑：根据音频和视频数据的长度与目标长度的差异，决定是否进行填充（用零填充）或截断（根据随机或居中截断）。

通过这种方式，DefaultCollator 类能够确保每个批次中的数据具有一致的长度，并且保留了原始音频和视频的时间步长信息，方便模型处理对比学习任务中的多模态数据。

（5）定义一个名为 UCFDataset 的类，用于处理和管理 UCF 数据集。类 UCFDataset 主要负责数据的加载、预处理和分割操作，并提供对不同类别数据和特征的访问。UCFDataset 包含了多种属性和方法，以便根据指定的零样本学习模式和数据集分割（训练、验证、测试）来动态获取和处理音频、视频和文本数据。它支持从预处理后的特征文件中读取数据，并确保所有特征和目标标签的一致性，以便于训练深度学习模型。

（6）定义一个名为 ActivityNetDataset 的类，用于处理和管理 ActivityNet 数据集。类 ActivityNetDataset 主要负责数据的预处理、加载和分割等操作，支持零样本学习模式。ActivityNetDataset 可以根据不同的数据集分割（训练、验证、测试）动态获取音频、视频和文本数据。类中包含多个属性和方法，用于读取预处理后的特征文件，并确保所有特征和目标标签的一致性，以便于训练深度学习模型。该类还可以根据不同的模式（如全体、已见、未见）提供相应类别的数据。

9.3.3 辅助函数

文件 utils.py 提供了用于实验设置、模型加载、评估和可视化等操作的辅助函数，包括读取特征数据、设置随机种子、保存实验配置、记录实验参数和指标到 TensorBoard、保存最佳模型、加载模型参数、评估数据集、计算最佳评估指标、获取类别名称、加载模型权重、绘制柱状图以及保存类别性能等功能。文件 utils.py 的具体实现流程如下所示。

（1）函数 read_features 的功能是从文件中读取特征数据，并返回特征数据、视频名称列表和帧率。

```
def read_features(path):
    with open(path, 'rb') as f:
        x = pickle.load(f)
    data = x['features']
    fps=x['fps']
    url = [str(u) for u in list(x['video_names'])]
    return data, url, fps
```

（2）函数 fix_seeds 的功能是设置随机种子，确保实验可复现性。

```
def fix_seeds(seed=42):

    np.random.seed(seed)
    os.environ["PYTHONHASHSEED"] = str(seed)
```

```
np.random.seed(seed)
torch.manual_seed(seed)
torch.cuda.manual_seed(seed)
torch.cuda.manual_seed_all(seed)
torch.backends.cudnn.benchmark = False
torch.backends.cudnn.deterministic = True
```

（3）函数 print_model_size 的功能是打印模型参数数量，用于评估模型复杂度。

```
def print_model_size(model, logger):
    num_params = 0
    for param in model.parameters():
        if param.requires_grad:
            num_params += param.numel()
    logger.info(
        "Created network [%s] with total number of parameters: %.1f million."
        % (type(model).__name__, num_params / 1000000)
    )
```

（4）函数 get_git_revision_hash 的功能是获取当前 Git 代码库的提交哈希值。

```
def get_git_revision_hash():
    try:
        hash_string = (
            subprocess.check_output(["git", "rev-parse", "HEAD"])
            .decode("ascii")
            .strip()
        )
    except:
        hash_string = ""
    return hash_string
```

（5）函数 dump_config_yaml 的功能是将实验配置保存为 YAML 格式的文件。

```
def dump_config_yaml(args, exp_dir):
    args_dict = deepcopy(vars(args))
    for k,v in args_dict.items():
        if isinstance(v, pathlib.PosixPath):
            args_dict[k] = v.as_posix()
    with open((exp_dir/"args.yaml"), "w") as f:
        yaml.safe_dump(args_dict, f)
```

（6）函数 log_hparams 的功能是将实验参数和指标记录到 TensorBoard 中。

```
def log_hparams(writer, args, metrics):
    args_dict = vars(args)
    for k,v in args_dict.items():
        if isinstance(v, pathlib.PosixPath):
            args_dict[k] = v.as_posix()
    del metrics["recall"]
    metrics = {"Eval/"+k: v for k,v in metrics.items()}
    del args_dict['audio_hip_blocks']
    del args_dict['video_hip_blocks']
    writer.add_hparams(args_dict, metrics)
```

（7）函数 setup_experiment 的功能是设置实验，创建实验目录并保存参数、日志和模型检查点。

```
def setup_experiment(args, *stats):
    if args.exp_name == "":
        exp_name = f"{datetime.now().strftime('%b%d_%H-%M-%S_%f')}_{socket.gethostname()}"
    else:
```

```
        exp_name = str(args.exp_name) + f"_{datetime.now().strftime('%b%d_%H-%M-%S_%f')}
            _{socket.gethostname()}"

    exp_dir = (args.log_dir / exp_name)
    exp_dir.mkdir(parents=True)

    (exp_dir / "checkpoints").mkdir()
    pickle.dump(args, (exp_dir / "args.pkl").open("wb"))

    dump_config_yaml(args, exp_dir)
    train_stats = PD_Stats(exp_dir / "train_stats.pkl", stats)
    val_stats = PD_Stats(exp_dir / "val_stats.pkl", stats)
    logger = create_logger(exp_dir / "train.log")
    logger.info(f"Start experiment {exp_name}")
    logger.info(
        "\n".join(f"{k}: {str(v)}" for k, v in sorted(dict(vars(args)).items()))
    )
    logger.info(f"The experiment will be stored in {exp_dir.resolve()}\n")
    logger.info("")
    writer = SummaryWriter(log_dir=exp_dir)
    return logger, exp_dir, writer, train_stats, val_stats
```

（8）函数 setup_evaluation 的功能是设置评估参数、加载评估模型并创建评估目录。

```
def setup_evaluation(args, *stats):

    eval_dir = args.load_path_stage_B
    assert eval_dir.exists()

    test_stats = PD_Stats(eval_dir / "test_stats.pkl", ['seen', 'unseen', 'hm', 'zsl'])
    logger = create_logger(eval_dir / "eval.log")

    logger.info(f"Start evaluation {eval_dir}")
    logger.info(
        "\n".join(f"{k}: {str(v)}" for k, v in sorted(dict(vars(args)).items()))
    )
    logger.info(f"Loaded configuration {args.load_path_stage_B / 'args.pkl'}")
    logger.info(
        "\n".join(f"{k}: {str(v)}" for k, v in sorted(dict(vars(load_args(args.load_path
            _stage_B))).items()))
    )
    logger.info(f"The evaluation will be stored in {eval_dir.resolve()}\n")
    logger.info("")
    # for Tensorboard hparam logging
    writer = SummaryWriter(log_dir=eval_dir)
    return logger, eval_dir, test_stats, writer
```

（9）函数 setup_visualizations 的功能是设置可视化参数、加载模型并创建可视化目录。

```
def setup_visualizations(args, *stats):

    eval_dir = args.load_path_stage_B
    assert eval_dir.exists()

    # test_stats = PD_Stats(eval_dir / "test_stats.pkl", list(sorted(stats)))
    test_stats = PD_Stats(eval_dir / "test_stats.pkl", ['seen', 'unseen', 'hm', 'zsl'])
    logger = create_logger(eval_dir / "visuals.log")

    logger.info(f"Start visualizing {eval_dir}")
    logger.info(
        "\n".join(f"{k}: {str(v)}" for k, v in sorted(dict(vars(args)).items()))
    )
    logger.info(f"Loaded configuration {args.load_path_stage_B / 'args.pkl'}")
```

```python
logger.info(
    "\n".join(f"{k}: {str(v)}" for k, v in sorted(dict(vars(load_args(args.load_path
    _stage_B))).items())))
)
logger.info(f"The evaluation will be stored in {eval_dir.resolve()}\n")
logger.info("")
# for Tensorboard hparam logging
writer = SummaryWriter(log_dir=eval_dir)

return logger, eval_dir, test_stats, writer
```

（10）函数 save_best_model 的功能是保存最佳模型的参数和优化器状态，并记录在日志中。

```python
def save_best_model(epoch, best_metric, model, optimizer, log_dir, args, metric="",
checkpoint=False):
    logger = logging.getLogger()
    logger.info(f"Saving model to {log_dir} with {metric} = {best_metric:.4f}")
    if optimizer is None:
        optimizer = model.optimizer_gen
    save_dict = {
        "epoch": epoch + 1,
        "model": model.state_dict() if args.data_parallel == False else model.module.
        state_dict(),
        "optimizer": optimizer.state_dict(),
        "metric": metric
    }
    if checkpoint:
        torch.save(
            save_dict,
            log_dir / f"{model.__class__.__name__}_{metric}_ckpt_{epoch}.pt"
        )
    else:
        torch.save(
            save_dict,
            log_dir / f"{model.__class__.__name__}_{metric}.pt"
        )
```

（11）函数 check_best_loss 的功能是检查并保存最佳损失值的模型。

```python
def check_best_loss(epoch, best_loss, best_epoch, val_loss, model, optimizer, log_dir, args):
    if not best_loss:
        save_best_model(epoch, val_loss, model, optimizer, log_dir, args, metric="loss")
        return val_loss, epoch
    if val_loss < best_loss:
        best_loss = val_loss
        best_epoch = epoch
        save_best_model(epoch, best_loss, model, optimizer, log_dir, args, metric="loss")
    return best_loss, best_epoch
```

（12）函数 check_best_score 的功能是检查并保存最佳评分的模型。

```python
def check_best_score(epoch, best_score, best_epoch, hm_score, model, optimizer, log_dir,
args):
    if not best_score:
        save_best_model(epoch, hm_score, model, optimizer, log_dir, args, metric="score")
        return hm_score, epoch
    if hm_score > best_score:
        best_score = hm_score
        best_epoch = epoch
        save_best_model(epoch, best_score, model, optimizer, log_dir, args, metric="score")
    return best_score, best_epoch
```

（13）函数 load_model_parameters 的功能是加载模型的参数。

```python
def load_model_parameters(model, model_weights):
    logger = logging.getLogger()
    loaded_state = model_weights
    self_state = model.state_dict()
    for name, param in loaded_state.items():
        param = param
        if 'module.' in name:
            name = name.replace('module.', '')
        if name in self_state.keys():
            self_state[name].copy_(param)
        else:
            logger.info("didnt load ", name)
```

（14）函数 cos_dist 的功能是计算两个张量之间的余弦相似度。

```python
def cos_dist(a, b):
    # https://stackoverflow.com/questions/50411191/how-to-compute-the-cosine-similarity-
    #in-pytorch-for-all-rows-in-a-matrix-with-re
    a_norm = a / a.norm(dim=1)[:, None]
    b_norm = b / b.norm(dim=1)[:, None]
    res = torch.mm(a_norm, b_norm.transpose(0, 1))
    return res
```

（15）函数 evaluate_dataset_baseline 的功能是实现基线模型在数据集上的评估。

```python
def evaluate_dataset_baseline(dataset_tuple, model, device, distance_fn, best_beta=None,
                              new_model_sequence=False,
                              args=None, save_performances=False):
    dataset=dataset_tuple[0]
    data_loader=dataset_tuple[1]
    data_t = torch.tensor(dataset.all_data['text']).to(device)
    accumulated_audio_emb=[]
    accumulated_video_emb=[]
    accumulated_data_num=[]
    for batch_idx, (data, target) in tqdm(enumerate(data_loader)):
        data_a = data['positive']["audio"].to(device)
        data_v = data['positive']["video"].to(device)
        data_num = target["positive"].to(device)
        masks = {}
        masks['positive'] = {'audio': data['positive']['audio_mask'], 'video': data[
        'positive']['video_mask']}
        timesteps = {}
        timesteps['positive'] = {'audio': data['positive']['timestep']['audio'], 'video':
        data['positive']['timestep']['video']}

        all_data = (
            data_a, data_v, data_num, data_t, masks['positive'], timesteps['positive']
        )
        try:
            if args.z_score_inputs:
                all_data = tuple([(x - torch.mean(x)) / torch.sqrt(torch.var(x)) for x in
                all_data])
        except AttributeError:
            print("Namespace has no fitting attribute. Continuing")

        model.eval()
        with torch.no_grad():
            audio_emb, video_emb, emb_cls = model.get_embeddings(all_data[0],all_data[1],
            all_data[3], all_data[4], all_data[5])
            accumulated_audio_emb.append(audio_emb)
            accumulated_video_emb.append(video_emb)
```

```
            outputs_all = (audio_emb, video_emb, emb_cls)
        accumulated_data_num.append(data_num)

    stacked_audio_emb=torch.cat(accumulated_audio_emb)
    stacked_video_emb=torch.cat(accumulated_video_emb)
    data_num=torch.cat(accumulated_data_num)
    emb_cls=outputs_all[2]
    outputs_all=(stacked_audio_emb, stacked_video_emb, emb_cls)
    a_p, v_p, t_p = outputs_all
    # a_p = None
    video_evaluation = get_best_evaluation(dataset, data_num, a_p, v_p, t_p, mode="video
    ", device=device,distance_fn=distance_fn, best_beta=best_beta,
    save_performances=save_performances,args=args)
    return {
        "audio": video_evaluation,
        "video": video_evaluation,
        "both": video_evaluation
    }
```

（16）函数 get_best_evaluation 的功能是计算在给定 beta 值下的最佳评估指标。

```
def get_best_evaluation(dataset, targets, a_p, v_p, t_p, mode, device, distance_fn, best_beta=
None, save_performances=False, args=None, attention_weights=None):
    seen_scores = []
    zsl_scores = []
    unseen_scores = []
    hm_scores = []
    per_class_recalls = []
    start = 0 # 0
    end = 5 # 3
    steps = (end - start) * 15 + 1  # steps = (end - start) * 5 + 1
    betas = torch.tensor([best_beta], dtype=torch.float, device=device) if best_beta
    else torch.linspace(start, end, steps,device=device)
    seen_label_array = torch.tensor(dataset.seen_class_ids, dtype=torch.long, device=device)
    unseen_label_array = torch.tensor(dataset.unseen_class_ids, dtype=torch.long, device
    =device)
    seen_unseen_array = torch.tensor(np.sort(np.concatenate((dataset.seen_class_ids,
    dataset.unseen_class_ids))), dtype=torch.long, device=device)
    classes_embeddings = t_p
    with torch.no_grad():
        for beta in betas:
            if a_p == None:
                distance_mat = torch.zeros((v_p.shape[0], len(dataset.all_class_ids)),
                dtype=torch.float,device=device) + 99999999999999
                distance_mat_zsl = torch.zeros((v_p.shape[0], len(dataset.all_class_ids)
                ), dtype=torch.float,device=device) + 99999999999999
            else:
                distance_mat = torch.zeros((a_p.shape[0], len(dataset.all_class_ids)),
                dtype=torch.float,device=device) + 99999999999999
                distance_mat_zsl = torch.zeros((a_p.shape[0], len(dataset.all_class_ids)
                ), dtype=torch.float,device=device) + 99999999999999
            if mode == "audio":
                distance_mat[:, seen_unseen_array] = torch.cdist(a_p, classes_embeddings
                )  # .pow(2)
                mask = torch.zeros(len(dataset.all_class_ids), dtype=torch.long, device=device)
                mask[seen_label_array] = 99999999999999
                distance_mat_zsl = distance_mat + mask
                if distance_fn == "SquaredL2Loss":
                    distance_mat[:, seen_unseen_array] = distance_mat[:,
                    seen_unseen_array].pow(2)
                    distance_mat_zsl[:, unseen_label_array] = distance_mat_zsl[:,
                    unseen_label_array].pow(2)
            elif mode == "video":
```

```python
            # L2
            v_p = v_p.type(torch.float32)
            distance_mat[:, seen_unseen_array] = torch.cdist(v_p, classes_embeddings
            )  # .pow(2)
            mask = torch.zeros(len(dataset.all_class_ids),dtype=torch.long,device=device)
            mask[seen_label_array] = 99999999999999
            distance_mat_zsl = distance_mat + mask
            if distance_fn == "SquaredL2Loss":
                distance_mat[:, seen_unseen_array] = distance_mat[:,
                seen_unseen_array].pow(2)
                distance_mat_zsl[:, unseen_label_array] = distance_mat_zsl[:,
                unseen_label_array].pow(2)
        elif mode == "both":
            # L2
            audio_distance = torch.cdist(a_p, classes_embeddings, p=2)  # .pow(2)
            video_distance = torch.cdist(v_p, classes_embeddings, p=2)  # .pow(2)

            if distance_fn == "SquaredL2Loss":
                audio_distance = audio_distance.pow(2)
                video_distance = video_distance.pow(2)
            # Sum
            distance_mat[:, seen_unseen_array] = (audio_distance + video_distance)

            mask = torch.zeros(len(dataset.all_class_ids), dtype=torch.long, device=
            device)
            mask[seen_label_array] = 99999999999999
            distance_mat_zsl = distance_mat + mask

    mask = torch.zeros(len(dataset.all_class_ids), dtype=torch.long, device=
    device) + beta
    mask[unseen_label_array] = 0
    neighbor_batch = torch.argmin(distance_mat + mask, dim=1)
    match_idx = neighbor_batch.eq(targets.int()).nonzero().flatten()
    match_counts = torch.bincount(neighbor_batch[match_idx], minlength=len
    (dataset.all_class_ids))[
        seen_unseen_array]
    target_counts = torch.bincount(targets, minlength=len(dataset.all_class_ids)
    )[seen_unseen_array]
    per_class_recall = torch.zeros(len(dataset.all_class_ids), dtype=torch.float
    , device=device)
    per_class_recall[seen_unseen_array] = match_counts / target_counts
    seen_recall_dict = per_class_recall[seen_label_array]
    unseen_recall_dict = per_class_recall[unseen_label_array]
    s = seen_recall_dict.mean()
    u = unseen_recall_dict.mean()

    if save_performances:
        seen_dict = {k: v for k, v in zip(np.array(dataset.all_class_names)
        [seen_label_array.cpu().numpy()], seen_recall_dict.cpu().numpy())}
        unseen_dict = {k: v for k, v in zip(np.array(dataset.all_class_names)
        [unseen_label_array.cpu().numpy()], unseen_recall_dict.cpu().numpy())}
        save_class_performances(seen_dict, unseen_dict, dataset.dataset_name)

    hm = (2 * u * s) / ((u + s) + np.finfo(float).eps)

    neighbor_batch_zsl = torch.argmin(distance_mat_zsl, dim=1)
    match_idx = neighbor_batch_zsl.eq(targets.int()).nonzero().flatten()
    match_counts = torch.bincount(neighbor_batch_zsl[match_idx], minlength=len
    (dataset.all_class_ids))[
        seen_unseen_array]
    target_counts = torch.bincount(targets, minlength=len(dataset.all_class_ids)
    )[seen_unseen_array]
    per_class_recall = torch.zeros(len(dataset.all_class_ids), dtype=torch.float
```

```
                , device=device)
            per_class_recall[seen_unseen_array] = match_counts / target_counts
            zsl = per_class_recall[unseen_label_array].mean()

            zsl_scores.append(zsl.item())
            seen_scores.append(s.item())
            unseen_scores.append(u.item())
            hm_scores.append(hm.item())
            per_class_recalls.append(per_class_recall.tolist())
    argmax_hm = np.argmax(hm_scores)
    max_seen = seen_scores[argmax_hm]
    max_zsl = zsl_scores[argmax_hm]
    max_unseen = unseen_scores[argmax_hm]
    max_hm = hm_scores[argmax_hm]
    max_recall = per_class_recalls[argmax_hm]
    best_beta = betas[argmax_hm].item()
    return {
        "seen": max_seen,
        "unseen": max_unseen,
        "hm": max_hm,
        "recall": max_recall,
        "zsl": max_zsl,
        "beta": best_beta
    }
```

（17）函数 get_class_names 的功能是从文件中获取类别名称。

```
def get_class_names(path):
    if isinstance(path, str):
        path = Path(path)
    with path.open("r") as f:
        classes = sorted([line.strip() for line in f])
    return classes
```

（18）函数 load_model_weights 的功能是加载模型的权重。

```
def load_model_weights(weights_path, model):
    logging.info(f"Loading model weights from {weights_path}")
    load_dict = torch.load(weights_path)
    model_weights = load_dict["model"]
    epoch = load_dict["epoch"]
    logging.info(f"Load from epoch: {epoch}")
    load_model_parameters(model, model_weights)
    return epoch
```

（19）函数 plot_hist_from_dict 的功能是根据字典数据绘制柱状图。

```
def plot_hist_from_dict(dict):
    plt.bar(range(len(dict)), list(dict.values()), align="center")
    plt.xticks(range(len(dict)), list(dict.keys()), rotation='vertical')
    plt.tight_layout()
    plt.show()
```

（20）函数 save_class_performances 的功能是保存类别性能到文件中。

```
def save_class_performances(seen_dict, unseen_dict, dataset_name, args=None):
    roor_path = args.load_path_stage_B
    seen_dir = os.path.join(roor_path, f'class_performance_{dataset_name}_seen.pkl')
    unseen_dir = os.path.join(roor_path, f'class_performance_{dataset_name}_unseen.pkl')
    seen_path = Path(seen_dir)
    unseen_path = Path(unseen_dir)
    with seen_path.open("wb") as f:
        pickle.dump(seen_dict, f)
```

```
        logging.info(f"Saving seen class performances to {seen_path}")
    with unseen_path.open("wb") as f:
        pickle.dump(unseen_dict, f)
        logging.info(f"Saving unseen class performances to {unseen_path}")
```

9.4 特征提取

在本项目的"clip_feature_extraction"目录中保存了特征提取功能的程序文件。特征提取的作用是将原始的视频数据转换为可以输入模型的图像和音频特征表示，以便于后续的模型训练和评估，同时降低计算复杂度，加速模型训练过程，并简化建模流程。具体来说，特征提取的主要目的如下。

- **降低计算复杂度**：通过提取视频的图像和音频特征，可以将原始的高维视频数据转换为更加紧凑且语义丰富的特征表示，从而降低后续模型训练和评估的计算复杂度。
- **加速模型训练**：使用预训练的模型提取特征可以节省大量的时间和计算资源，因为预训练模型已经学习到了丰富的视觉和音频特征，无须从头开始训练。
- **方便建模**：提取的特征可以作为模型的输入，简化了模型的设计和实现过程，使模型训练更加方便和高效。

9.4.1 从 ActivityNet 数据集提取特征

文件 get_clip_features_activitynet.py 的功能是从 ActivityNet 数据集的视频中提取视觉和音频特征，并保存为.pkl 文件。首先加载预训练的 CLIP 模型和 WavCaps 音频编码器模型，然后对指定目录下的视频文件进行处理：从每个视频中提取中间帧的图像特征和音频特征，然后将这些特征与类别标签一起保存到.pkl 文件中，用于后续的任务或分析。文件 get_clip_features_activitynet.py 的具体代码如下所示。

```
def read_prepare_audio(audio_path, device):
    audio, _ = librosa.load(audio_path, sr=32000, mono=True)
    audio = torch.tensor(audio).unsqueeze(0).to(device)
    # pad
    if audio.shape[-1] < 32000 * 10:
        pad_length = 32000 * 10 - audio.shape[-1]
        audio = F.pad(audio, [0, pad_length], "constant", 0.0)
    # crop
    elif audio.shape[-1] > 32000 * 10:
        center = audio.shape[-1] // 2
        start  = center - (32000 * 5)
        end    = center + (32000 * 5)
        audio = audio[:, start:end]
    return audio

parser = argparse.ArgumentParser(description='Create the datasets.')
parser.add_argument('--finetuned_model', type=str_to_bool, default=True,
                    help='Whether to load original Clip and WavCaps models or the
                    finetuned verisons')

parser.add_argument('--index', default=0, type=int)
parser.add_argument('--gpu', default=0, type=int)
args = parser.parse_args()

# device = 'cuda:4'
device = 'cuda:'+str(args.gpu)
model, preprocess = clip.load("ViT-B/32", device=device)
```

```python
if args.finetuned_model == True:
    model_path = '/home/aoq234/dev/ClipClap/logs/activitynet_clip_finetuning/epoch1_lr06
_activity_Sep14_13-30-51_901727_callisto/checkpoints/Clip_ActivityNet_finetuned.pt'
    save_path = '/home/aoq234/akata-shared/aoq234/mnt/activitynet_features_finetuned_
clip_wavcaps'+str(args.index)
    checkpoint = torch.load(model_path)
    model.load_state_dict(checkpoint['model_state_dict'])
else:
    model_path = "ViT-B/32"
    save_path = '/home/aoq234/akata-shared/aoq234/mnt/activitynet_features_original_clip
_wavcaps'+str(args.index)

model = model.to(device)
model.eval()
input_resolution = model.visual.input_resolution
context_length = model.context_length
vocab_size = model.vocab_size

print("Model parameters:", f"{np.sum([int(np.prod(p.shape)) for p in model.parameters()]
):,}")
print("Input resolution:", input_resolution)
print("Context length:", context_length)
print("Vocab size:", vocab_size)
output_list_no_average=[]
path=Path("/home/aoq234/akata-shared/datasets/ActivityNet/v1-2-trim") # path to search
for videos

dict_csv={}
list_classes=[]
count=0
dict_classes_ids={}
list_of_files=[]

for f in tqdm(path.glob("**/*.mp4")):
    class_name=str(f).split("/")[-2]
    list_of_files.append(f)
    if class_name not in list_classes:
        list_classes.append(class_name)

path=Path("/home/aoq234/akata-shared/datasets/ActivityNet/v1-3-trim")

for f in tqdm(path.glob("**/*.mp4")):
    class_name=str(f).split("/")[-2]
    list_of_files.append(f)
    if class_name not in list_classes:
        list_classes.append(class_name)

chunk=int(len(list_of_files)/3)+1
list_of_files=list_of_files[args.index*chunk:(args.index+1)*chunk]
list_classes.sort()

for index,val in enumerate(sorted(list_classes)):
    dict_classes_ids[val]=index

with open("/home/aoq234/dev/CLIP-GZSL/WavCaps/retrieval/settings/inference.yaml", "r") as f:
    config = yaml.safe_load(f)

wavcaps_model = ASE(config)
wavcaps_model.to(device)
if args.finetuned_model == True:
    cp_path = '/home/aoq234/dev/ClipClap/logs/activitynet_wavcaps_finetuning/epoch20_
lr1e6_activity_Sep18_08-11-22_070278_callisto/checkpoints/WavCaps_ActivityNet_finetuned.pt'
```

```python
        state_dict_key = 'model_state_dict'
    else:
        cp_path = '/home/aoq234/dev/CLIP-GZSL/WavCaps/retrieval/pretrained_models/audio_encoders/HTSAT_BERT_zero_shot.pt'
        state_dict_key = 'model'
cp = torch.load(cp_path)
wavcaps_model.load_state_dict(cp[state_dict_key])
wavcaps_model.eval()
print("Model weights loaded from {}".format(cp_path))

counter=0
for f in tqdm(list_of_files):
    counter+=1
    if counter%1000==0:
        with open(save_path, 'wb') as handle:
            pickle.dump(output_list_no_average, handle, protocol=pickle.HIGHEST_PROTOCOL)
    try:
        # audio
        mp4_version = AudioSegment.from_file(str(f), "mp4")
        mp4_version.export("/home/aoq234/akata-shared/aoq234/mnt/activity_dummy"+str(args.index)+".wav", format="wav")
        audio = read_prepare_audio("/home/aoq234/akata-shared/aoq234/mnt/activity_dummy"+str(args.index)+".wav", device)
        with torch.no_grad():
            audio_emb = wavcaps_model.encode_audio(audio).squeeze()
        audio_emb = audio_emb.cpu().detach().numpy() # (1024,)
        #获取中间索引
        cap = cv2.VideoCapture(str(f))
        frameCount = int(cap.get(cv2.CAP_PROP_FRAME_COUNT))
        fc = 0
        ret = True
        while (fc < frameCount and ret):
            ret, image=cap.read()
            if ret==True:
                fc += 1
        mid_frame_idx = (fc // 2)
        cap.release()
        cap = cv2.VideoCapture(str(f))
        frameCount = int(cap.get(cv2.CAP_PROP_FRAME_COUNT))
        frameWidth = int(cap.get(cv2.CAP_PROP_FRAME_WIDTH))
        frameHeight = int(cap.get(cv2.CAP_PROP_FRAME_HEIGHT))
        video = torch.zeros((1, 3, 224, 224), dtype=torch.float32)
        fc = 0
        ret = True
        p= torchvision.transforms.Compose([
            torchvision.transforms.ToPILImage()
        ])
        while (fc != mid_frame_idx and ret):

            ret, image=cap.read()
            if ret==True:
                fc += 1
                if fc == mid_frame_idx:
                    torch_image=torch.from_numpy(image)
                    torch_image=torch_image.permute(2,0,1)
                    torch_image=p(torch_image)
                    torch_image=preprocess(torch_image) # 返回 (3, 224, 224)
                    video[0]=torch_image
        cap.release()
        list_clips=[]
        frame = video[0]
        frame = frame.unsqueeze(0).to(device)   #返回的形状是 (1, 3, 224, 224)，等同于 (批量大
                                                #小，通道数，高度，宽度)
```

```
                    if fc == 0:
image_features = np.array([])#在ActivityNet的一些视频中，当使用read()函数读取时，在第一帧返回了
#false，并且帧计数器（fc）始终为0。在这种情况下，TCAF代码库会返回一个空列表作为时间特征，并且对于平均特
#征返回NaN值。
                    else:
                        with torch.no_grad():
                            image_features = model.encode_image(frame) # shape torch.Size([1, 512])
                            image_features /= image_features.norm(dim=-1, keepdim=True)
                            image_features = image_features.squeeze()
                            image_features=image_features.cpu().detach().numpy() # shape (512,)
                except Exception as e:
                    print(e)
                    print(f)
                    continue
                name_file=str(f).split("/")[-1]
                class_name=str(f).split("/")[-2]
                class_id=dict_classes_ids[class_name]
                result_list=[image_features, class_id, audio_emb, name_file]
                output_list_no_average.append(result_list)

with open(save_path, 'wb') as handle:
    pickle.dump(output_list_no_average, handle, protocol=pickle.HIGHEST_PROTOCOL)
```

9.4.2 从 UCF101 数据集提取特征

文件 get_clip_features_ucf.py 用于从 UCF101 数据集的视频中提取视觉和音频特征，其中包括通过 CLIP 模型提取视觉特征和通过 WavCaps 模型提取音频特征，然后将这些特征保存到文件中。文件 get_clip_features_ucf.py 的具体代码如下。

```
def read_prepare_audio(audio_path, device):
    audio, _ = librosa.load(audio_path, sr=32000, mono=True)
    audio = torch.tensor(audio).unsqueeze(0).to(device)
    # pad
    if audio.shape[-1] < 32000 * 10:
        pad_length = 32000 * 10 - audio.shape[-1]
        audio = F.pad(audio, [0, pad_length], "constant", 0.0)
    # crop
    elif audio.shape[-1] > 32000 * 10:
        center = audio.shape[-1] // 2
        start  = center - (32000 * 5)
        end    = center + (32000 * 5)
        audio = audio[:, start:end]
    return audio

parser = argparse.ArgumentParser(description='Create the datasets.')
parser.add_argument('--finetuned_model', type=str_to_bool, default=True,
                    help='Whether to load original Clip and WavCaps models or the
finetuned verisons')

args = parser.parse_args()

device = 'cuda:4'
model, preprocess = clip.load("ViT-B/32", device=device)
if args.finetuned_model == True:
    model_path = '/home/aoq234/dev/ClipClap/logs/clip_finetuning/second_try_Aug25_19-53-
14_475888_callisto/checkpoints/clip_finetuned.pt'
    save_path = '/home/aoq234/akata-shared/aoq234/mnt/ucf_features_finetuned_clip_wavcaps'
    checkpoint = torch.load(model_path)
    model.load_state_dict(checkpoint['model_state_dict'])
else:
    model_path = "ViT-B/32"
```

9.4 特征提取

```python
        save_path = '/home/aoq234/akata-shared/aoq234/mnt/ucf_features_original_clip_wavcaps'
model = model.to(device)
model.eval()
input_resolution = model.visual.input_resolution
context_length = model.context_length
vocab_size = model.vocab_size

print("Model parameters:", f"{np.sum([int(np.prod(p.shape)) for p in model.parameters()]):,}")
print("Input resolution:", input_resolution)
print("Context length:", context_length)
print("Vocab size:", vocab_size)
output_list_no_average=[]

path=Path("/home/aoq234/akata-shared/datasets/UCF101/UCF-101")  #搜索视频的路径
dict_csv={}
list_classes=[]
count=0
dict_classes_ids={}

for f in tqdm(path.glob("**/*.avi")):
    class_name=str(f).split("/")[-2]
    if class_name not in list_classes:
        list_classes.append(class_name)
list_classes.sort()

for index,val in enumerate(sorted(list_classes)):
    dict_classes_ids[val]=index
with open("/home/aoq234/dev/CLIP-GZSL/WavCaps/retrieval/settings/inference.yaml", "r") as f:
    config = yaml.safe_load(f)
wavcaps_model = ASE(config)
wavcaps_model.to(device)

if args.finetuned_model == True:
    cp_path = '/home/aoq234/dev/ClipClap/logs/wavcaps_finetuning/first_try_Aug29_07-25-0
    5_613403_callisto/checkpoints/WavCaps_finetuned.pt'
    state_dict_key = 'model_state_dict'
else:
    cp_path = '/home/aoq234/dev/CLIP-GZSL/WavCaps/retrieval/pretrained_models/
    audio_encoders/HTSAT_BERT_zero_shot.pt'
    state_dict_key = 'model'

cp = torch.load(cp_path)
wavcaps_model.load_state_dict(cp[state_dict_key])
wavcaps_model.eval()
print("Model weights loaded from {}".format(cp_path))

counter=0

for f in tqdm(path.glob("**/*.avi")):
    counter+=1
    if counter%3000==0:
        with open(save_path, 'wb') as handle:
            pickle.dump(output_list_no_average, handle, protocol=pickle.HIGHEST_PROTOCOL)
    try:
        # audio
        mp4_version = AudioSegment.from_file(str(f), "avi")
        mp4_version.export("/home/aoq234/akata-shared/aoq234/mnt/ucf_dummy_tmp.wav", format="wav")
        audio = read_prepare_audio("/home/aoq234/akata-shared/aoq234/mnt/ucf_dummy_tmp.wav", device)
        with torch.no_grad():
```

```python
            audio_emb = wavcaps_model.encode_audio(audio).squeeze()
            audio_emb = audio_emb.cpu().detach().numpy() # (1024,)

            cap = cv2.VideoCapture(str(f))
            frameCount = int(cap.get(cv2.CAP_PROP_FRAME_COUNT))
            frameWidth = int(cap.get(cv2.CAP_PROP_FRAME_WIDTH))
            frameHeight = int(cap.get(cv2.CAP_PROP_FRAME_HEIGHT))
            video = torch.zeros((frameCount, frameHeight,frameWidth, 3), dtype=torch.float32)
              # torch.Size([164, 240, 320, 3])
            fc = 0
            ret = True
            while (fc < frameCount and ret):
                ret, image=cap.read()
                if ret==True:
                    torch_image=torch.from_numpy(image)
                    video[fc]=torch_image
                    fc += 1
            cap.release()
            list_clips=[]

            p= torchvision.transforms.Compose([
                torchvision.transforms.ToPILImage() # requires tensor of shape  C x H x W
            ])
            mid_frame_idx = video.shape[0] // 2
            frame = video[mid_frame_idx]
            frame=frame.permute(2, 0, 1)
            frame=p(frame)
            # preprocess()调用toTensor 转换,该转换返回一个形状为 (C x H x W) 的 torch.FloatTensor
            # 返回shape  (1, 3, 224, 224) = (bs, channels, height, width)
            frame = preprocess(frame).unsqueeze(0).to(device) # preprocess and add batch
            dimension
            with torch.no_grad():
                image_features = model.encode_image(frame) # shape torch.Size([1, 512])
            image_features /= image_features.norm(dim=-1, keepdim=True)
            image_features = image_features.squeeze()
            image_features=image_features.cpu().detach().numpy() # shape (512,)

        except Exception as e:
            print(e)
            print(f)
            continue

        name_file=str(f).split("/")[-1]
        class_name=str(f).split("/")[-2]
        class_id=dict_classes_ids[class_name]
        result_list=[image_features, class_id, audio_emb, name_file]
        output_list_no_average.append(result_list)

with open(save_path, 'wb') as handle:
    pickle.dump(output_list_no_average, handle, protocol=pickle.HIGHEST_PROTOCOL)
```

9.4.3 从 VGGSound 数据集提取特征

文件 get_clip_features_vggsound.py 用于从 VGGSound 数据集的视频中提取视觉和音频特征,首先加载了 CLIP 和 WavCaps 模型,然后对视频进行特征提取,最后将提取的特征保存到文件中。文件 get_clip_features_vggsound.py 的具体代码如下所示。

```python
def read_prepare_audio(audio_path, device):
    audio, _ = librosa.load(audio_path, sr=32000, mono=True)
    audio = torch.tensor(audio).unsqueeze(0).to(device)
    # pad
```

```python
        if audio.shape[-1] < 32000 * 10:
            pad_length = 32000 * 10 - audio.shape[-1]
            audio = F.pad(audio, [0, pad_length], "constant", 0.0)
        # crop
        elif audio.shape[-1] > 32000 * 10:
            center = audio.shape[-1] // 2
            start = center - (32000 * 5)
            end   = center + (32000 * 5)
            audio = audio[:, start:end]
        return audio

parser=argparse.ArgumentParser(description="GZSL with ESZSL")
parser.add_argument('--finetuned_model', type=str_to_bool, default=False,
                    help='Whether to load original Clip and WavCaps models or the finetuned verisons')
parser.add_argument('--index', default=0, type=int)
parser.add_argument('--gpu', default=0, type=int)
args=parser.parse_args()
# device = 'cuda:4'
device = 'cuda:'+str(args.gpu)
model, preprocess = clip.load("ViT-B/32", device=device)
if args.finetuned_model == True:
    raise NotImplementedError()

else:
    model_path = "ViT-B/32"
    save_path = '/home/aoq234/akata-shared/aoq234/mnt/vggsound_features_original_clip_wavcaps'+str(args.index)
model = model.to(device)
model.eval()
input_resolution = model.visual.input_resolution
context_length = model.context_length
vocab_size = model.vocab_size
print("Model parameters:", f"{np.sum([int(np.prod(p.shape)) for p in model.parameters()]):,}")
print("Input resolution:", input_resolution)
print("Context length:", context_length)
print("Vocab size:", vocab_size)

with open("/home/aoq234/dev/CLIP-GZSL/WavCaps/retrieval/settings/inference.yaml", "r") as f:
    config = yaml.safe_load(f)

wavcaps_model = ASE(config)
wavcaps_model.to(device)
if args.finetuned_model == True:
    raise NotImplementedError()
else:
    cp_path = '/home/aoq234/dev/CLIP-GZSL/WavCaps/retrieval/pretrained_models/audio_encoders/HTSAT_BERT_zero_shot.pt'
    state_dict_key = 'model'

cp = torch.load(cp_path)
wavcaps_model.load_state_dict(cp[state_dict_key])
wavcaps_model.eval()
print("Model weights loaded from {}".format(cp_path))
output_list_no_average=[]

path=Path("/home/aoq234/akata-shared/datasets/vggsound/video")
dict_csv={}
list_classes=[]
count=0
with open('/home/aoq234/akata-shared/datasets/vggsound/metadata/vggsound.csv', 'r') as read_obj:
```

```python
    csv_reader = reader(read_obj)

    for row in csv_reader:
        key=str(row[0])+"_"+str(row[1])
        if row[2] not in list_classes:
            list_classes.append(row[2])
        dict_csv[key]=[row[2],row[3]]
list_classes.sort()
dict_classes_ids={}
for index,val in enumerate(sorted(list_classes)):
    dict_classes_ids[val]=index

list_of_files=[]
for f in tqdm(path.glob("**/*.mp4")):
        list_of_files.append(f)
chunk=int(len(list_of_files)/3)+1
list_of_files=list_of_files[args.index*chunk:(args.index+1)*chunk]
counter=0

for f in tqdm(list_of_files):
    counter+=1
    if counter%3000==0:
        with open(save_path, 'wb') as handle:
            pickle.dump(output_list_no_average, handle, protocol=pickle.HIGHEST_PROTOCOL)
    try:
        # 音频
        mp4_version = AudioSegment.from_file(str(f), "mp4")
        mp4_version.export("/home/aoq234/akata-shared/aoq234/mnt/vggsound_dummy"+str(args.index)+".wav", format="wav")
        audio = read_prepare_audio("/home/aoq234/akata-shared/aoq234/mnt/vggsound_dummy"+str(args.index)+".wav", device)
        with torch.no_grad():
            audio_emb = wavcaps_model.encode_audio(audio).squeeze()
        audio_emb = audio_emb.cpu().detach().numpy() # (1024,)
        p= torchvision.transforms.Compose([
            torchvision.transforms.ToPILImage()
        ])
        #获取中间索引
        while True:
            cap = cv2.VideoCapture(str(f))
            frameCount = int(cap.get(cv2.CAP_PROP_FRAME_COUNT))
            fc = 0
            ret = True
            while (fc < frameCount and ret):
                ret, image=cap.read()
                if ret==True:
                    fc += 1
            mid_frame_idx = (fc // 2)
            cap.release()
            if fc!=0:
                break
        while True:
            cap = cv2.VideoCapture(str(f))
            frameCount = int(cap.get(cv2.CAP_PROP_FRAME_COUNT))
            frameWidth = int(cap.get(cv2.CAP_PROP_FRAME_WIDTH))
            frameHeight = int(cap.get(cv2.CAP_PROP_FRAME_HEIGHT))
            video = torch.zeros((1, 3, 224, 224), dtype=torch.float32)
            fc = 0
            ret = True
            while (fc < frameCount and ret):
                ret, image=cap.read()
                if ret==True:
```

```
                        fc += 1
                        if fc == mid_frame_idx:
                            torch_image=torch.from_numpy(image)
                            torch_image=torch_image.permute(2,0,1)
                            torch_image=p(torch_image)
                            torch_image=preprocess(torch_image) # 返回 (3, 224, 224)
                            video[0]=torch_image
                    cap.release()
                    if fc!=0:
                        break
                list_clips=[]
                frame = video[0]
                frame = frame.unsqueeze(0).to(device)   #返回的形状为 (1, 3, 224, 224) ;表示(bs,
                channels, height, width)
                with torch.no_grad():
                    image_features = model.encode_image(frame) # shape torch.Size([1, 512])
                    image_features /= image_features.norm(dim=-1, keepdim=True)
                    image_features = image_features.squeeze()
                    image_features=image_features.cpu().detach().numpy() # shape (512,)
            except Exception as e:
                print(e)
                print(f)
                continue
            name_file=str(f).split("/")[-1]
            splitted_path=name_file.rsplit('_', 1)
            class_id=splitted_path[0]
            number=int(splitted_path[1].split(".")[0])
            search_name=class_id+"_"+str(number)
            class_name=dict_csv[search_name][0]
            class_id=dict_classes_ids[class_name]
            result_list=[image_features, class_id, audio_emb, name_file]
            output_list_no_average.append(result_list)

with open(save_path, 'wb') as handle:
    pickle.dump(output_list_no_average, handle, protocol=pickle.HIGHEST_PROTOCOL)
```

9.5 多模态模型

本项目实现了一个多模态模型，包括模型定义、训练过程、评估指标、优化器定义和评估过程，通过整合这些部分实现了从多模态数据处理、模型训练到评估的完整流程。

9.5.1 多模态数据学习模型

文件 clipclap_model.py 实现了多模态数据嵌入学习和对比学习模型，首先实现了一个名为 ClipClap_model 的模型类和一个名为 EmbeddingNet 的嵌入网络类。然后实现了一些辅助函数 disable_running_stats 和 enable_running_stats，用于在训练过程中启用或禁用批归一化层的 running statistics。文件 clipclap_model.py 的主要功能是实现一个用于多模态数据嵌入学习和对比学习的模型，并提供了灵活的损失计算和参数优化策略。其具体实现流程如下。

（1）类 EmbeddingNet 是一个通用的嵌入网络类，用于定义嵌入层的结构，包括线性层、批归一化层和激活函数等。

```
class EmbeddingNet(nn.Module):
    def __init__(self, input_size, output_size, dropout, use_bn, hidden_size=-1):
        super(EmbeddingNet, self).__init__()
        modules = []
```

```python
            if hidden_size > 0:
                modules.append(nn.Linear(in_features=input_size, out_features=hidden_size))
                if use_bn:
                    modules.append(nn.BatchNorm1d(num_features=hidden_size))
                modules.append(nn.ReLU())
                modules.append(nn.Dropout(dropout))
                modules.append(nn.Linear(in_features=hidden_size, out_features=output_size))
                modules.append(nn.BatchNorm1d(num_features=output_size))
                modules.append(nn.ReLU())
                modules.append(nn.Dropout(dropout))
            else:
                modules.append(nn.Linear(in_features=input_size, out_features=output_size))
                modules.append(nn.BatchNorm1d(num_features=output_size))
                modules.append(nn.ReLU())
                modules.append(nn.Dropout(dropout))
            self.fc = nn.Sequential(*modules)

    def forward(self, x):
        output = self.fc(x)
        return output

    def get_embedding(self, x):
        return self.forward(x)
```

（2）类 ClipClap_model 是一个深度学习模型，用于多模态数据（音频、视频、文本）的嵌入学习和对比学习任务，主要包括以下功能。

- 初始化模型参数和优化器：定义了模型的结构和参数，包括编码器、解码器、损失函数、优化器等。
- 前向传播：实现了模型的前向传播逻辑，根据输入的音频、视频和文本数据，计算对应的嵌入向量。
- 损失计算：计算了重构损失、正交损失和交叉熵损失，并返回总损失和各个损失项。
- 参数优化：根据损失函数进行参数优化，支持使用不同的优化器进行训练，包括常规的 Adam 优化器和 SAM（Sharpness-Aware Minimization）优化器。
- 获取嵌入向量：提供了获取嵌入向量的方法，用于对数据进行嵌入表示。

```python
class ClipClap_model(nn.Module):
    def __init__(self, params_model, input_size_audio, input_size_video):
        super(ClipClap_model, self).__init__()

        print('Initializing model variables...', end='')
        #嵌入的维度
        self.dim_out = params_model['dim_out']
        self.input_dim_audio = input_size_audio
        self.input_dim_video = input_size_video

        self.hidden_size_decoder=params_model['decoder_hidden_size']
        self.drop_proj_o=params_model['dropout_decoder']
        self.drop_proj_w=params_model['additional_dropout']
        self.reg_loss=params_model['reg_loss']
        self.cross_entropy_loss=params_model['cross_entropy_loss']
        self.hidden_size_encoder=params_model['encoder_hidden_size']
        self.drop_enc=params_model['dropout_encoder']
        self.rec_loss = params_model['rec_loss']

        self.lr_scheduler = params_model['lr_scheduler']

        print('Initializing trainable models...', end='')
```

```python
        self.modality = params_model['modality']
        self.word_embeddings = params_model['word_embeddings']

        if self.modality == 'audio':
            self.O_enc = EmbeddingNet(
                input_size=1024,
                output_size=512,
                dropout=0.1,
                use_bn=True
            )
            self.W_enc = EmbeddingNet(
                input_size=1024,
                output_size=512,
                dropout=0.1,
                use_bn=True
            )
        elif self.modality == 'video':
            self.O_enc = EmbeddingNet(
                input_size=512,
                output_size=512,
                dropout=0.1,
                use_bn=True
            )
            self.W_enc = EmbeddingNet(
                input_size=512,
                output_size=512,
                dropout=0.1,
                use_bn=True
            )
        else:
            self.O_enc = EmbeddingNet(
                input_size=1536,
                output_size=512,
                dropout=0.1,
                use_bn=True
            )
            w_in_dim = 1536
            if self.word_embeddings == 'wavcaps':
                w_in_dim = 1024
            elif self.word_embeddings == 'clip':
                w_in_dim = 512

            self.W_enc = EmbeddingNet(
                input_size=w_in_dim,
                output_size=512,
                dropout=0.1,
                use_bn=True
            )
        word_embedding_dim = 512
        self.O_proj = EmbeddingNet(
            input_size=512,
            hidden_size=self.hidden_size_decoder,
            output_size=self.dim_out,
            dropout=self.drop_proj_o,
            use_bn=params_model['embeddings_batch_norm']
        )
        self.D_o = EmbeddingNet(
            input_size=self.dim_out,
            hidden_size=self.hidden_size_decoder,
            output_size=word_embedding_dim,
            dropout=self.drop_proj_o,
            use_bn=params_model['embeddings_batch_norm']
```

```python
        )
        self.W_proj= EmbeddingNet(
            input_size=word_embedding_dim,
            output_size=self.dim_out,
            dropout=self.drop_proj_w,
            use_bn=params_model['embeddings_batch_norm']
        )
        self.D_w = EmbeddingNet(
            input_size=self.dim_out,
            output_size=word_embedding_dim,
            dropout=self.drop_proj_w,
            use_bn=params_model['embeddings_batch_norm']
        )
        #优化器
        print('Defining optimizers...', end='')
        self.lr = params_model['lr']

        optimizer = params_model['optimizer']
        self.is_sam_optim = False
        if optimizer == 'adam':
            self.optimizer_gen = optim.Adam(
                self.parameters(),
                lr=self.lr, weight_decay=1e-5
            )
            if self.lr_scheduler:
                self.scheduler_learning_rate = optim.lr_scheduler.ReduceLROnPlateau
                (self.optimizer_gen, 'max', patience=3, verbose=True)

        elif optimizer == 'adam-sam':
            self.optimizer_gen = SAM(self.parameters(), optim.Adam, lr=self.lr,
            weight_decay=1e-5)
            self.is_sam_optim = True
            if self.lr_scheduler:
                #在基础优化器上进行学习率调度处理
                self.scheduler_learning_rate = optim.lr_scheduler.ReduceLROnPlateau
                (self.optimizer_gen.base_optimizer, 'max', patience=3, verbose=True)
        else:
            raise NotImplementedError

        print('Done')

        #损失函数
        print('Defining losses...', end='')
        self.criterion_cyc = nn.MSELoss()
        self.criterion_cls = nn.CrossEntropyLoss()
        self.MSE_loss = nn.MSELoss()
        print('Done')

    def optimize_scheduler(self, value):
        if self.lr_scheduler:
            self.scheduler_learning_rate.step(value)

    def forward(self, a, v, w, masks, timesteps):
        b, _ = a.shape
        device = a.device
        v = v.type(torch.float32)
        if self.modality == 'audio':
            w = w[:,512:]
            model_input = a

        elif self.modality == 'video':
            w = w[:,:512]
            model_input = v
```

```python
        else:
            if self.word_embeddings == 'wavcaps':
                w = w[:,512:]
            elif self.word_embeddings == 'clip':
                w = w[:,:512]
        model_input = torch.cat((v, a), dim=1)


    o = self.O_enc(model_input)

    w = self.W_enc(w)
    theta_o = self.O_proj(o)
    rho_o = self.D_o(theta_o)
    theta_w = self.W_proj(w)
    rho_w=self.D_w(theta_w)
    output = {
        "theta_w": theta_w,
        "w": w,
        "rho_w": rho_w,
        "theta_o": theta_o,
        "rho_o": rho_o,
    }
    return output

def compute_loss(self, outputs, embeddings_crossentropy, gt_cross_entropy):
    theta_w = outputs['theta_w']
    w = outputs['w']
    rho_w = outputs['rho_w']
    theta_o = outputs['theta_o']
    rho_o = outputs['rho_o']
    device = theta_w.device

    if self.cross_entropy_loss==True:
        if self.modality == 'audio':
            embeddings_crossentropy = embeddings_crossentropy[:,512:]
        elif self.modality == 'video':
            embeddings_crossentropy = embeddings_crossentropy[:,:512]
        else:
            if self.word_embeddings == 'wavcaps':
                embeddings_crossentropy = embeddings_crossentropy[:,512:]
            elif self.word_embeddings == 'clip':
                embeddings_crossentropy = embeddings_crossentropy[:,:512]

        embedding_cross_entropy=self.W_proj(self.W_enc(embeddings_crossentropy))
        Cross_loss=nn.CrossEntropyLoss()
        scores=torch.matmul(theta_o, embedding_cross_entropy.t()) # (bs, 64) x (K_seen
        , 64).T = (bs, 64) x (64, K_seen) = (bs, K_seen)
        # gt_cross_entropy = [1, 3, 2, 55, 97, 45, ...] list of gt class labels ->
        #shape (bs,)
        l_ce=Cross_loss(scores, gt_cross_entropy)
    else:
        l_ce = torch.tensor(0., device=device)

    if self.reg_loss==True:
        l_reg = (
            self.MSE_loss(theta_o, theta_w)
        )
    else:
        l_reg = torch.tensor(0., device=device)


    if self.rec_loss == True:
        l_rec = (
```

```python
                    self.MSE_loss(w, rho_o) +
                    self.MSE_loss(w, rho_w)
            )
        else:
            l_rec = torch.tensor(0., device=device)


        loss_total = l_rec+l_reg+l_ce
        loss_dict = {
            "Loss/total_loss": loss_total.detach().cpu(),
            "Loss/loss_reg": l_reg.detach().cpu(),
            "Loss/loss_cmd_rec": l_rec.detach().cpu(),
            "Loss/cross_entropy": l_ce.detach().cpu()

        }
        return loss_total, loss_dict

# cls_numeric = 类别索引
# cls_embedding = 目标的 Word2Vec 嵌入
def optimize_params(self, audio, video, cls_numeric, cls_embedding, masks, timesteps
, embedding_crossentropy, optimize=False):
    if not self.is_sam_optim:
        # 前向传播
        outputs = self.forward(audio, video, cls_embedding, masks, timesteps)

        # 反向传播
        loss_numeric, loss = self.compute_loss(outputs, embedding_crossentropy, cls_numeric)

        if optimize == True:
            self.optimizer_gen.zero_grad()
            loss_numeric.backward()
            self.optimizer_gen.step()

    else:
        # SAM 优化器需要两次 forward / backward

        enable_running_stats(self)
        outputs = self.forward(audio, video, cls_embedding, masks, timesteps)
        loss_numeric, loss = self.compute_loss(outputs, embedding_crossentropy, cls_numeric)

        if optimize:
            #第一次 forward-backward 步骤
            # self.optimizer_gen.zero_grad()
            loss_numeric.backward()
            self.optimizer_gen.first_step(zero_grad=True)

            #第二次 forward-backward 步骤
            disable_running_stats(self)
            outputs_second = self.forward(audio, video, cls_embedding, masks, timesteps)
            second_loss, _ = self.compute_loss(outputs_second,
                embedding_crossentropy,  cls_numeric)
            second_loss.backward()
            self.optimizer_gen.second_step(zero_grad=True)

    return loss_numeric, loss

def get_embeddings(self, a, v, w, masks, timesteps):
    b, _ = a.shape
    device = a.device
    v = v.type(torch.float32)
    if self.modality == 'audio':
```

```
            w = w[:,512:]
            model_input = a
        elif self.modality == 'video':
            w = w[:,:512]
            model_input = v
        else:
            if self.word_embeddings == 'wavcaps':
                w = w[:,512:]
            elif self.word_embeddings == 'clip':
                w = w[:,:512]
            model_input = torch.cat((v, a), dim=1)
        o = self.O_enc(model_input)

        w = self.W_enc(w)
        theta_o = self.O_proj(o)
        theta_w=self.W_proj(w)
        return theta_o, theta_o, theta_w
```

9.5.2 性能评估指标

文件 metrics.py 定义了用于评估模型性能的多个指标类，包括计算平均非零三元组数量、类别重叠百分比、详细损失、目标难度、平均类别准确率等。这些指标用于衡量模型在训练和评估过程中的表现，并提供了各种不同层面的性能评估。文件 metrics.py 的具体实现流程如下。

（1）基类 Metric 定义了评估指标类的通用接口，包括计算指标值、重置指标值、获取当前指标值和获取指标名称等功能。子类需要继承该基类并实现具体的指标计算逻辑。

```
class Metric:
    def __init__(self):
        pass

    def __call__(self, outputs, target, loss):
        """计算指标值，需要在子类中实现。

        Args:
            outputs (tensor): 模型输出.
            target (tensor): 目标标签.
            loss (tensor): 损失值.

        Raises:
            NotImplementedError: 需要在子类中实现具体的计算逻辑.
        """
        raise NotImplementedError

    def reset(self):
        """重置指标值."""
        raise NotImplementedError

    def value(self):
        """获取当前指标值."""
        raise NotImplementedError

    def name(self):
        """获取指标名称."""
        raise NotImplementedError
```

（2）类 AverageNonzeroTripletsMetric 用于计算小批量数据中非零三元组的平均数量，通常用于评估训练过程中的三元组损失。

```python
class AverageNonzeroTripletsMetric(Metric):
    '''
    统计小批量中非零三元组的平均数量
    '''
    def __init__(self):
        super(AverageNonzeroTripletsMetric, self).__init__()
        self.values = []

    def __call__(self, outputs, target, loss):
        """计算非零三元组的平均数量,并将结果存储在列表中。

        Args:
            outputs (tensor): 模型输出.
            target (tensor): 目标标签.
            loss (tensor): 损失值.

        Returns:
            float: 平均非零三元组的数量.
        """
        self.values.append(loss[1])
        return self.value()

    def reset(self):
        """重置值列表."""
        self.values = []

    def value(self):
        """计算并返回平均非零三元组的数量."""
        return np.mean(self.values)

    def name(self):
        """获取指标名称."""
        return 'Average nonzero triplets'
```

（3）类 PercentOverlappingClasses 用于计算两个标签列表之间的类别重叠百分比,并返回平均的重叠百分比作为指标。

```python
class PercentOverlappingClasses(Metric):
    def __init__(self):
        super(PercentOverlappingClasses, self).__init__()
        self.values = []

    def __call__(self, outputs, target, loss):
        """计算两个标签列表之间的类别重叠百分比,并将结果存储在列表中。

        Args:
            outputs (tensor): 模型输出.
            target (tuple): 包含两个张量的元组,表示两个标签列表.
            loss (tensor): 损失值.

        """
        labels1, labels2 = target
        assert len(labels1) == len(labels2)
        percent_overlap = len(torch.where(labels1.eq(labels2))[0]) / len(labels1)
        self.values.append(percent_overlap)

    def reset(self):
        """重置值列表."""
        self.values = []

    def value(self):
        """计算并返回类别重叠的平均百分比."""
```

```
        value = np.mean(self.values)
        assert value == 0.
        return {"class_overlap": value}

    def name(self):
        """获取指标名称."""
        return "Average p,q class overlap [%]"
```

（4）类 DetailedLosses 用于计算和跟踪模型训练过程中的各种详细损失，并计算 CMD 和 CT 损失的总和，用于调试和监视训练过程中的损失变化。

```
class DetailedLosses(Metric):
    def __init__(self):
        super(DetailedLosses, self).__init__()
        self.cmd = []          # 存储 CMD 损失
        self.ct = []           # 存储 CT 损失
        self.l_rec = []        # 存储 Reconstruction 损失
        self.l_cta = []        # 存储 Cross-modal Tripet Alignment 损失
        self.l_ctv = []        # 存储 Cross-modal Tripet Alignment 损失
        self.l_ta = []         # 存储 Tripet Alignment 损失
        self.l_at = []         # 存储 Alignment 损失
        self.l_tv = []         # 存储 Triplet Visuo-Tactile 损失
        self.l_vt = []         # 存储 Visuo-Tactile 损失

    def __call__(self, outputs, target, loss):
        """计算并存储各种详细损失，并计算 CMD 和 CT 损失的总和。

        Args:
            outputs: 模型输出.
            target: 目标.
            loss: 损失值.
        """
        self.l_rec.append(loss[1]["cmd"]["l_rec"].item())
        self.l_cta.append(loss[1]["cmd"]["l_cta"].item())
        self.l_ctv.append(loss[1]["cmd"]["l_ctv"].item())
        self.l_ta.append(loss[1]["ct"]["l_ta"].item())
        self.l_at.append(loss[1]["ct"]["l_at"].item())
        self.l_tv.append(loss[1]["ct"]["l_tv"].item())
        self.l_vt.append(loss[1]["ct"]["l_vt"].item())
        self.cmd.append(self.l_rec[-1] + self.l_cta[-1] + self.l_ctv[-1])
        self.ct.append(self.l_ta[-1] + self.l_at[-1] + self.l_tv[-1] + self.l_vt[-1])

    def reset(self):
        """重置所有损失列表."""
        self.cmd = []
        self.ct = []
        self.l_rec = []
        self.l_cta = []
        self.l_ctv = []
        self.l_ta = []
        self.l_at = []
        self.l_tv = []
        self.l_vt = []

    def value(self):
        """计算并返回各种损失的均值，以及 CMD 和 CT 损失的总和."""
        return {
            "cmd": np.mean(self.cmd),
            "ct": np.mean(self.ct),
            "l_rec": np.mean(self.l_rec),
            "l_cta": np.mean(self.l_cta),
            "l_ctv": np.mean(self.l_ctv),
```

```python
            "l_ta": np.mean(self.l_ta),
            "l_at": np.mean(self.l_at),
            "l_tv": np.mean(self.l_tv),
            "l_vt": np.mean(self.l_vt)
        }

    def name(self):
        """获取指标名称."""
        return "Debug losses"
```

（5）类 TargetDifficulty 用于计算音频和视频样本中不同难度级别的三元组比例，以帮助评估模型在处理难度不同的样本时的表现。

```python
class TargetDifficulty(Metric):
    def __init__(self, margin, distance_fn):
        super(TargetDifficulty, self).__init__()
        self.margin = margin    # 三元组难度的边界值
        self.distance_fn = distance_fn    # 距离函数，用于计算样本之间的距离
        self.easy_audio = []      # 存储音频样本中易样本的比例
        self.hard_audio = []      # 存储音频样本中难样本的比例
        self.semi_hard_audio = []     # 存储音频样本中半难样本的比例
        self.easy_video = []      # 存储视频样本中易样本的比例
        self.hard_video = []      # 存储视频样本中难样本的比例
        self.semi_hard_video = []     # 存储视频样本中半难样本的比例

    def __call__(self, outputs, target, loss):
        # 获取模型输出
        _, a1, v1, t1, a2, v2, _, _, _, _, _, _ = outputs
        # 计算音频样本的三元组难度
        easy_audio, hard_audio, semi_hard_audio = self._get_triplet_difficulty(anchor=t1
        , positive=a1, negative=a2,margin=self.margin)
        # 计算视频样本的三元组难度
        easy_video, hard_video, semi_hard_video = self._get_triplet_difficulty(anchor=t1
        , positive=v1, negative=v2,margin=self.margin)
        # 将计算结果存储起来
        self.easy_audio.append(easy_audio)
        self.hard_audio.append(hard_audio)
        self.semi_hard_audio.append(semi_hard_audio)
        self.easy_video.append(easy_video)
        self.hard_video.append(hard_video)
        self.semi_hard_video.append(semi_hard_video)

    def reset(self):
        # 重置存储的难度比例列表
        self.easy_audio = []
        self.hard_audio = []
        self.semi_hard_audio = []
        self.easy_video = []
        self.hard_video = []
        self.semi_hard_video = []

    def value(self):
        # 返回各类难度样本的平均比例
        return {
            "easy_audio": np.mean(self.easy_audio),
            "hard_audio": np.mean(self.hard_audio),
            "semi_hard_audio": np.mean(self.semi_hard_audio),
            "easy_video": np.mean(self.easy_video),
            "hard_video": np.mean(self.hard_video),
            "semi_hard_video": np.mean(self.semi_hard_video),
        }
```

```python
    def name(self):
        # 获取指标名称
        return "Target difficulties"

    def _get_triplet_difficulty(self, anchor, positive, negative, margin):
        # 计算三元组的难度
        distance_positive = self.distance_fn(anchor, positive)
        distance_negative = self.distance_fn(anchor, negative)
        easy_targets = distance_negative > distance_positive + margin
        hard_targets = distance_negative < distance_positive
        semi_hard_targets = distance_negative < distance_positive + margin
        return (
            np.mean(easy_targets.cpu().numpy()),
            np.mean(hard_targets.cpu().numpy()),
            np.mean(semi_hard_targets.cpu().numpy())
        )
```

（6）类 MeanClassAccuracy 用于计算每种模态（音频、视频、同时包含音频和视频）的平均类别准确率，评估模型在不同模态下的分类性能。

```python
class MeanClassAccuracy(Metric):
    def __init__(self, model, dataset, device, distance_fn, new_model_sequence, args):
        super(MeanClassAccuracy, self).__init__()
        self.model = model  # 模型
        self.new_model_sequence = new_model_sequence  # 新模型序列
        self.dataset = dataset  # 数据集
        self.device = device  # 设备
        self.distance_fn = distance_fn  # 距离函数
        self.args = args  # 参数
        # 存储各个指标值的列表
        self.audio_seen = []
        self.audio_unseen = []
        self.audio_hm = []
        self.audio_recall = []
        self.audio_beta = []
        self.audio_zsl = []
        self.video_seen = []
        self.video_unseen = []
        self.video_hm = []
        self.video_recall = []
        self.video_beta = []
        self.video_zsl = []
        self.both_seen = []
        self.both_unseen = []
        self.both_hm = []
        self.both_recall = []
        self.both_beta = []
        self.both_zsl = []

    def __call__(self, outputs, target, loss_outputs):
        # 对数据集进行评估
        evaluation = evaluate_dataset_baseline(dataset_tuple=self.dataset, model=self.
        model, device=self.device,distance_fn=self.distance_fn,new_model_sequence=self.
        new_model_sequence,args=self.args)
        # 将评估结果存储到对应的列表中
        self.audio_seen.append(evaluation["audio"]["seen"])
        self.audio_unseen.append(evaluation["audio"]["unseen"])
        self.audio_hm.append(evaluation["audio"]["hm"])
        self.audio_recall.append(evaluation["audio"]["recall"])
        self.audio_beta.append(evaluation["audio"]["beta"])
        self.audio_zsl.append(evaluation["audio"]["zsl"])
```

```python
            self.video_seen.append(evaluation["video"]["seen"])
            self.video_unseen.append(evaluation["video"]["unseen"])
            self.video_hm.append(evaluation["video"]["hm"])
            self.video_recall.append(evaluation["video"]["recall"])
            self.video_beta.append(evaluation["video"]["beta"])
            self.video_zsl.append(evaluation["video"]["zsl"])

            self.both_seen.append(evaluation["both"]["seen"])
            self.both_unseen.append(evaluation["both"]["unseen"])
            self.both_hm.append(evaluation["both"]["hm"])
            self.both_recall.append(evaluation["both"]["recall"])
            self.both_beta.append(evaluation["both"]["beta"])
            self.both_zsl.append(evaluation["both"]["zsl"])

    def reset(self):
        # 重置存储指标值的列表
        self.audio_seen = []
        self.audio_unseen = []
        self.audio_hm = []
        self.audio_recall = []
        self.audio_beta = []
        self.audio_zsl = []

        self.video_seen = []
        self.video_unseen = []
        self.video_hm = []
        self.video_recall = []
        self.video_beta = []
        self.video_zsl = []

        self.both_seen = []
        self.both_unseen = []
        self.both_hm = []
        self.both_recall = []
        self.both_beta = []
        self.both_zsl = []

    def value(self):
        # 返回各个指标的平均值
        return {
            "audio_seen": np.mean(self.audio_seen),
            "audio_unseen": np.mean(self.audio_unseen),
            "audio_hm": np.mean(self.audio_hm),
            "audio_recall": np.mean(self.audio_recall, axis=0),
            "audio_beta": np.mean(self.audio_beta),
            "audio_zsl": np.mean(self.audio_zsl),

            "video_seen": np.mean(self.video_seen),
            "video_unseen": np.mean(self.video_unseen),
            "video_hm": np.mean(self.video_hm),
            "video_recall": np.mean(self.video_recall, axis=0),
            "video_beta": np.mean(self.video_beta),
            "video_zsl": np.mean(self.video_zsl),

            "both_seen": np.mean(self.both_seen),
            "both_unseen": np.mean(self.both_unseen),
            "both_hm": np.mean(self.both_hm),
            "both_recall": np.mean(self.both_recall, axis=0),
            "both_beta": np.mean(self.both_beta),
            "both_zsl": np.mean(self.both_zsl)
        }

    def name(self):
```

```
        # 获取指标名称
        return "Mean class accuracies per modality"
```

9.5.3 模型优化器

文件 optimizer.py 实现了 SAM 优化器，这是一种自适应调整学习率的优化算法。SAM 优化器是针对多模态模型的优化器，用于优化模型的参数以最小化损失函数。SAM 优化器通过对参数梯度的方向和幅度进行自适应调整，提高了模型训练的鲁棒性和收敛速度。文件 optimizer.py 的具体代码如下所示。

```python
import torch

class SAM(torch.optim.Optimizer):
    def __init__(self, params, base_optimizer, rho=0.05, adaptive=False, **kwargs):
        assert rho >= 0.0, f"Invalid rho, should be non-negative: {rho}"
        defaults = dict(rho=rho, adaptive=adaptive, **kwargs)
        super(SAM, self).__init__(params, defaults)
        self.base_optimizer = base_optimizer(self.param_groups, **kwargs)
        self.param_groups = self.base_optimizer.param_groups

    @torch.no_grad()
    def first_step(self, zero_grad=False):
        grad_norm = self._grad_norm()
        for group in self.param_groups:
            scale = group["rho"] / (grad_norm + 1e-12)

            for p in group["params"]:
                if p.grad is None: continue
                self.state[p]["old_p"] = p.data.clone()
                e_w = (torch.pow(p, 2) if group["adaptive"] else 1.0) * p.grad * scale.to(p)
                p.add_(e_w)  # 上升到局部最大值 "w + e(w)"
        if zero_grad: self.zero_grad()

    @torch.no_grad()
    def second_step(self, zero_grad=False):
        for group in self.param_groups:
            for p in group["params"]:
                if p.grad is None: continue
                p.data = self.state[p]["old_p"]  # 从 "w + e(w)" 返回到 "w"
        self.base_optimizer.step()  # 执行实际的 "sharpness-aware" 更新
        if zero_grad: self.zero_grad()

    @torch.no_grad()
    def step(self, closure=None):
        assert closure is not None, "Sharpness Aware Minimization requires closure, but it was not provided"
        closure = torch.enable_grad()(closure)  # 闭包应执行完整的前向-后向传播
        self.first_step(zero_grad=True)
        closure()
        self.second_step()

    def _grad_norm(self):
        shared_device = self.param_groups[0]["params"][0].device
        # 将所有内容放在同一个设备上，以防模型并行
        norm = torch.norm(
                    torch.stack([
                        ((torch.abs(p) if group["adaptive"] else 1.0) * p.grad).norm(p=2)
                        .to(shared_device)
                        for group in self.param_groups for p in group["params"]
                        if p.grad is not None
```

```
                    ]),
                    p=2
                )
        return norm

    def load_state_dict(self, state_dict):
        super().load_state_dict(state_dict)
        self.base_optimizer.param_groups = self.param_groups
```

SAM 通过两个步骤来提高优化的鲁棒性，第一步根据梯度方向更新参数，第二步在局部最大值处更新参数。文件 optimizer.py 实现了 SAM 优化器的两个步骤，并提供了加载状态字典的功能。在上述代码中，SAM 优化器的 first_step 方法执行了梯度方向的更新，并根据梯度的大小自适应地调整学习率，而 second_step 方法在局部最大值处更新参数。step 方法结合了这两个步骤，并要求提供一个闭包函数进行完整的前向传播和反向传播。

9.5.4　模型训练和验证

文件 train.py 实现了模型的训练和验证功能，用于训练多模态模型并评估其性能，包括训练过程中的参数更新、损失计算和性能评估，以及验证过程中的性能指标计算和记录。通过这些函数，可以对多模态模型进行训练，并监控其在训练集和验证集上的表现，最终返回最佳模型的损失和得分。文件 train.py 的具体实现流程如下。

（1）函数 train 的功能是对模型进行训练的一个步骤，包括迭代数据集中的每个批次，计算损失并更新模型参数，同时记录训练过程中的损失值和指标，并将统计信息写入日志和 TensorBoard。

```
def train(train_loader, val_loader, model, criterion, optimizer, lr_scheduler, epochs,
device, writer, metrics,
          train_stats, val_stats, log_dir, new_model_sequence, args):
    best_loss = None
    best_score = None
    best_epoch = None
    best_loss_epoch = None
    for epoch in range(epochs):
        train_loss = train_step(train_loader, model, criterion, optimizer, epoch, epochs,
            writer, device, metrics, train_stats,  args)
        val_loss, val_hm = val_step(val_loader, model, criterion, epoch, epochs, writer,
            device, metrics, val_stats, args)
        best_loss, best_loss_epoch = check_best_loss(epoch, best_loss, best_loss_epoch,
            val_loss, model, optimizer, log_dir, args)
        best_score, best_epoch = check_best_score(epoch, best_score, best_epoch, val_hm,
            model, optimizer, log_dir, args)
        if args.save_checkpoints:
            save_best_model(epoch, val_hm, model, optimizer, log_dir / "checkpoints",
                args, metric=args.best_model_criterion, checkpoint=True)
        model.optimize_scheduler(val_hm)
    if args.best_model_criterion == 'loss':
        return best_loss, best_score, best_loss_epoch
    elif args.best_model_criterion == 'score':
        return best_loss, best_score, best_epoch
```

（2）函数 add_loss_details 的功能是将批次损失的详细信息添加到当前的损失详情中，用于在训练或验证过程中累积每个损失组件的值。

```
def add_loss_details(current_loss_details, batch_loss_details):
    for key, value in current_loss_details.items():
        if key not in batch_loss_details:
            batch_loss_details[key] = value
```

```
        else:
            batch_loss_details[key] += value
    return batch_loss_details
```

（3）函数 add_logs_tensorboard 的功能是将损失详情记录到 TensorBoard 中，用于可视化训练和验证过程中的损失值。

```
def add_logs_tensorboard(batch_loss_details, writer, batch_idx, step, which_stage):
    writer.add_scalar(f"Loss/total_loss_" + which_stage, batch_loss_details['Loss/
    total_loss'] / (batch_idx), step)
    writer.add_scalar(f"Loss/loss_reg_" + which_stage, batch_loss_details['Loss/loss_reg
    '] / (batch_idx), step)
    writer.add_scalar(f"Loss/loss_cmd_rec_" + which_stage, batch_loss_details['Loss/
    loss_cmd_rec'] / (batch_idx), step)
    writer.add_scalar(f"Loss/cross_entropy_" + which_stage, batch_loss_details['Loss/
    cross_entropy'] / (batch_idx), step)
```

（4）函数 train_step 是用于训练模型的一个步骤，包括计算训练损失、更新模型参数、记录指标，并将统计信息写入日志和 TensorBoard。

```
def train_step(data_loader, model, criterion, optimizer, epoch, epochs, writer, device,
metrics, stats, args):
    logger = logging.getLogger()
    model.train()

    for metric in metrics:
        metric.reset()
    embeddings, mapping_dict = data_loader.dataset.zsl_dataset.map_embeddings_target
    batch_loss = 0
    batch_loss_details = {}
    for batch_idx, (data, target) in tqdm(enumerate(data_loader)):
        p = data["positive"]
        q = data["negative"]
        x_p_a = p["audio"].to(device)
        x_p_v = p["video"].to(device)
        x_p_t = p["text"].to(device)
        x_p_num = target["positive"].to(device)

        masks = {}
        masks['positive'] = {'audio': p['audio_mask'], 'video': p['video_mask']}

        timesteps = {}
        timesteps['positive'] = {'audio': p['timestep']['audio'], 'video': p['timestep']
        ['video']}
        inputs = (
            x_p_a, x_p_v, x_p_num, x_p_t, masks['positive'], timesteps['positive']
        )
        if args.z_score_inputs:
            inputs = tuple([(x - torch.mean(x)) / torch.sqrt(torch.var(x)) for x in inputs])

        if args.cross_entropy_loss == True:
            for i in range(inputs[2].shape[0]):
                inputs[2][i] = mapping_dict[(inputs[2][[i]]).item()]
        if args.cross_entropy_loss == True:
            loss, loss_details = model.optimize_params(*inputs, embedding_crossentropy=
            embeddings, optimize=True)
        else:
            loss, loss_details = model.optimize_params(*inputs, embedding_crossentropy=
            None, optimize=True)
        batch_loss_details = add_loss_details(loss_details, batch_loss_details)
        audio_emb, video_emb, emb_cls = model.get_embeddings(inputs[0], inputs[1],
        inputs[3], inputs[4], inputs[5])
```

```python
            outputs = torch.stack([video_emb, emb_cls], dim=0)
            batch_loss += loss.item()
            p_target = target["positive"].to(device)
            q_target = target["negative"].to(device)

            #统计信息
            iteration = len(data_loader) * epoch + batch_idx
            if iteration % len(data_loader) == 0:
                for metric in metrics:
                    if isinstance(metric, MeanClassAccuracy):
                        continue
                    metric(outputs, (p_target, q_target), (loss, loss_details))
                    for key, value in metric.value().items():
                        if "recall" in key:
                            continue
                        writer.add_scalar(
                            f"train_{key}", value, iteration
                        )
        batch_loss /= (batch_idx + 1)
        stats.update((epoch, batch_loss, None))
        add_logs_tensorboard(batch_loss_details, writer, (batch_idx + 1), len(data_loader) *
(epoch + 1), "train")
        logger.info(
            f"TRAIN\t"
            f"Epoch: {epoch}/{epochs}\t"
            f"Iteration: {iteration}\t"
            f"Loss: {batch_loss:.4f}\t"
        )
        return batch_loss
```

（5）函数 val_step 的功能是对模型进行验证的一个步骤，用于评估模型在验证集上的性能，计算验证损失和指标，记录验证过程中的损失值、指标和评分，并将统计信息写入日志和 TensorBoard。

```python
def val_step(data_loader, model, criterion, epoch, epochs, writer, device, metrics,
stats, args=None):
    logger = logging.getLogger()
    model.eval()
    for metric in metrics:
        metric.reset()
    embeddings, mapping_dict = data_loader.dataset.zsl_dataset.map_embeddings_target
    with torch.no_grad():
        batch_loss = 0
        hm_score = 0
        seen_score = 0
        unseen_score = 0
        batch_loss_details = {}
        for batch_idx, (data, target) in tqdm(enumerate(data_loader)):
            p = data["positive"]
            q = data["negative"]
            x_p_a = p["audio"].to(device)
            x_p_v = p["video"].to(device)
            x_p_t = p["text"].to(device)
            x_p_num = target["positive"].to(device)

            masks = {}
            masks['positive'] = {'audio': p['audio_mask'], 'video': p['video_mask']}

            timesteps = {}
            timesteps['positive'] = {'audio': p['timestep']['audio'], 'video': p['
timestep']['video']}
```

```python
        inputs = (
            x_p_a, x_p_v, x_p_num, x_p_t, masks['positive'], timesteps['positive']
        )
        if args.z_score_inputs:
            inputs = tuple([(x - torch.mean(x)) / torch.sqrt(torch.var(x)) for x in inputs])

        if args.cross_entropy_loss == True:
            for i in range(inputs[2].shape[0]):
                inputs[2][i] = mapping_dict[(inputs[2][[i]]).item()]
        if args.cross_entropy_loss == True:
            loss, loss_details = model.optimize_params(*inputs,
                embedding_crossentropy=embeddings, optimize=False)
        else:
            loss, loss_details = model.optimize_params(*inputs,
                embedding_crossentropy=None, optimize=False)
        batch_loss_details = add_loss_details(loss_details, batch_loss_details)
        audio_emb, video_emb, emb_cls = model.get_embeddings(inputs[0], inputs[1],
        inputs[3], inputs[4], inputs[5])
        outputs = (video_emb, emb_cls)
        batch_loss += loss.item()
        p_target = target["positive"].to(device)
        q_target = target["negative"].to(device)

        #统计信息
        iteration = len(data_loader) * epoch + batch_idx
        if iteration % len(data_loader) == 0:
            for metric in metrics:
                metric(outputs, (p_target, q_target), (loss, loss_details))
                for key, value in metric.value().items():
                    if "recall" in key:
                        continue
                    if "both_hm" in key:
                        hm_score = value
                        writer.add_scalar(
                            f"metric_val/{key}", value, iteration
                        )
                    if "both_zsl" in key:
                        zsl_score = value
                        writer.add_scalar(
                            f"metric_val/{key}", value, iteration
                        )
                    if "both_seen" in key:
                        seen_score = value
                        writer.add_scalar(
                            f"metric_val/{key}", value, iteration
                        )
                    if "both_unseen" in key:
                        unseen_score = value
                        writer.add_scalar(
                            f"metric_val/{key}", value, iteration
                        )
batch_loss /= (batch_idx + 1)
stats.update((epoch, batch_loss, hm_score))
add_logs_tensorboard(batch_loss_details, writer, (batch_idx + 1), len
(data_loader) * (epoch + 1), "val")
logger.info(
    f"VALID\t"
    f"Epoch: {epoch}/{epochs}\t"
    f"Iteration: {iteration}\t"
    f"Loss: {batch_loss:.4f}\t"
    f"ZSL score: {zsl_score:.4f}\t"
```

```
            f"Seen score: {seen_score:.4f}\t"
            f"Unseen score: {unseen_score:.4f}\t"
            f"HM: {hm_score:.4f}"
        )
    return batch_loss, hm_score
```

9.5.5 模型的评估

文件 get_evaluation.py 的功能是定义函数 get_evaluation,通过此函数执行模型的评估过程,包括加载模型和数据集、进行模型推断、在验证集和测试集上计算性能指标,最后将评估结果记录到日志和 TensorBoard 中。文件 get_evaluation.py 的具体代码如下所示。

```
def get_evaluation(args):

    config = load_args(args.load_path_stage_B)
    config.root_dir = args.root_dir
    if config.input_size is not None:
        config.input_size_audio = config.input_size
        config.input_size_video = config.input_size
    assert config.retrain_all, f"--retrain_all flag is not set in load_path_stage_B. Are you sure this is the correct path?. {args.load_path_stage_B}"
    fix_seeds(config.seed)
    logger, eval_dir, test_stats, tb_writer = setup_evaluation(args, config.__dict__.keys())

    if args.dataset_name == "AudioSetZSL":
        val_all_dataset = AudioSetZSLDataset(
            args=config,
            dataset_split="val",
            zero_shot_mode="all",
        )
        test_dataset = AudioSetZSLDataset(
            args=config,
            dataset_split="test",
            zero_shot_mode="all",
        )
    elif args.dataset_name == "VGGSound":
        val_all_dataset = VGGSoundDataset(
            args=config,
            dataset_split="val",
            zero_shot_mode=None,
        )
        test_dataset = VGGSoundDataset(
            args=config,
            dataset_split="test",
            zero_shot_mode=None,
        )
    elif args.dataset_name == "UCF":
        val_all_dataset = UCFDataset(
            args=config,
            dataset_split="val",
            zero_shot_mode=None,
        )
        test_dataset = UCFDataset(
            args=config,
            dataset_split="test",
            zero_shot_mode=None,
        )
    elif args.dataset_name == "ActivityNet":
        val_all_dataset = ActivityNetDataset(
            args=config,
            dataset_split="val",
```

9.5 多模态模型

```python
            zero_shot_mode=None,
        )
        test_dataset = ActivityNetDataset(
            args=config,
            dataset_split="test",
            zero_shot_mode=None,
        )
    else:
        raise NotImplementedError()

    contrastive_val_dataset = ContrastiveDataset(val_all_dataset)
    contrastive_test_dataset = ContrastiveDataset(test_dataset)

    if config.selavi == False:
        collator_test = DefaultCollator(mode=args.batch_seqlen_test, max_len=args.batch_
            seqlen_test_maxlen, trim=args.batch_seqlen_test_trim)
    elif config.selavi==True:
        collator_test = DefaultCollator(mode=args.batch_seqlen_test, max_len=args.batch_
            seqlen_test_maxlen, trim=args.batch_seqlen_test_trim,rate_video=1, rate_audio=1)
    final_val_loader = data.DataLoader(
        dataset=contrastive_val_dataset,
        collate_fn=collator_test,
        batch_size=args.eval_bs,
        num_workers=args.eval_num_workers,
    )

    final_test_loader = data.DataLoader(
        dataset=contrastive_test_dataset,
        collate_fn=collator_test,
        batch_size=args.eval_bs,
        num_workers=args.eval_num_workers,
    )

    model_params = get_model_params(
        config.lr, config.reg_loss, config.embedding_dropout, config.decoder_dropout,
        config.additional_dropout,config.embeddings_hidden_size, config.decoder_hidden_size,
        config.embeddings_batch_norm, config.rec_loss,config.cross_entropy_loss,
        config.transformer_use_embedding_net, config.transformer_dim, config.transformer
        _depth, config.transformer_heads,
        config.transformer_dim_head, config.transformer_mlp_dim, config.transformer_dropout,
        config.transformer_embedding_dim, config.transformer_embedding_time_len, config.
        transformer_embedding_dropout,
        config.transformer_embedding_time_embed_type, config.
        transformer_embedding_fourier_scale, config.transformer_embedding_embed_augment_
        position,
        config.lr_scheduler, config.optimizer, config.use_self_attention,
        config.use_cross_attention, config.transformer_average_features,
        config.audio_only, config.video_only, config.transformer_use_class_token,
        config.transformer_embedding_modality,
        config.modality, config.word_embeddings
    )
    if config.new_model_sequence==True:
        model_A = ClipClap_model(params_model=model_params, input_size_audio=config.
        input_size_audio, input_size_video=config.input_size_video)
    else:
        raise AttributeError("No correct model_A name.")
print_model_size(model_A, logger)
logging.info(model_A)
model_B = copy.deepcopy(model_A)
# weights_path_stage_A = list(args.load_path_stage_A.glob("*_score.pt"))[0]
weights_path_stage_A = list(args.load_path_stage_A.glob(f"*_{config.
best_model_criterion}.pt"))[0]
epoch_A = load_model_weights(weights_path_stage_A, model_A)
```

```python
    weights_path_stage_B = list((args.load_path_stage_B / "checkpoints").glob
    (f"*_ckpt_{epoch_A - 1}.pt"))[0]
    _ = load_model_weights(weights_path_stage_B, model_B)
    model_A.to(config.device)
    model_B.to(config.device)
    results = test(
        eval_name=args.eval_name,
        val_dataset=(val_all_dataset, final_val_loader),
        test_dataset=(test_dataset, final_test_loader),
        model_A=model_A,
        model_B=model_B,
        device=args.device,
        distance_fn=config.distance_fn,
        test_stats=test_stats,
        eval_dir=eval_dir,
        new_model_sequence=config.new_model_sequence,
        args=config,
        save_performances=args.eval_save_performances
    )
    # Tensorboard 超参数记录
    log_hparams(tb_writer, config, results['both'])

    logger.info("FINISHED")
    # return results['both']

if __name__ == "__main__":
    args, eval_args = args_main()
    get_evaluation(eval_args)
```

9.5.6 主文件

文件 main.py 是整个模型训练和评估功能的主程序，包括根据参数运行不同阶段的训练和评估、加载数据集和模型、设置训练过程中的优化器和学习率调度器、执行训练和评估，以及记录日志和模型性能指标。文件 main.py 的具体实现流程如下。

（1）函数 run 的功能是根据指定的运行模式（stage-1、stage-2、eval、all），依次执行对应的训练阶段和评估过程。stage-1 阶段负责模型的第一次训练，stage-2 阶段在第一阶段的基础上进行微调，eval 阶段对训练好的模型进行评估。

```python
def run():
    # 获取命令行参数和评估参数
    args, eval_args = args_main()
    # 运行模式（stage-1、stage-2、eval、all）
    run_mode = args.run
    best_epoch = None
    # 如果是 stage-1 或者 all 模式
    if run_mode == 'stage-1' or run_mode == 'all':
        # 设定为不重新训练全部模型，并且不保存检查点
        args.retrain_all = False
        args.save_checkpoints = False
        # 执行第一阶段训练，并获取最佳 epoch
        path_stage_1, best_epoch = main(args)
        # 设置评估参数的 stage A 路径
        eval_args.load_path_stage_A = path_stage_1

    # 如果是 stage-2 或者 all 模式
    if run_mode == 'stage-2' or run_mode == 'all':
        # 如果还未设置 stage-2 的参数，则设置
        args.retrain_all = True
        args.save_checkpoints = True
```

```
        print('best_epoch after stage1: ', best_epoch)

        # 如果有最佳 epoch,则只训练所需的 epoch
        if best_epoch is not None:
            args.epochs = best_epoch + 1
        print('args.epochs after stage1: ', args.epochs)

        # 执行第二阶段训练,并获取路径
        path_stage_2, _ = main(args)
        # 设置评估参数的 stage B 路径
        eval_args.load_path_stage_B = path_stage_2

    # 如果是 eval 或者 all 模式
    if run_mode == 'eval' or run_mode == 'all':
        # 确保加载了 stage A 和 stage B 的路径
        assert eval_args.load_path_stage_A != None
        assert eval_args.load_path_stage_B != None
        # 进行评估
        get_evaluation(eval_args)
```

（2）函数 main(args) 的功能是根据传入的参数配置进行模型训练的主控制逻辑，包括设置实验日志、加载数据集、初始化模型，进行训练并记录训练过程中的指标和日志信息，最后返回最佳模型的保存路径和最佳轮次。

```
def main(args):
    # 如果指定了输入大小,则将音频和视频输入大小设置为相同
    if args.input_size is not None:
        args.input_size_audio = args.input_size
        args.input_size_video = args.input_size
    # 设置随机种子
    fix_seeds(args.seed)

    # 设置实验日志记录器、日志目录、Tensorboard 写入器以及训练和验证统计信息
    logger, log_dir, writer, train_stats, val_stats = setup_experiment(args, "epoch",
    "loss", "hm")

    logger.info("Git commit hash: " + get_git_revision_hash())

    # 根据数据集名称加载训练集、验证集和用于全部类别的验证集
    if args.dataset_name == "AudioSetZSL":
        train_dataset = AudioSetZSLDataset(
            args=args,
            dataset_split="train",
            zero_shot_mode="seen",
        )

        val_dataset = AudioSetZSLDataset(
            args=args,
            dataset_split="val",
            zero_shot_mode="seen",
        )

        train_val_dataset = AudioSetZSLDataset(
            args=args,
            dataset_split="train_val",
            zero_shot_mode="seen",
        )

        val_all_dataset = AudioSetZSLDataset(
            args=args,
            dataset_split="val",
            zero_shot_mode="all",
```

```python
            )
        elif args.dataset_name == "VGGSound":
            if args.retrain_all==False:
                train_dataset = VGGSoundDataset(
                    args=args,
                    dataset_split="train",
                    zero_shot_mode="train",
                )

            if args.retrain_all==True:
                train_val_dataset = VGGSoundDataset(
                    args=args,
                    dataset_split="train_val",
                    zero_shot_mode=None,
                )

            val_all_dataset = VGGSoundDataset(
                args=args,
                dataset_split="val",
                zero_shot_mode=None,
            )
        elif args.dataset_name == "UCF":
            if args.retrain_all==False:
                train_dataset = UCFDataset(
                    args=args,
                    dataset_split="train",
                    zero_shot_mode="train",
                )
            # 如果在第二阶段，则在训练和验证集合上进行训练
            if args.retrain_all==True:
                train_val_dataset = UCFDataset(
                    args=args,
                    dataset_split="train_val",
                    zero_shot_mode=None,
                )

            val_all_dataset = UCFDataset(
                args=args,
                dataset_split="val",
                zero_shot_mode=None,
            )
        elif args.dataset_name == "ActivityNet":
            if args.retrain_all==False:
                train_dataset = ActivityNetDataset(
                    args=args,
                    dataset_split="train",
                    zero_shot_mode="train",
                )

            if args.retrain_all==True:
                train_val_dataset = ActivityNetDataset(
                    args=args,
                    dataset_split="train_val",
                    zero_shot_mode=None,
                )
            val_all_dataset = ActivityNetDataset(
                args=args,
                dataset_split="val",
                zero_shot_mode=None,
            )
        else:
            raise NotImplementedError()
```

```python
# 如果不是第二阶段,则对训练集进行对比学习数据集的处理
if args.retrain_all==False:
    contrastive_train_dataset = ContrastiveDataset(train_dataset)
# 如果是第二阶段,则对训练+验证集进行对比学习数据集的处理
if args.retrain_all==True:
    contrastive_train_val_dataset = ContrastiveDataset(train_val_dataset)
# 对用于全部类别的验证集进行对比学习数据集的处理
contrastive_val_all_dataset = ContrastiveDataset(val_all_dataset)

# 如果不是第二阶段,则设置训练集的采样器
if args.retrain_all==False:
    train_sampler = SamplerFactory(logger).get(
        class_idxs=list(contrastive_train_dataset.target_to_indices.values()),
        batch_size=args.bs,
        n_batches=args.n_batches,
        alpha=1, # 使每个批次中的类别均匀分布
        kind='random'
    )

# 如果是第二阶段,则设置"训练+验证集"的采样器
if args.retrain_all==True:
    train_val_sampler = SamplerFactory(logger).get(
        class_idxs=list(contrastive_train_val_dataset.target_to_indices.values()),
        batch_size=args.bs,
        n_batches=args.n_batches,
        alpha=1,
        kind='random'
    )

# 设置全部类别的验证集的采样器
val_all_sampler = SamplerFactory(logger).get(
    class_idxs=list(contrastive_val_all_dataset.target_to_indices.values()),
    batch_size=args.bs,
    n_batches=args.n_batches,
    alpha=1,
    kind='random'
)
# 如果不是自注意力模型,则设置训练和测试数据的默认处理器
if args.selavi==False:
    collator_train = DefaultCollator(mode=args.batch_seqlen_train, max_len=args.
        batch_seqlen_train_maxlen, trim=args.batch_seqlen_train_trim,)
    collator_test = DefaultCollator(mode=args.batch_seqlen_test, max_len=args.
        batch_seqlen_test_maxlen, trim=args.batch_seqlen_test_trim)
# 如果是自注意力模型,则设置自注意力模型的数据处理器
elif args.selavi==True:
    collator_train = DefaultCollator(mode=args.batch_seqlen_train, max_len=args.
        batch_seqlen_train_maxlen,trim=args.batch_seqlen_train_trim,rate_video=1, rate_audio=1)
    collator_test = DefaultCollator(mode=args.batch_seqlen_test, max_len=args.
        batch_seqlen_test_maxlen,trim=args.batch_seqlen_test_trim,rate_video=1, rate_audio=1)

# 如果不是第二阶段,则设置训练数据加载器
if args.retrain_all==False:
    train_loader = data.DataLoader(
        dataset=contrastive_train_dataset,
        batch_sampler=train_sampler,
        collate_fn=collator_train,
        num_workers=4
    )

# 设置全部类别的验证集的测试数据加载器
final_test_loader=data.DataLoader(
    dataset=contrastive_val_all_dataset,
```

```python
        collate_fn=collator_test,
        batch_size=args.bs,
        num_workers=4
    )

    # 如果是第二阶段,则设置"训练+验证数据"加载器
    if args.retrain_all==True:
        train_val_loader = data.DataLoader(
            dataset=contrastive_train_val_dataset,
            batch_sampler=train_val_sampler,
            collate_fn=collator_train,
            num_workers=4
        )

    # 设置全部类别的验证集的验证数据加载器
    val_all_loader = data.DataLoader(
        dataset=contrastive_val_all_dataset,
        batch_sampler=val_all_sampler,
        collate_fn=collator_test,
        num_workers=4
    )
    # 返回模型参数的字典
    model_params = get_model_params(
        args.lr, args.reg_loss, args.embedding_dropout,
        args.decoder_dropout, args.additional_dropout, args.embeddings_hidden_size, args
        .decoder_hidden_size,
        args.embeddings_batch_norm, args.rec_loss, args.cross_entropy_loss,
        args.transformer_use_embedding_net, args.transformer_dim, args.transformer_depth
        , args.transformer_heads,
        args.transformer_dim_head, args.transformer_mlp_dim, args.transformer_dropout,
        args.transformer_embedding_dim, args.transformer_embedding_time_len, args.
        transformer_embedding_dropout,
        args.transformer_embedding_time_embed_type, args.
        transformer_embedding_fourier_scale, args.transformer_embedding_embed_augment_
        position,
        args.lr_scheduler, args.optimizer, args.use_self_attention,
        args.use_cross_attention, args.transformer_average_features,
        args.audio_only, args.video_only, args.transformer_use_class_token, args.
        transformer_embedding_modality,
        args.modality, args.word_embeddings
    )
    # 如果是新的模型序列,则创建模型
    if args.new_model_sequence==True:
        model = ClipClap_model(model_params, input_size_audio=args.input_size_audio,
        input_size_video=args.input_size_video)
    else:
        raise AttributeError("No correct model name.")
    # 打印模型大小信息
    print_model_size(model, logger)
    model.to(args.device)

    # 根据参数中指定的距离函数获取距离函数对象
    distance_fn = getattr(sys.modules[__name__], args.distance_fn)()
    # 定义评估指标列表
    metrics = [
        MeanClassAccuracy(model=model, dataset=(val_all_dataset, final_test_loader),
        device=args.device, distance_fn=distance_fn, new_model_sequence=args.new_model_
        sequence, args=args)
                ]
    logger.info(model)
    logger.info(None)
    logger.info(None)
    logger.info(None)
```

```
    logger.info([metric.__class__.__name__ for metric in metrics])
    optimizer = None
    lr_scheduler = None

    # 进行训练并获取最佳损失、最佳分数和最佳轮次
    best_loss, best_score, best_epoch = train(
        train_loader=train_val_loader if args.retrain_all else train_loader,
        val_loader=val_all_loader,
        model=model,
        criterion=None,
        optimizer=optimizer,
        lr_scheduler=None,
        epochs=args.epochs,
        device=args.device,
        writer=writer,
        metrics=metrics,
        train_stats=train_stats,
        val_stats=val_stats,
        log_dir=log_dir,
        new_model_sequence=args.new_model_sequence,
        args=args
    )
    logger.info(f"FINISHED. Run is stored at {log_dir}")
    return log_dir, best_epoch
```

9.6 调试运行

至此，本项目的主要源码介绍完毕，接下来开始运行本项目。简单来说，本项目的调试运行步骤如下。

（1）安装依赖，具体命令如下所示。

```
conda env create -f clipclap.yml
conda activate clipclap
```

（2）下载特征。下载 CLIP 和 CLAP 特征数据并解压缩，具体命令如下所示。

```
unzip data.zip
```

（3）训练模型。

- 对于 UCF-GZSL 数据集，使用以下命令进行训练。

```
nohup python3 main.py --cfg config/clipclap.yaml \
                --device cuda:6 \
                --root_dir /path/to/UCF \
                --log_dir logs/ClipClap_UCF \
                --dataset_name UCF \
                --epochs 20 \
                --lr 0.00007 \
                --use_wavcaps_embeddings True \
                --modality both \
                --word_embeddings both \
                --run all > logs/ClipClap_UCF.log &
```

- 对于 ActivityNet-GZSL 数据集，使用以下命令进行训练。

```
nohup python3 main.py --cfg config/clipclap.yaml \
                --device cuda:6 \
                --root_dir /path/to/ActivityNet \
                --log_dir logs/ClipClap_ActivityNet \
```

```
                        --dataset_name ActivityNet \
                        --epochs 15 \
                        --lr 0.0001 \
                        --use_wavcaps_embeddings True \
                        --modality both   \
                        --word_embeddings both   \
                        --run all > logs/ClipClap_ActivityNet.log &
```

- 对于 VGGSound-GZSL 数据集，使用以下命令进行训练。

```
nohup python3 main.py --cfg config/clipclap.yaml \
                        --device cuda:5 \
                        --root_dir /path/to/VGGSound   \
                        --log_dir logs/ClipClap_VGGSound \
                        --dataset_name VGGSound \
                        --epochs 15 \
                        --lr 0.0001 \
                        --use_wavcaps_embeddings True \
                        --modality both   \
                        --word_embeddings both   \
                        --run all > logs/ClipClap_VGGSound.log &
```

（4）评估模型。评估可以在训练时自动进行，也可以手动进行。对于手动评估，可以运行以下命令实现。

```
python3 get_evaluation.py --cfg config/clipclap.yaml \
                        --load_path_stage_A PATH_STAGE_A \
                        --load_path_stage_B PATH_STAGE_B \
                        --dataset_name DATASET_NAME \
                        --root_dir ROOT_DIR
```

（5）提取特征。如果需要从头开始提取特征，首先安装依赖并下载 WavCaps 的模型权重，然后运行以下命令提取 CLIP/CLAP 特征。

```
python3 clip_feature_extraction/get_clip_features_activitynet.py
python3 clip_feature_extraction/get_clip_features_ucf.py
python3 clip_feature_extraction/get_clip_features_vggsound.py
```

（6）获取类别嵌入。运行以下脚本以获取类别嵌入。

```
python3 clip_embeddings_extraction/get_clip_embeddings_activitynet.py
python3 clip_embeddings_extraction/get_clip_embeddings_ucf.py
python3 clip_embeddings_extraction/get_clip_embeddings_vggsound.py
```

上述步骤将实现多模态模型的训练、评估以及特征提取功能，其中基于 UCF-GZSL 数据集模型的输入音频特征、输入视觉特征、学习到的"音频-视觉"嵌入的可视化结果如图 9-1 所示。

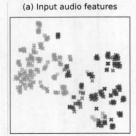
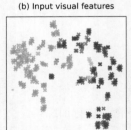
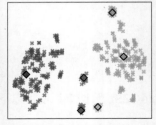

图 9-1　基于 UCF-GZSL 数据集模型的可视化图

基于 ActivityNet-GZSL 数据集模型的输入音频特征、输入视觉特征、学习到的"音频-视觉"

嵌入的可视化结果如图 9-2 所示。

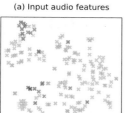
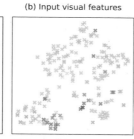
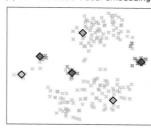

图 9-2　基于 ActivityNet-GZSL 数据集模型的可视化图

第10章 基于 Diffusion Transformer 的文生图系统

文生图大模型是一种通过自然语言生成图像的人工智能技术，基于大规模的深度学习模型，如生成对抗网络（GAN）或变分自编码器（VAE）。文生图大模型通过学习大量文本和图像数据之间的对应关系，能够理解和生成与给定文本描述相符的图像。其基础包括多模态数据处理、文本和图像的联合嵌入表示、模型训练、图像生成算法以及评估和优化技术。文生图大模型技术在内容创作、设计、虚拟现实等领域具有广泛的应用前景。本章将详细讲解文生图大模型的基础知识。

10.1 Diffusion Transformer 介绍

Diffusion Transformer（DiT）是一种结合了扩散模型与 Transformer 架构的生成模型，其核心思想是利用扩散过程和 Transformer 的自注意力机制来生成高质量的图像。Diffusion Transformer 模型主要用于图像生成任务，旨在通过更强大的特征提取和建模能力来提升生成效果。

10.1.1 Diffusion Transformer 的特点

Diffusion Transformer 主要应用于高质量图像生成、图像修复、图像超分辨率等领域，在需要生成逼真且细腻的图像时，DiT 展现出了强大的潜力。DiT 的主要特点如下所示。

- Transformer 架构：DiT 使用了 Transformer 作为其主要架构。Transformer 原本广泛用于自然语言处理任务，现在被应用于图像生成中。Transformer 的自注意力机制能够有效捕捉全局信息，使得生成的图像具有更高的分辨率和细节。
- 扩散模型：扩散模型是一种逐步将噪声引导到目标分布的生成模型。在 DiT 中，扩散过程用于逐步生成图像，这一过程通过模型多次迭代，从随机噪声中引导生成清晰的图像。
- 高效的生成能力：相比传统的卷积神经网络或者其他生成模型，DiT 结合了 Transformer 的长程依赖捕捉能力，使得其在生成多样化和高分辨率图像方面表现出色。
- 可扩展性：DiT 通过使用更大的模型和更多的数据集，可以扩展到更大规模的图像生成任务，能够处理更复杂的图像生成需求。

10.1.2 Stable Diffusion 和 Diffusion Transformer 的区别和联系

Stable Diffusion 和 Diffusion Transformer 是两种不同的模型架构和技术概念，但它们都涉及扩散模型（Diffusion Model）这一类生成模型。下面将详细讲解两者的主要区别和关系。

1. Stable Diffusion

Stable Diffusion 是一种基于扩散模型的图像生成方法，它通过逐步将噪声添加到数据中，然后训练一个模型来逆转这个过程，最终生成高质量的图像。Stable Diffusion 是一种特定类型的扩散模型，强调稳定性和控制生成过程。Stable Diffusion 的主要特点如下。

- 高分辨率图像生成：Stable Diffusion 能够生成高分辨率的图像，并且生成过程是稳定的，不容易陷入局部最优。
- 使用潜在空间（Latent Space）：Stable Diffusion 通过在潜在空间中执行扩散过程，提高了计算效率，并且能够更好地控制生成结果。

2. Diffusion Transformer

Diffusion Transformer 是一种将 Transformer 与扩散模型相结合的架构，旨在利用 Transformer 强大的序列建模能力来处理扩散过程中的时间步。该模型可以用于文本生成、图像生成或其他任务中的扩散过程。Diffusion Transformer 的主要特点如下。

- Transformer 架构：利用 Transformer 的自注意力机制来捕捉扩散过程中的复杂依赖关系。
- 多模态生成：可以应用于处理文本、图像等多模态数据的生成任务。

3. 区别

（1）模型架构

- Stable Diffusion 主要基于扩散模型中的经典架构，它通常使用 UNet 结构来处理图像生成任务。
- Diffusion Transformer 则结合了 Transformer 和扩散模型，采用了 Transformer 的自注意力机制来建模扩散过程中的依赖关系。

（2）应用场景

- Stable Diffusion 主要应用于高分辨率图像生成，强调图像质量和控制生成过程。
- Diffusion Transformer 则是一个更为通用的架构，可以用于多种生成任务，包括文本、图像和多模态数据的生成。

4. 联系

二者都属于扩散模型的一种实现方式，都是通过逐步还原噪声来生成数据。其中，Stable Diffusion 是对传统扩散模型的一种改进，专注于图像生成的稳定性和效率；而 Diffusion Transformer 则将扩散模型与 Transformer 结合，探索在不同模态数据上的应用。

总的来说，Stable Diffusion 和 Diffusion Transformer 都是基于扩散模型的生成技术，但它们有着不同的架构和应用场景。Stable Diffusion 专注于图像生成的稳定性，而 Diffusion Transformer 则试图将 Transformer 引入扩散过程，应用于更广泛的生成任务中。

10.2 项目介绍

本项目旨在实现一个基于扩散模型和 Transformer 架构的 DiT 生成模型，主要用于生成高质量图像和实现特征提取功能。

- 生成高质量图像：项目通过使用扩散模型和 Transformer 架构，可以根据输入的条件或噪声生成清晰、细节丰富的图像。这意味着用户可以利用该模型来创造新的视觉内容，例如艺术作品、照片等。
- 特征提取：项目还可以从输入图像中提取有用的特征表示。这些特征可以用于后续的任务，如分类、检索或其他计算机视觉应用。这种特征提取能力使得模型在理解和分析图像方面具有更强的表现。

本项目包含多个核心功能模块，包括模型训练、特征提取、模型下载、数据处理等，支持在分布式环境下高效训练和应用预训练模型。本项目的功能模块主要包括以下几个部分。

- 模型训练模块：通过多个训练脚本（如 train_amp.py、train_baseline.py、train_features.py 等）实现对 DiT 模型的训练。每个脚本可能针对不同的训练配置或技术进行优化，例如 AMP（自动混合精度）和标准训练。
- 特征提取模块：从输入图像中提取特征，利用预训练的 DiT 模型将图像映射到潜在空间，并保存提取的特征和标签。

- 模型下载模块：实现预训练模型的下载功能，确保用户能够获取所需的模型权重，支持从本地加载自定义模型。
- 数据处理模块：数据处理功能包括图像的预处理、中心裁剪、归一化等，通过 torchvision 库提供的转换功能进行操作，确保数据符合模型输入要求。
- 分布式训练支持模块：使用 PyTorch 的 DDP（Distributed Data Parallel，分布式数据并行）功能，支持在多个 GPU 上进行高效的分布式训练，提高模型训练速度和效率。
- 日志记录模块：创建并管理日志记录，跟踪训练过程中的重要信息，帮助开发者进行调试和性能分析。
- 辅助函数模块：提供了各种辅助函数，如更新 EMA（Exponential Moving Average，指数移动平均）模型、设置模型参数的梯度标志、清理分布式训练等函数，便于管理训练过程中的各项任务。

整体而言，这些功能模块协同工作，形成一个完整的深度学习训练和推理流程，旨在生成高质量图像和有效提取特征。

10.3 准备预训练模型

本项目会采用不同的训练策略、参数或设置，以实现对 DiT 模型的训练和优化。文件 download.py 用于管理和下载预训练的 DiT 模型，如果指定的模型名称在预定义的预训练模型集合中，则会从网络下载该模型；否则，会尝试从指定的本地路径加载模型。如果模型不在本地路径中，脚本将会抛出错误提示。此外，文件 download.py 还可以下载所有列出的预训练模型到本地。

```python
from torchvision.datasets.utils import download_url
import torch
import os

# 预定义的预训练模型集合
pretrained_models = {'DiT-XL-2-512x512.pt', 'DiT-XL-2-256x256.pt'}

def find_model(model_name):
    """
    查找预训练的 DiT 模型, 如果必要则下载它。
    另外, 也可以从本地路径加载模型。
    """
    if model_name in pretrained_models:  # 查找/下载预训练的 DiT 模型
        return download_model(model_name)
    else:  # 从自定义路径加载 DiT 模型
        assert os.path.isfile(model_name), f'无法在{model_name}找到 DiT 检查点'
        checkpoint = torch.load(model_name, map_location=lambda storage, loc: storage)
        if "ema" in checkpoint:  # 支持来自 train.py 的检查点
            checkpoint = checkpoint["ema"]
        return checkpoint

def download_model(model_name):
    """
    从网络下载预训练的 DiT 模型。
    """
    assert model_name in pretrained_models
    local_path = f'pretrained_models/{model_name}'
    if not os.path.isfile(local_path):
        os.makedirs('pretrained_models', exist_ok=True)  # 创建模型存储目录
        web_path = f'https://dl.fbaipublicfiles.com/DiT/models/{model_name}'
        download_url(web_path, 'pretrained_models')  # 下载模型文件
```

```
        model = torch.load(local_path, map_location=lambda storage, loc: storage)
        # 加载下载的模型
        return model

if __name__ == "__main__":
    # 下载所有 DiT 检查点
    for model in pretrained_models:
        download_model(model)
    print('完成。')
```

10.4 扩散模型核心模块

本项目的"diffusion"目录包含了与扩散模型相关的核心功能模块，主要用于实现和优化扩散过程的采样与训练。这些模块支持高效的扩散算法、时间步的重新采样及多种采样策略，以减小训练过程中的方差并提高模型性能。整体上，这些功能促进了生成任务的有效执行，支持扩散模型的构建、训练和优化。

10.4.1 计算高斯分布概率

文件 diffusion_utils.py 用于计算高斯分布的相关概率，包括 Kullback-Leibler（KL）散度、连续高斯分布的对数似然和离散高斯分布的对数似然。

```
import torch as th
import numpy as np
def normal_kl(mean1, logvar1, mean2, logvar2):
    """
    计算两个高斯分布之间的 KL 散度。
    参数:
        mean1: 第一个高斯分布的均值
        logvar1: 第一个高斯分布的对数方差
        mean2: 第二个高斯分布的均值
        logvar2: 第二个高斯分布的对数方差
    返回: 两个高斯分布之间的 KL 散度
    """
    tensor = None
    for obj in (mean1, logvar1, mean2, logvar2):
        if isinstance(obj, th.Tensor):
            tensor = obj
            break
    assert tensor is not None, "至少一个参数必须是 Tensor"

    # 强制将方差转换为 Tensor。广播可以将标量转换为 Tensor,
    # 但对 th.exp() 不适用。
    logvar1, logvar2 = [
        x if isinstance(x, th.Tensor) else th.tensor(x).to(tensor)
        for x in (logvar1, logvar2)
    ]

    return 0.5 * (
        -1.0
        + logvar2
        - logvar1
        + th.exp(logvar1 - logvar2)
        + ((mean1 - mean2) ** 2) * th.exp(-logvar2)
    )


def approx_standard_normal_cdf(x):
```

```
    """
    计算标准正态分布的累积分布函数的快速近似值。
    参数：
        x：输入值
    返回：标准正态分布的累积分布函数值
    """
    return 0.5 * (1.0 + th.tanh(np.sqrt(2.0 / np.pi) * (x + 0.044715 * th.pow(x, 3))))


def continuous_gaussian_log_likelihood(x, *, means, log_scales):
    """
    计算连续高斯分布的对数似然。
    参数：
        x：目标值
        means：高斯均值 Tensor
        log_scales：高斯对数标准差 Tensor
    返回：与 x 形状相同的对数概率 Tensor（以 nats 为单位）
    """
    centered_x = x - means
    inv_stdv = th.exp(-log_scales)
    normalized_x = centered_x * inv_stdv
    log_probs = th.distributions.Normal(th.zeros_like(x), th.ones_like(x)).log_prob(normalized_x)
    return log_probs


def discretized_gaussian_log_likelihood(x, *, means, log_scales):
    """
    计算离散高斯分布的对数似然，适用于给定图像的离散化。
    参数：
        x：目标图像，假定为 uint8 值，已缩放到 [-1, 1] 范围
        means：高斯均值 Tensor
        log_scales：高斯对数标准差 Tensor
    返回：与 x 形状相同的对数概率 Tensor（以 nats 为单位）
    """
    assert x.shape == means.shape == log_scales.shape
    centered_x = x - means
    inv_stdv = th.exp(-log_scales)
    plus_in = inv_stdv * (centered_x + 1.0 / 255.0)
    cdf_plus = approx_standard_normal_cdf(plus_in)
    min_in = inv_stdv * (centered_x - 1.0 / 255.0)
    cdf_min = approx_standard_normal_cdf(min_in)
    log_cdf_plus = th.log(cdf_plus.clamp(min=1e-12))
    log_one_minus_cdf_min = th.log((1.0 - cdf_min).clamp(min=1e-12))
    cdf_delta = cdf_plus - cdf_min
    log_probs = th.where(
        x < -0.999,
        log_cdf_plus,
        th.where(x > 0.999, log_one_minus_cdf_min, th.log(cdf_delta.clamp(min=1e-12))),
    )
    assert log_probs.shape == x.shape
    return log_probs
```

上述代码主要用于处理与高斯分布相关的数学计算，具体如下。
- KL 散度：用于衡量两个高斯分布之间的差异。
- 连续高斯对数似然：用于计算某个值在连续高斯分布下的对数似然。
- 离散高斯对数似然：用于计算离散化后的图像在高斯分布下的对数似然。

10.4.2 实现扩散模型

文件 gaussian_diffusion.py 实现了扩散模型的核心功能，包括采样和训练过程。此文件定义了

多种方法来生成样本，使用逐步反向扩散过程（如 DDIM 和其他变体），并计算变分下界（VLB）以优化模型性能。通过这些功能，文件支持从高维数据中生成样本并评估模型的训练损失。

（1）ModelMeanType 是一个枚举类，用于定义扩散模型中不同类型的输出均值。这些均值决定了模型在生成过程中所预测的内容。

```
class ModelMeanType(enum.Enum):

    PREVIOUS_X = enum.auto()
    START_X = enum.auto()
    EPSILON = enum.auto()
```

对上述代码的具体说明如下。

- PREVIOUS_X：这个选项表示模型的输出是对上一个时间步 x_{t-1} 的预测。也就是说，模型会根据当前的噪声图像预测出前一步的图像。这种方式在逐步去噪的过程中非常常见。
- START_X：这个选项表示模型的输出是对初始图像 x_0 的预测。这种方式通常用于在扩散过程中重建原始图像，可能在一些特定的采样任务中使用。
- EPSILON：这个选项表示模型输出的是当前图像中的噪声部分（通常表示为 ϵ）。模型试图预测加到图像上的噪声，这样在去噪时可以更准确地恢复原始图像。

（2）ModelVarType 是一个枚举类，用于定义扩散模型中不同类型的方差处理方式。方差在扩散模型中扮演着重要角色，它影响生成过程中的噪声强度和生成图像的多样性。

```
class ModelVarType(enum.Enum):

    LEARNED = enum.auto()
    FIXED_SMALL = enum.auto()
    FIXED_LARGE = enum.auto()
    LEARNED_RANGE = enum.auto()
```

（3）下面代码定义了一个枚举类 LossType，用于表示不同的损失函数类型。它包括均方误差（MSE）损失、重缩放的均方误差损失以及变分下界损失等。该枚举类还提供了一个方法 is_vb()，用于判断当前损失类型是否属于变分损失。

```
class LossType(enum.Enum):
    MSE = enum.auto()    # 使用原始均方误差损失（学习方差时也使用 KL 损失）
    RESCALED_MSE = (
        enum.auto()
    )    # 使用原始均方误差损失（学习方差时使用重缩放的 KL 损失）
    KL = enum.auto()    # 使用变分下界损失
    RESCALED_KL = enum.auto()    # 类似于 KL，但重缩放以估计完整的变分下界

    def is_vb(self):
        # 判断当前损失类型是否为变分损失
        return self == LossType.KL or self == LossType.RESCALED_KL
```

（4）下面代码定义了一个名为 _warmup_beta 的函数，用于生成一个线性增加的 beta 值数组，主要用于扩散模型训练过程中的 beta 参数热身（warm-up）。该函数接受起始 beta 值、结束 beta 值、扩散时间步数和热身比例，并根据这些参数生成一个包含 beta 值的数组。

```
def _warmup_beta(beta_start, beta_end, num_diffusion_timesteps, warmup_frac):
    # 创建一个全为 beta_end 的数组，长度为 num_diffusion_timesteps
    betas = beta_end * np.ones(num_diffusion_timesteps, dtype=np.float64)

    # 计算热身阶段的时间步数
    warmup_time = int(num_diffusion_timesteps * warmup_frac)
```

```python
    # 用线性插值填充热身阶段的 beta 值
    betas[:warmup_time] = np.linspace(beta_start, beta_end, warmup_time, dtype=np.float64)

    return betas
```

（5）下面代码定义了一个名为 get_beta_schedule 的函数，用于根据给定的调度方式生成 beta 值数组。该函数根据不同的调度类型（如线性、平方等）计算并返回 beta 值，以便在扩散模型的训练过程中使用。

```python
def get_beta_schedule(beta_schedule, *, beta_start, beta_end, num_diffusion_timesteps):
    # 根据指定的调度类型生成 beta 值
    if beta_schedule == "quad":
        # 生成平方调度的 beta 值
        betas = (
            np.linspace(
                beta_start ** 0.5,
                beta_end ** 0.5,
                num_diffusion_timesteps,
                dtype=np.float64,
            )
            ** 2
        )
    elif beta_schedule == "linear":
        # 生成线性调度的 beta 值
        betas = np.linspace(beta_start, beta_end, num_diffusion_timesteps, dtype=np.float64)
    elif beta_schedule == "warmup10":
        # 生成热身比例为 0.1 的 beta 值
        betas = _warmup_beta(beta_start, beta_end, num_diffusion_timesteps, 0.1)
    elif beta_schedule == "warmup50":
        # 生成热身比例为 0.5 的 beta 值
        betas = _warmup_beta(beta_start, beta_end, num_diffusion_timesteps, 0.5)
    elif beta_schedule == "const":
        # 生成常数 beta 值
        betas = beta_end * np.ones(num_diffusion_timesteps, dtype=np.float64)
    elif beta_schedule == "jsd":  # 1/T, 1/(T-1), 1/(T-2), ..., 1
        # 生成 JSD 调度的 beta 值
        betas = 1.0 / np.linspace(
            num_diffusion_timesteps, 1, num_diffusion_timesteps, dtype=np.float64
        )
    else:
        # 如果指定的调度类型未实现，抛出错误
        raise NotImplementedError(beta_schedule)

    # 确保返回的 beta 值数组的形状正确
    assert betas.shape == (num_diffusion_timesteps,)
    return betas
```

（6）定义函数 get_named_beta_schedule，根据指定的调度名称返回相应的 beta 值。这些 beta 值用于扩散模型中的训练过程，函数支持不同的调度方式，例如线性调度和平方余弦调度。

```python
def get_named_beta_schedule(schedule_name, num_diffusion_timesteps):
    if schedule_name == "linear":
        # Linear schedule from Ho et al, extended to work for any number of
        # diffusion steps.
        scale = 1000 / num_diffusion_timesteps
        return get_beta_schedule(
            "linear",
            beta_start=scale * 0.0001,
            beta_end=scale * 0.02,
            num_diffusion_timesteps=num_diffusion_timesteps,
        )
```

```
        elif schedule_name == "squaredcos_cap_v2":
            return betas_for_alpha_bar(
                num_diffusion_timesteps,
                lambda t: math.cos((t + 0.008) / 1.008 * math.pi / 2) ** 2,
            )
        else:
            raise NotImplementedError(f"unknown beta schedule: {schedule_name}")
```

（7）下面代码定义了一个名为 betas_for_alpha_bar 的函数，用于计算在给定的 alpha_bar 函数下的 beta 值序列。该序列用于扩散模型的训练过程，确保 beta 值不超过最大限制 max_beta。

```
def betas_for_alpha_bar(num_diffusion_timesteps, alpha_bar, max_beta=0.999):
    betas = []
    for i in range(num_diffusion_timesteps):
        t1 = i / num_diffusion_timesteps
        t2 = (i + 1) / num_diffusion_timesteps
        betas.append(min(1 - alpha_bar(t2) / alpha_bar(t1), max_beta))
    return np.array(betas)
```

（8）下面代码定义了一个名为 GaussianDiffusion 的类，用于实现高斯扩散过程的相关计算。它初始化了一系列属性，以便在后续的模型训练和推理中使用。

```
class GaussianDiffusion:
    def __init__(
        self,
        *,
        betas,
        model_mean_type,
        model_var_type,
        loss_type
    ):
        # 初始化模型的均值类型、方差类型和损失类型
        self.model_mean_type = model_mean_type
        self.model_var_type = model_var_type
        self.loss_type = loss_type

        # 将 beta 转换为 float64 类型以提高精度
        betas = np.array(betas, dtype=np.float64)
        self.betas = betas
        assert len(betas.shape) == 1, "betas must be 1-D"  # 确保 betas 是一维数组
        assert (betas > 0).all() and (betas <= 1).all()  # 确保 beta 值在(0, 1]范围内

        self.num_timesteps = int(betas.shape[0])  # 记录时间步数

        # 计算 alpha 值和累积乘积
        alphas = 1.0 - betas
        self.alphas_cumprod = np.cumprod(alphas, axis=0)  # 计算 alpha 的累积乘积
        self.alphas_cumprod_prev = np.append(1.0, self.alphas_cumprod[:-1])  # 前一个累积乘积
        self.alphas_cumprod_next = np.append(self.alphas_cumprod[1:], 0.0)  # 后一个累积乘积
        assert self.alphas_cumprod_prev.shape == (self.num_timesteps,)

        # 针对扩散过程的计算
        self.sqrt_alphas_cumprod = np.sqrt(self.alphas_cumprod)  # sqrt(α_t)
        self.sqrt_one_minus_alphas_cumprod = np.sqrt(1.0 - self.alphas_cumprod)
        # sqrt(1-α_t)
        self.log_one_minus_alphas_cumprod = np.log(1.0 - self.alphas_cumprod)  # log(1-α_t)
        self.sqrt_recip_alphas_cumprod = np.sqrt(1.0 / self.alphas_cumprod)  # sqrt(1/α_t)
        self.sqrt_recipm1_alphas_cumprod = np.sqrt(1.0 / self.alphas_cumprod - 1)
        # sqrt(1/α_t - 1)

        # 计算后验分布 q(x_{t-1} | x_t, x_0)
        self.posterior_variance = (
```

```python
        betas * (1.0 - self.alphas_cumprod_prev) / (1.0 - self.alphas_cumprod)
    )  # 后验方差
    # 计算后验方差的对数（进行了截断处理，避免开头的方差为 0）
    self.posterior_log_variance_clipped = np.log(
        np.append(self.posterior_variance[1], self.posterior_variance[1:])
    ) if len(self.posterior_variance) > 1 else np.array([])

    self.posterior_mean_coef1 = (
        betas * np.sqrt(self.alphas_cumprod_prev) / (1.0 - self.alphas_cumprod)
    )  # 后验均值系数 1
    self.posterior_mean_coef2 = (
        (1.0 - self.alphas_cumprod_prev) * np.sqrt(alphas) / (1.0 - self.alphas_cumprod)
    )  # 后验均值系数 2
```

（9）下面代码定义了类 GaussianDiffusion 中的一个方法 q_mean_variance，用于计算扩散过程中的均值和方差。

```python
def q_mean_variance(self, x_start, t):
    # 计算在时间步 t 时，给定起始样本 x_start 的均值和方差
    mean = _extract_into_tensor(self.sqrt_alphas_cumprod, t, x_start.shape) * x_start
    # 计算均值
    variance = _extract_into_tensor(1.0 - self.alphas_cumprod, t, x_start.shape)
    # 计算方差
    log_variance = _extract_into_tensor(self.log_one_minus_alphas_cumprod, t, x_start.shape)  # 计算对数方差
    return mean, variance, log_variance  # 返回均值、方差和对数方差
```

（10）下面代码定义了类 GaussianDiffusion 中的一个方法 q_sample，用于根据扩散过程中的样本和噪声生成新的样本。

```python
def q_sample(self, x_start, t, noise=None):
    if noise is None:
        noise = th.randn_like(x_start)
    assert noise.shape == x_start.shape
    return (
        _extract_into_tensor(self.sqrt_alphas_cumprod, t, x_start.shape) * x_start
        + _extract_into_tensor(self.sqrt_one_minus_alphas_cumprod, t, x_start.shape)
        * noise
    )
```

（11）下面代码定义了类 GaussianDiffusion 中的方法 q_posterior_mean_variance，用于计算后验分布的均值和方差。

```python
def q_posterior_mean_variance(self, x_start, x_t, t):
    # 确保 x_start 和 x_t 的形状一致
    assert x_start.shape == x_t.shape
    # 计算后验均值
    posterior_mean = (
        _extract_into_tensor(self.posterior_mean_coef1, t, x_t.shape) * x_start
        + _extract_into_tensor(self.posterior_mean_coef2, t, x_t.shape) * x_t
    )
    # 计算后验方差
    posterior_variance = _extract_into_tensor(self.posterior_variance, t, x_t.shape)
    # 计算裁剪后的后验对数方差
    posterior_log_variance_clipped = _extract_into_tensor(
        self.posterior_log_variance_clipped, t, x_t.shape
    )
    # 确保返回的均值、方差和对数方差的形状一致
    assert (
        posterior_mean.shape[0]
        == posterior_variance.shape[0]
```

10.4 扩散模型核心模块

```
        == posterior_log_variance_clipped.shape[0]
        == x_start.shape[0]
    )
    return posterior_mean, posterior_variance, posterior_log_variance_clipped
```

（12）定义方法 P_mean_Variance 的函数，它的作用是计算模型的均值和方差。

```
def p_mean_variance(self, model, x, t, clip_denoised=True, denoised_fn=None,
model_kwargs=None):
    if model_kwargs is None:
        model_kwargs = {}

    B, C = x.shape[:2]
    assert t.shape == (B,)
    model_output = model(x, t, **model_kwargs)
    if isinstance(model_output, tuple):
        model_output, extra = model_output
    else:
        extra = None

    if self.model_var_type in [ModelVarType.LEARNED, ModelVarType.LEARNED_RANGE]:
        assert model_output.shape == (B, C * 2, *x.shape[2:])
        model_output, model_var_values = th.split(model_output, C, dim=1)
        min_log = _extract_into_tensor(self.posterior_log_variance_clipped, t, x.shape)
        max_log = _extract_into_tensor(np.log(self.betas), t, x.shape)
        # The model_var_values is [-1, 1] for [min_var, max_var].
        frac = (model_var_values + 1) / 2
        model_log_variance = frac * max_log + (1 - frac) * min_log
        model_variance = th.exp(model_log_variance)
    else:
        model_variance, model_log_variance = {
            ModelVarType.FIXED_LARGE: (
                np.append(self.posterior_variance[1], self.betas[1:]),
                np.log(np.append(self.posterior_variance[1], self.betas[1:])),
            ),
            ModelVarType.FIXED_SMALL: (
                self.posterior_variance,
                self.posterior_log_variance_clipped,
            ),
        }[self.model_var_type]
        model_variance = _extract_into_tensor(model_variance, t, x.shape)
        model_log_variance = _extract_into_tensor(model_log_variance, t, x.shape)
```

（13）定义方法 p_mean_variance，用于计算扩散模型的预测均值和方差。

```
    def process_xstart(x):
        if denoised_fn is not None:
            x = denoised_fn(x)
        if clip_denoised:
            return x.clamp(-1, 1)
        return x

    if self.model_mean_type == ModelMeanType.START_X:
        pred_xstart = process_xstart(model_output)
    else:
        pred_xstart = process_xstart(
            self._predict_xstart_from_eps(x_t=x, t=t, eps=model_output)
        )
    model_mean, _, _ = self.q_posterior_mean_variance(x_start=pred_xstart, x_t=x, t=t)

    assert model_mean.shape == model_log_variance.shape == pred_xstart.shape == x.shape
    return {
        "mean": model_mean,
```

319

```
        "variance": model_variance,
        "log_variance": model_log_variance,
        "pred_xstart": pred_xstart,
        "extra": extra,
    }
```

（14）定义私有方法_predict_xstart_from_eps，用于从扩散模型的噪声预测起始样本。

```
def _predict_xstart_from_eps(self, x_t, t, eps):
    assert x_t.shape == eps.shape
    return (
        _extract_into_tensor(self.sqrt_recip_alphas_cumprod, t, x_t.shape) * x_t
        - _extract_into_tensor(self.sqrt_recipm1_alphas_cumprod, t, x_t.shape) * eps
    )
```

（15）定义私有方法_predict_eps_from_xstart，用于根据预测的起始样本计算噪声。

```
def _predict_eps_from_xstart(self, x_t, t, pred_xstart):
    return (
        _extract_into_tensor(self.sqrt_recip_alphas_cumprod, t, x_t.shape) *
        x_t - pred_xstart
    ) / _extract_into_tensor(self.sqrt_recipm1_alphas_cumprod, t, x_t.shape)
```

（16）定义方法 condition_mean，用于根据给定的条件函数调整均值。

```
def condition_mean(self, cond_fn, p_mean_var, x, t, model_kwargs=None):
    gradient = cond_fn(x, t, **model_kwargs)
    new_mean = p_mean_var["mean"].float() + p_mean_var["variance"] * gradient.float()
    return new_mean
```

（17）定义方法 condition_score，用于根据条件函数调整模型的预测。

```
def condition_score(self, cond_fn, p_mean_var, x, t, model_kwargs=None):
    alpha_bar = _extract_into_tensor(self.alphas_cumprod, t, x.shape)

    eps = self._predict_eps_from_xstart(x, t, p_mean_var["pred_xstart"])
    eps = eps - (1 - alpha_bar).sqrt() * cond_fn(x, t, **model_kwargs)

    out = p_mean_var.copy()
    out["pred_xstart"] = self._predict_xstart_from_eps(x, t, eps)
    out["mean"], _, _ = self.q_posterior_mean_variance(x_start=out["pred_xstart"], x_t=x, t=t)
    return out
```

（18）下面代码定义了类 GaussianDiffusion 中的 p_sample 方法，用于根据模型的预测生成样本。

```
def p_sample(
    self,
    model,
    x,
    t,
    clip_denoised=True,
    denoised_fn=None,
    cond_fn=None,
    model_kwargs=None,
):
    out = self.p_mean_variance(
        model,
        x,
        t,
        clip_denoised=clip_denoised,
```

```
            denoised_fn=denoised_fn,
            model_kwargs=model_kwargs,
        )
        noise = th.randn_like(x)
        nonzero_mask = (
            (t != 0).float().view(-1, *([1] * (len(x.shape) - 1)))
        )  # no noise when t == 0
        if cond_fn is not None:
            out["mean"] = self.condition_mean(cond_fn, out, x, t,
                model_kwargs=model_kwargs)
        sample = out["mean"] + nonzero_mask * th.exp(0.5 * out["log_variance"]) * noise
        return {"sample": sample, "pred_xstart": out["pred_xstart"]}
```

（19）下面代码定义了类 GaussianDiffusion 中的方法 p_sample_loop，用于执行一个完整的采样循环。

```
    def p_sample_loop(
        self,
        model,
        shape,
        noise=None,
        clip_denoised=True,
        denoised_fn=None,
        cond_fn=None,
        model_kwargs=None,
        device=None,
        progress=False,
    ):
        final = None
        for sample in self.p_sample_loop_progressive(
            model,
            shape,
            noise=noise,
            clip_denoised=clip_denoised,
            denoised_fn=denoised_fn,
            cond_fn=cond_fn,
            model_kwargs=model_kwargs,
            device=device,
            progress=progress,
        ):
            final = sample
        return final["sample"]
```

（20）下面代码定义了 p_sample_loop_progressive 方法，用于逐步生成样本。该方法接受一个模型和目标形状，并可以选择性地使用初始噪声。它通过反向时间步的方式，逐步调用 p_sample 方法进行采样，同时支持裁剪去噪结果和应用条件函数。方法中还提供了进度条选项，方便监控采样过程，最终返回生成的样本。

```
    def p_sample_loop_progressive(
        self,
        model,
        shape,
        noise=None,
        clip_denoised=True,
        denoised_fn=None,
        cond_fn=None,
        model_kwargs=None,
        device=None,
        progress=False,
    ):
        if device is None:
```

```python
        device = next(model.parameters()).device
assert isinstance(shape, (tuple, list))
if noise is not None:
    img = noise
else:
    img = th.randn(*shape, device=device)
indices = list(range(self.num_timesteps))[::-1]

if progress:
    # Lazy import so that we don't depend on tqdm.
    from tqdm.auto import tqdm

    indices = tqdm(indices)

for i in indices:
    t = th.tensor([i] * shape[0], device=device)
    with th.no_grad():
        out = self.p_sample(
            model,
            img,
            t,
            clip_denoised=clip_denoised,
            denoised_fn=denoised_fn,
            cond_fn=cond_fn,
            model_kwargs=model_kwargs,
        )
        yield out
        img = out["sample"]
```

（21）下面代码实现了 ddim_sample 方法，用于通过确定性扩散过程生成样本。该方法接收模型和输入数据，并计算当前时间步的均值和方差。若提供了条件函数，则会根据条件更新输出。随后，它使用预测的起始图像和噪声计算最终样本。在计算中，考虑了扩散过程中的先前累积系数和噪声，通过公式确定最终样本的均值，并在必要时添加噪声。最后返回生成的样本和预测的起始图像。

```python
def ddim_sample(
    self,
    model,
    x,
    t,
    clip_denoised=True,
    denoised_fn=None,
    cond_fn=None,
    model_kwargs=None,
    eta=0.0,
):
    out = self.p_mean_variance(
        model,
        x,
        t,
        clip_denoised=clip_denoised,
        denoised_fn=denoised_fn,
        model_kwargs=model_kwargs,
    )
    if cond_fn is not None:
        out = self.condition_score(cond_fn, out, x, t, model_kwargs=model_kwargs)
    eps = self._predict_eps_from_xstart(x, t, out["pred_xstart"])

    alpha_bar = _extract_into_tensor(self.alphas_cumprod, t, x.shape)
    alpha_bar_prev = _extract_into_tensor(self.alphas_cumprod_prev, t, x.shape)
    sigma = (
```

```
            eta
            * th.sqrt((1 - alpha_bar_prev) / (1 - alpha_bar))
            * th.sqrt(1 - alpha_bar / alpha_bar_prev)
        )
        # Equation 12.
        noise = th.randn_like(x)
        mean_pred = (
            out["pred_xstart"] * th.sqrt(alpha_bar_prev)
            + th.sqrt(1 - alpha_bar_prev - sigma ** 2) * eps
        )
        nonzero_mask = (
            (t != 0).float().view(-1, *([1] * (len(x.shape) - 1)))
        )  # no noise when t == 0
        sample = mean_pred + nonzero_mask * sigma * noise
        return {"sample": sample, "pred_xstart": out["pred_xstart"]}
```

（22）下面代码实现了 ddim_reverse_sample 方法，用于在确定性路径下生成反向扩散样本。该方法首先确保 eta 参数为 0，以符合反向常微分方程的要求。然后，它调用 p_mean_variance 计算当前时间步的均值和方差，并在必要时根据条件函数进行更新。接着，计算从当前样本推导出的噪声，并利用下一个时间步的累积系数计算预测的均值。最终返回生成的样本和预测的起始图像。

```
    def ddim_reverse_sample(
        self,
        model,
        x,
        t,
        clip_denoised=True,
        denoised_fn=None,
        cond_fn=None,
        model_kwargs=None,
        eta=0.0,
    ):
        assert eta == 0.0, "Reverse ODE only for deterministic path"
        out = self.p_mean_variance(
            model,
            x,
            t,
            clip_denoised=clip_denoised,
            denoised_fn=denoised_fn,
            model_kwargs=model_kwargs,
        )
        if cond_fn is not None:
            out = self.condition_score(cond_fn, out, x, t, model_kwargs=model_kwargs)
        eps = (
            _extract_into_tensor(self.sqrt_recip_alphas_cumprod, t, x.shape) * x
            - out["pred_xstart"]
        ) / _extract_into_tensor(self.sqrt_recipm1_alphas_cumprod, t, x.shape)
        alpha_bar_next = _extract_into_tensor(self.alphas_cumprod_next, t, x.shape)

        mean_pred = out["pred_xstart"] * th.sqrt(alpha_bar_next) + th.sqrt(
        1 - alpha_bar_next) * eps

        return {"sample": mean_pred, "pred_xstart": out["pred_xstart"]}
```

（23）下面代码实现了 ddim_sample_loop 方法，用于生成一系列样本。该方法接受模型、目标形状、噪声、是否剪裁去噪声、去噪函数、条件函数及其他参数。它调用 ddim_sample_loop_progressive 方法逐步生成样本，直到所有时间步完成。最后，返回生成的最终样本。此方法主要用于在扩散模型的 DDIM 采样过程中提供便利的接口。

```python
    def ddim_sample_loop(
        self,
        model,
        shape,
        noise=None,
        clip_denoised=True,
        denoised_fn=None,
        cond_fn=None,
        model_kwargs=None,
        device=None,
        progress=False,
        eta=0.0,
    ):
        final = None
        for sample in self.ddim_sample_loop_progressive(
            model,
            shape,
            noise=noise,
            clip_denoised=clip_denoised,
            denoised_fn=denoised_fn,
            cond_fn=cond_fn,
            model_kwargs=model_kwargs,
            device=device,
            progress=progress,
            eta=eta,
        ):
            final = sample
        return final["sample"]
```

（24）下面代码实现了 ddim_sample_loop_progressive 方法，用于逐步生成样本。它接受多个参数，包括模型、目标形状、噪声、去噪剪裁标志、去噪函数、条件函数及设备等。该方法首先确定设备并初始化输入图像（如果未提供噪声则使用随机噪声），然后按时间步反向遍历扩散过程。通过调用 ddim_sample 方法获取每个时间步的样本，并在生成的样本中更新图像。使用 yield 语句返回每个时间步的输出，以便实现逐步采样。这种方式可以在生成过程中显示进度条，便于用户跟踪生成进度。

```python
    def ddim_sample_loop_progressive(
        self,
        model,
        shape,
        noise=None,
        clip_denoised=True,
        denoised_fn=None,
        cond_fn=None,
        model_kwargs=None,
        device=None,
        progress=False,
        eta=0.0,
    ):
        if device is None:
            device = next(model.parameters()).device
        assert isinstance(shape, (tuple, list))
        if noise is not None:
            img = noise
        else:
            img = th.randn(*shape, device=device)
        indices = list(range(self.num_timesteps))[::-1]

        if progress:
            # 延迟导入，这样就不必依赖于 tqdm 库
```

```
                from tqdm.auto import tqdm

                indices = tqdm(indices)

            for i in indices:
                t = th.tensor([i] * shape[0], device=device)
                with th.no_grad():
                    out = self.ddim_sample(
                        model,
                        img,
                        t,
                        clip_denoised=clip_denoised,
                        denoised_fn=denoised_fn,
                        cond_fn=cond_fn,
                        model_kwargs=model_kwargs,
                        eta=eta,
                    )
                    yield out
                    img = out["sample"]
```

（25）下面这段代码定义了_vb_terms_bpd 方法，用于计算变分下界中的每个时间步的目标值。首先，它通过 q_posterior_mean_variance 方法获取真实后验分布的均值和对数方差，然后调用 p_mean_variance 方法得到模型预测的均值和方差。接着，计算了真实均值与模型预测均值之间的 KL 散度，并将其归一化以得到以比特为单位的值。随后，使用 discretized_gaussian_log_likelihood 方法计算解码器的负对数似然（NLL），并同样归一化。最终，输出包含 KL 散度和解码器 NLL，根据时间步 t 的值选择相应的值。输出的字典包含计算结果和模型预测的起始样本（pred_xstart）。

```
    def _vb_terms_bpd(
        self, model, x_start, x_t, t, clip_denoised=True, model_kwargs=None
    ):
        true_mean, _, true_log_variance_clipped = self.q_posterior_mean_variance(
            x_start=x_start, x_t=x_t, t=t
        )
        out = self.p_mean_variance(
            model, x_t, t, clip_denoised=clip_denoised, model_kwargs=model_kwargs
        )
        kl = normal_kl(
            true_mean, true_log_variance_clipped, out["mean"], out["log_variance"]
        )
        kl = mean_flat(kl) / np.log(2.0)

        decoder_nll = -discretized_gaussian_log_likelihood(
            x_start, means=out["mean"], log_scales=0.5 * out["log_variance"]
        )
        assert decoder_nll.shape == x_start.shape
        decoder_nll = mean_flat(decoder_nll) / np.log(2.0)

        output = th.where((t == 0), decoder_nll, kl)
        return {"output": output, "pred_xstart": out["pred_xstart"]}
```

（26）下面代码定义了 training_losses 方法，主要用于计算模型在训练过程中的损失。首先，它生成一个经过扩散过程的样本 x_t。接着，根据不同的损失类型（如 KL 散度或均方误差），计算相应的损失值。对于 KL 散度损失，training_losses 方法通过调用_vb_terms_bpd 方法来计算变分下界并存储在 terms 字典中。如果损失类型为均方误差，代码会首先通过模型获取输出，然后根据不同的均值类型（如前一帧的样本或原始样本）确定目标值，并计算均方误差损失。如果存在变分边界项，则将其与均方误差损失相加，最终返回一个包含损失信息的字典。

```python
def training_losses(self, model, x_start, t, model_kwargs=None, noise=None):
    if model_kwargs is None:
        model_kwargs = {}
    if noise is None:
        noise = th.randn_like(x_start)
    x_t = self.q_sample(x_start, t, noise=noise)

    terms = {}

    if self.loss_type == LossType.KL or self.loss_type == LossType.RESCALED_KL:
        terms["loss"] = self._vb_terms_bpd(
            model=model,
            x_start=x_start,
            x_t=x_t,
            t=t,
            clip_denoised=False,
            model_kwargs=model_kwargs,
        )["output"]
        if self.loss_type == LossType.RESCALED_KL:
            terms["loss"] *= self.num_timesteps
    elif self.loss_type == LossType.MSE or self.loss_type == LossType.RESCALED_MSE:
        model_output = model(x_t, t, **model_kwargs)

        if self.model_var_type in [
            ModelVarType.LEARNED,
            ModelVarType.LEARNED_RANGE,
        ]:
            B, C = x_t.shape[:2]
            assert model_output.shape == (B, C * 2, *x_t.shape[2:])
            model_output, model_var_values = th.split(model_output, C, dim=1)
            # Learn the variance using the variational bound, but don't let
            # it affect our mean prediction.
            frozen_out = th.cat([model_output.detach(), model_var_values], dim=1)
            terms["vb"] = self._vb_terms_bpd(
                model=lambda *args, r=frozen_out: r,
                x_start=x_start,
                x_t=x_t,
                t=t,
                clip_denoised=False,
            )["output"]
            if self.loss_type == LossType.RESCALED_MSE:
                # Divide by 1000 for equivalence with initial implementation.
                # Without a factor of 1/1000, the VB term hurts the MSE term.
                terms["vb"] *= self.num_timesteps / 1000.0

        target = {
            ModelMeanType.PREVIOUS_X: self.q_posterior_mean_variance(
                x_start=x_start, x_t=x_t, t=t
            )[0],
            ModelMeanType.START_X: x_start,
            ModelMeanType.EPSILON: noise,
        }[self.model_mean_type]
        assert model_output.shape == target.shape == x_start.shape
        terms["mse"] = mean_flat((target - model_output) ** 2)
        if "vb" in terms:
            terms["loss"] = terms["mse"] + terms["vb"]
        else:
            terms["loss"] = terms["mse"]
    else:
        raise NotImplementedError(self.loss_type)

    return terms
```

（27）定义 _prior_bpd 方法，用于计算给定样本 x_start 的先验分布的比特每维度（BPD）。它通过获取在特定时间步的均值和对数方差，计算样本与标准正态分布之间的 KL 散度，并将结果以比特为单位归一化，最终返回每维度的比特数。

```python
def _prior_bpd(self, x_start):
    batch_size = x_start.shape[0]
    t = th.tensor([self.num_timesteps - 1] * batch_size, device=x_start.device)
    qt_mean, _, qt_log_variance = self.q_mean_variance(x_start, t)
    kl_prior = normal_kl(
        mean1=qt_mean, logvar1=qt_log_variance, mean2=0.0, logvar2=0.0
    )
    return mean_flat(kl_prior) / np.log(2.0)
```

（28）定义 calc_bpd_loop 方法，用于计算指定输入样本 x_start 的总比特每维度（BPD）。它通过遍历每个时间步，生成带噪声的样本 x_t，并计算变分下界（VLB）和其他损失项。对于每个时间步，首先生成噪声并采样 x_t，然后计算当前时间步的 VLB 及其对应的均方误差（MSE）。最后，计算先验 BPD，并将所有时间步的 VLB 和先验 BPD 相加以获得总 BPD，返回总 BPD、先验 BPD、各时间步的 VLB，以及其他 MSE 相关信息。

```python
def calc_bpd_loop(self, model, x_start, clip_denoised=True, model_kwargs=None):
    device = x_start.device
    batch_size = x_start.shape[0]

    vb = []
    xstart_mse = []
    mse = []
    for t in list(range(self.num_timesteps))[::-1]:
        t_batch = th.tensor([t] * batch_size, device=device)
        noise = th.randn_like(x_start)
        x_t = self.q_sample(x_start=x_start, t=t_batch, noise=noise)
        # Calculate VLB term at the current timestep
        with th.no_grad():
            out = self._vb_terms_bpd(
                model,
                x_start=x_start,
                x_t=x_t,
                t=t_batch,
                clip_denoised=clip_denoised,
                model_kwargs=model_kwargs,
            )
        vb.append(out["output"])
        xstart_mse.append(mean_flat((out["pred_xstart"] - x_start) ** 2))
        eps = self._predict_eps_from_xstart(x_t, t_batch, out["pred_xstart"])
        mse.append(mean_flat((eps - noise) ** 2))

    vb = th.stack(vb, dim=1)
    xstart_mse = th.stack(xstart_mse, dim=1)
    mse = th.stack(mse, dim=1)

    prior_bpd = self._prior_bpd(x_start)
    total_bpd = vb.sum(dim=1) + prior_bpd
    return {
        "total_bpd": total_bpd,
        "prior_bpd": prior_bpd,
        "vb": vb,
        "xstart_mse": xstart_mse,
        "mse": mse,
    }
```

（29）定义函数_extract_into_tensor，用于从一个 NumPy 数组 arr 中提取特定时间步的值并转换为 PyTorch 张量。函数的输入参数包括时间步 timesteps 和希望广播到的形状 broadcast_shape。

```python
def _extract_into_tensor(arr, timesteps, broadcast_shape):
    res = th.from_numpy(arr).to(device=timesteps.device)[timesteps].float()
    while len(res.shape) < len(broadcast_shape):
        res = res[..., None]
    return res + th.zeros(broadcast_shape, device=timesteps.device)
```

10.4.3 模型扩展

文件 respace.py 定义了一个扩展自 GaussianDiffusion 的 SpacedDiffusion 类，允许在基本扩散过程中跳过某些时间步。

```python
import numpy as np
import torch as th

from .gaussian_diffusion import GaussianDiffusion

def space_timesteps(num_timesteps, section_counts):
    
    if isinstance(section_counts, str):
        if section_counts.startswith("ddim"):
            desired_count = int(section_counts[len("ddim") :])
            for i in range(1, num_timesteps):
                if len(range(0, num_timesteps, i)) == desired_count:
                    return set(range(0, num_timesteps, i))
            raise ValueError(
                f"cannot create exactly {num_timesteps} steps with an integer stride"
            )
        section_counts = [int(x) for x in section_counts.split(",")]
    size_per = num_timesteps // len(section_counts)
    extra = num_timesteps % len(section_counts)
    start_idx = 0
    all_steps = []
    for i, section_count in enumerate(section_counts):
        size = size_per + (1 if i < extra else 0)
        if size < section_count:
            raise ValueError(
                f"cannot divide section of {size} steps into {section_count}"
            )
        if section_count <= 1:
            frac_stride = 1
        else:
            frac_stride = (size - 1) / (section_count - 1)
        cur_idx = 0.0
        taken_steps = []
        for _ in range(section_count):
            taken_steps.append(start_idx + round(cur_idx))
            cur_idx += frac_stride
        all_steps += taken_steps
        start_idx += size
    return set(all_steps)


class SpacedDiffusion(GaussianDiffusion):
    
    def __init__(self, use_timesteps, **kwargs):
        self.use_timesteps = set(use_timesteps)
        self.timestep_map = []
```

```python
        self.original_num_steps = len(kwargs["betas"])

        base_diffusion = GaussianDiffusion(**kwargs)  # pylint: disable=missing-kwoa
        last_alpha_cumprod = 1.0
        new_betas = []
        for i, alpha_cumprod in enumerate(base_diffusion.alphas_cumprod):
            if i in self.use_timesteps:
                new_betas.append(1 - alpha_cumprod / last_alpha_cumprod)
                last_alpha_cumprod = alpha_cumprod
                self.timestep_map.append(i)
        kwargs["betas"] = np.array(new_betas)
        super().__init__(**kwargs)

    def p_mean_variance(
        self, model, *args, **kwargs
    ):  # pylint: disable=signature-differs
        return super().p_mean_variance(self._wrap_model(model), *args, **kwargs)

    def training_losses(
        self, model, *args, **kwargs
    ):
        return super().training_losses(self._wrap_model(model), *args, **kwargs)

    def condition_mean(self, cond_fn, *args, **kwargs):
        return super().condition_mean(self._wrap_model(cond_fn), *args, **kwargs)

    def condition_score(self, cond_fn, *args, **kwargs):
        return super().condition_score(self._wrap_model(cond_fn), *args, **kwargs)

    def _wrap_model(self, model):
        if isinstance(model, _WrappedModel):
            return model
        return _WrappedModel(
            model, self.timestep_map, self.original_num_steps
        )

    def _scale_timesteps(self, t):
        return t


class _WrappedModel:
    def __init__(self, model, timestep_map, original_num_steps):
        self.model = model
        self.timestep_map = timestep_map
        self.original_num_steps = original_num_steps

    def __call__(self, x, ts, **kwargs):
        map_tensor = th.tensor(self.timestep_map, device=ts.device, dtype=ts.dtype)
        new_ts = map_tensor[ts]
        return self.model(x, new_ts, **kwargs)
```

上述代码的主要功能如下。
- 时间步生成：space_timesteps 函数根据给定的时间步数和部分计数生成要使用的时间步。可以根据不同的区间划分策略返回一组时间步。
- 初始化：在 SpacedDiffusion 的初始化方法中，创建了一个 GaussianDiffusion 的实例，并根据指定的时间步更新了模型的 beta 值。这确保了扩散过程只在所需的时间步上进行计算。
- 模型包装：通过 _wrap_model 方法包装给定的模型，以适应新的时间步映射。
- 重写方法：重写了几个方法（如 p_mean_variance 和 training_losses），使它们调用包装后的模型。这些方法在扩散模型中负责计算均值、方差和训练损失等。

- _WrappedModel 类：定义了一个内部类_WrappedModel，用于存储原始模型和时间步映射，并在调用时将输入的时间步映射到实际使用的时间步。

通过这些功能，SpacedDiffusion 能够灵活地处理扩散过程中的时间步，允许在训练和采样中只关注关键的时间步，从而提高计算效率。

10.4.4 采样器调度

文件 timestep_sampler.py 实现了一组用于扩散过程时间步采样的类，主要包括采样器调度（ScheduleSampler）及其具体实现。

（1）下面代码定义了一个名为 create_named_schedule_sampler 的工厂函数，用于根据给定名称创建不同类型的调度采样器。

```python
def create_named_schedule_sampler(name, diffusion):
    """
    创建一个调度采样器，从预定义采样器的库中选择。
    :param name: 采样器的名称。
    :param diffusion: 用于采样的扩散对象。
    """
    if name == "uniform":
        return UniformSampler(diffusion)
        # 如果名称为"uniform"，返回均匀采样器
    elif name == "loss-second-moment":
        return LossSecondMomentResampler(diffusion)
        # 如果名称为"loss-second-moment"，返回基于损失的二次矩采样器
    else:
        raise NotImplementedError(f"unknown schedule sampler: {name}")
        # 如果名称未知，抛出异常
```

（2）下面代码定义了一个名为 ScheduleSampler 的抽象基类，用于在扩散过程中对时间步进行抽样。类 ScheduleSampler 的主要功能是提供时间步的分布，目的是减少目标函数的方差。

```python
class ScheduleSampler(ABC):
    """
    在扩散过程中的时间步分布，用于减少目标方差。
    默认情况下，采样器执行无偏重要性采样，目标的均值保持不变。
    然而，子类可以重写 sample() 方法，以更改采样项的权重，从而允许目标发生实际变化。
    """

    @abstractmethod
    def weights(self):
        """
        获取每个扩散步骤的权重的 numpy 数组。
        权重不必归一化，但必须是正值。
        """

    def sample(self, batch_size, device):
        """
        为一批样本进行重要性采样的时间步。
        :param batch_size: 时间步的数量。
        :param device: 用于保存的 PyTorch 设备。
        :return: 一个元组 (timesteps, weights)：
                 - timesteps: 时间步索引的张量。
                 - weights: 用于缩放结果损失的权重张量。
        """
        w = self.weights()        # 获取权重
        p = w / np.sum(w)         # 计算概率分布
        indices_np = np.random.choice(len(p), size=(batch_size,), p=p)    # 按照概率进行采样
```

10.4 扩散模型核心模块

```
        indices = th.from_numpy(indices_np).long().to(device)    # 转换为张量并移动到指定设备
        weights_np = 1 / (len(p) * p[indices_np])                # 计算权重
        weights = th.from_numpy(weights_np).float().to(device)   # 转换为张量并移动到指定设备
        return indices, weights    # 返回时间步索引和权重
```

（3）下面代码定义了一个名为 UniformSampler 的类，它是 ScheduleSampler 的子类，专门用于实现均匀采样的策略。

```
class UniformSampler(ScheduleSampler):
    def __init__(self, diffusion):
        self.diffusion = diffusion    # 存储扩散对象
        self._weights = np.ones([diffusion.num_timesteps])    # 初始化每个时间步的权重为1

    def weights(self):
        return self._weights    # 返回权重
```

（4）下面代码定义了一个名为 LossAwareSampler 的类，它是 ScheduleSampler 的子类，旨在根据模型的损失动态更新时间步的权重。

```
class LossAwareSampler(ScheduleSampler):
    def update_with_local_losses(self, local_ts, local_losses):
        """
        使用模型的损失更新权重。
        从每个进程调用此方法，传入一批时间步及其对应的损失。
        该方法将执行同步，以确保所有进程保持相同的重加权。
        :param local_ts: 包含时间步的整数张量。
        :param local_losses: 包含损失的 1D 张量。
        """
        batch_sizes = [
            th.tensor([0], dtype=th.int32, device=local_ts.device)
            for _ in range(dist.get_world_size())
        ]
        dist.all_gather(
            batch_sizes,
            th.tensor([len(local_ts)], dtype=th.int32, device=local_ts.device),
        )

        # 填充 all_gather 的批次为最大批次大小
        batch_sizes = [x.item() for x in batch_sizes]
        max_bs = max(batch_sizes)

        timestep_batches = [th.zeros(max_bs).to(local_ts) for bs in batch_sizes]
        loss_batches = [th.zeros(max_bs).to(local_losses) for bs in batch_sizes]
        dist.all_gather(timestep_batches, local_ts)
        dist.all_gather(loss_batches, local_losses)
        timesteps = [
            x.item() for y, bs in zip(timestep_batches, batch_sizes) for x in y[:bs]
        ]
        losses = [x.item() for y, bs in zip(loss_batches, batch_sizes) for x in y[:bs]]
        self.update_with_all_losses(timesteps, losses)

    @abstractmethod
    def update_with_all_losses(self, ts, losses):
        """
        使用模型的损失更新权重。
        子类应重写此方法，根据模型的损失更新权重。
        此方法直接更新权重，而不在工作线程之间同步。
        该方法由所有进程使用相同参数调用，因此应具有确定性行为，以保持跨工作进程的状态。
        :param ts: 时间步的整数列表。
        :param losses: 每个时间步的浮点损失列表。
        """
```

对上述代码的具体说明如下。
- **损失感知采样器**：LossAwareSampler 用于根据模型的损失动态更新时间步的权重，从而实现更智能的采样策略。
- **本地损失更新**：update_with_local_losses 方法接收本地的时间步和损失，进行同步以确保所有进程使用相同的时间步和损失数据。该方法使用 torch.distributed 中的 all_gather 来收集各个进程的损失和时间步信息。
- **权重更新**：方法的最后，调用 update_with_all_losses 来处理实际的权重更新逻辑，这个方法需要在子类中实现，以便根据具体模型的损失自定义权重的更新方式。
- **抽象方法**：update_with_all_losses 是一个抽象方法，要求子类实现特定的损失更新逻辑，以保证不同进程的一致性和确定性。

（5）下面的代码定义了一个名为 LossSecondMomentResampler 的类，它是 LossAwareSampler 的子类，旨在通过利用损失的二阶矩来动态调整时间步的权重。

```python
class LossSecondMomentResampler(LossAwareSampler):
    def __init__(self, diffusion, history_per_term=10, uniform_prob=0.001):
        # 初始化参数
        self.diffusion = diffusion  # 扩散对象
        self.history_per_term = history_per_term  # 每个时间步的历史记录数量
        self.uniform_prob = uniform_prob  # 均匀概率，控制采样的多样性
        # 初始化损失历史记录数组，形状为 (时间步数量，每个时间步的历史记录数量)
        self._loss_history = np.zeros(
            [diffusion.num_timesteps, history_per_term], dtype=np.float64
        )
        # 初始化每个时间步的损失计数
        self._loss_counts = np.zeros([diffusion.num_timesteps], dtype=np.int)

    def weights(self):
        """
        计算当前时间步的权重。
        如果尚未达到预热状态，则返回均匀权重。
        否则，计算基于损失历史的权重。
        """
        if not self._warmed_up():
            return np.ones([self.diffusion.num_timesteps], dtype=np.float64)
        # 计算权重为损失历史的二阶矩的平方根
        weights = np.sqrt(np.mean(self._loss_history ** 2, axis=-1))
        weights /= np.sum(weights)  # 归一化
        weights *= 1 - self.uniform_prob  # 考虑均匀概率
        weights += self.uniform_prob / len(weights)  # 添加均匀采样的影响
        return weights

    def update_with_all_losses(self, ts, losses):
        """
        使用传入的时间步和损失更新损失历史。
        :param ts: 时间步的列表。
        :param losses: 对应的损失列表。
        """
        for t, loss in zip(ts, losses):
            if self._loss_counts[t] == self.history_per_term:
                # 如果已达到历史记录限制，移除最旧的损失值
                self._loss_history[t, :-1] = self._loss_history[t, 1:]
                self._loss_history[t, -1] = loss
            else:
                # 记录新的损失值
                self._loss_history[t, self._loss_counts[t]] = loss
                self._loss_counts[t] += 1  # 更新计数
```

```python
def _warmed_up(self):
    """
    检查所有时间步是否都已达到历史记录的预热状态。
    """
    return (self._loss_counts == self.history_per_term).all()
```

10.5 训练模型

本项目的模型训练工作涉及多种训练策略和优化设置操作，以提高 DiT 模型的性能。在训练过程中，首先初始化分布式训练环境，并创建模型和优化器。然后，通过数据加载和预处理，将图像输入模型进行训练。在训练过程中，使用了不同的精度设置和动态损失计算方法来优化计算效率。每种策略都采用了适当的模型参数更新机制，并在训练期间保存模型检查点，便于后续使用和评估。最终，模型在多个训练周期内不断迭代，以降低损失并提升生成图像的质量。

10.5.1 定义不同配置的 DiT 模型

文件 models.py 实现了一个基于 Transformer 架构的扩散模型（DiT），主要用于图像生成任务。代码包含多种嵌入层，用于将时间步和类别标签转换为向量表示，以及多个核心组件，如 DiT 块和最终层，支持自适应层归一化调制。模型通过分块处理输入图像，结合时间和类别信息，生成高质量的输出图像。此外，文件 models.py 还提供了不同规模的 DiT 模型配置，以适应不同的计算需求和任务场景。

（1）函数 modulate 用于对输入张量 x 进行调制。具体来说，根据给定的 shift 和 scale 参数，按以下公式对 x 进行变换：首先通过 scale 调整 x 的幅度，然后通过 shift 偏移 x。这个操作在神经网络中常用于对特征进行动态调整，以提高模型的灵活性和表达能力。

```python
def modulate(x, shift, scale):
    return x * (1 + scale.unsqueeze(1)) + shift.unsqueeze(1)
```

（2）类 TimestepEmbedder 用于将标量时间步（timesteps）嵌入为向量表示。它的主要功能是生成时间步的正弦波嵌入，并将其传递给一个多层感知机（Multilayer Perceptron，MLP）进行进一步处理。

- 构造函数 __init__：初始化嵌入层，创建一个包含两层线性变换和激活函数 SiLU 的 MLP。
- 静态方法 timestep_embedding：生成正弦波嵌入。给定时间步 t 和输出维度 dim，它根据最大周期 max_period 计算正弦和余弦值，返回一个形状为(N, D)的张量，其中 N 是批量大小，D 是输出维度。
- forward 方法：接收时间步 t 作为输入，计算其频率嵌入，并将结果传递给 MLP，最终返回嵌入后的向量表示。

```python
class TimestepEmbedder(nn.Module):
    """
    将标量时间步嵌入为向量表示的类。
    """
    def __init__(self, hidden_size, frequency_embedding_size=256):
        super().__init__()
        self.mlp = nn.Sequential(
            nn.Linear(frequency_embedding_size, hidden_size, bias=True),  # 线性变换
            nn.SiLU(),  # SiLU 激活函数
            nn.Linear(hidden_size, hidden_size, bias=True),  # 线性变换
        )
        self.frequency_embedding_size = frequency_embedding_size
```

```python
    @staticmethod
    def timestep_embedding(t, dim, max_period=10000):
        """
        创建正弦波时间步嵌入。
        :param t: 一维张量 N 的索引，每个批量元素一个。可以是小数。
        :param dim: 输出的维度。
        :param max_period: 控制嵌入的最小频率。
        :return: 形状为(N, D)的正位置嵌入张量。
        """
        half = dim // 2
        freqs = torch.exp(
            -math.log(max_period) * torch.arange(start=0, end=half, dtype=torch.float32)
            / half
        ).to(device=t.device)  # 计算频率
        args = t[:, None].float() * freqs[None]  # 计算正弦波的输入参数
        embedding = torch.cat([torch.cos(args), torch.sin(args)], dim=-1)  # 计算正弦和余弦
        if dim % 2:
            embedding = torch.cat([embedding, torch.zeros_like(embedding[:, :1])],
                dim=-1)  # 添加零填充
        return embedding

    def forward(self, t):
        t_freq = self.timestep_embedding(t, self.frequency_embedding_size)  # 生成频率嵌入
        t_emb = self.mlp(t_freq)  # 通过 MLP 处理
        return t_emb  # 返回嵌入后的向量表示
```

（3）类 LabelEmbedder 用于将类别标签嵌入为向量表示，并处理标签丢弃，以支持无分类器引导（classifier-free guidance）。类 LabelEmbedder 中的成员如下。

- 构造函数 __init__：初始化嵌入层，创建一个大小为(num_classes + 1, hidden_size)的嵌入表，用于存储类别标签的嵌入向量。use_cfg_embedding 用于确定是否需要添加额外的标签（用于丢弃）。dropout_prob 是丢弃概率，用于控制标签丢弃的频率。
- 方法 token_drop：用于在训练过程中丢弃标签以实现无分类器引导。如果没有提供 force_drop_ids，则随机决定哪些标签被丢弃；否则，使用 force_drop_ids 中的值决定丢弃哪些标签。
- forward 方法：接收标签、训练标志和强制丢弃的 ID 作为输入。根据 train 标志和 dropout_prob 确定是否进行标签丢弃，然后通过嵌入表获取标签的嵌入向量，最后返回嵌入结果。

```python
class LabelEmbedder(nn.Module):
    """
    将类别标签嵌入为向量表示的类，同时处理标签丢弃以实现无分类器引导。
    """
    def __init__(self, num_classes, hidden_size, dropout_prob):
        super().__init__()
        use_cfg_embedding = dropout_prob > 0  # 判断是否使用分类器引导嵌入
        self.embedding_table = nn.Embedding(num_classes + use_cfg_embedding, hidden_size)
        # 创建嵌入表
        self.num_classes = num_classes  # 类别数量
        self.dropout_prob = dropout_prob  # 丢弃概率

    def token_drop(self, labels, force_drop_ids=None):
        """
        丢弃标签以启用无分类器引导。
        """
        if force_drop_ids is None:
            drop_ids = torch.rand(labels.shape[0], device=labels.device) <
            self.dropout_prob  # 随机决定丢弃的标签
```

```
        else:
            drop_ids = force_drop_ids == 1   # 使用 ID 强制丢弃
        labels = torch.where(drop_ids, self.num_classes, labels)
        # 将丢弃的标签替换为 num_classes
        return labels

    def forward(self, labels, train, force_drop_ids=None):
        use_dropout = self.dropout_prob > 0   # 判断是否使用丢弃
        if (train and use_dropout) or (force_drop_ids is not None):   # 训练状态下进行标签丢弃
            labels = self.token_drop(labels, force_drop_ids)
        embeddings = self.embedding_table(labels)   # 获取标签的嵌入向量
        return embeddings   # 返回嵌入结果
```

（4）类 DiTBlock 实现了一个具有自适应层归一化零（adaLN-Zero）条件的 DiT 块。

```
class DiTBlock(nn.Module):
    """
    具有自适应层归一化零（adaLN-Zero）条件的 DiT 块。
    """
    def __init__(self, hidden_size, num_heads, mlp_ratio=4.0, **block_kwargs):
        super().__init__()
        self.norm1 = nn.LayerNorm(hidden_size, elementwise_affine=False, eps=1e-6)
        # 第一层归一化
        self.attn = Attention(hidden_size, num_heads=num_heads, qkv_bias=True,
        **block_kwargs)   # 注意力机制
        self.norm2 = nn.LayerNorm(hidden_size, elementwise_affine=False, eps=1e-6)
        # 第二层归一化
        mlp_hidden_dim = int(hidden_size * mlp_ratio)   # 计算多层感知机的隐藏层维度
        approx_gelu = lambda: nn.GELU(approximate="tanh")   # 使用近似 GELU 激活函数
        self.mlp = Mlp(in_features=hidden_size, hidden_features=mlp_hidden_dim,
        act_layer=approx_gelu, drop=0)   # 多层感知机
        self.adaLN_modulation = nn.Sequential(   # 自适应层归一化调制
            nn.SiLU(),
            nn.Linear(hidden_size, 6 * hidden_size, bias=True)
        )

    def forward(self, x, c):
        """
        前向传播，处理输入特征 x 和条件 c。
        """
        # 使用自适应调制生成调整因子
        shift_msa, scale_msa, gate_msa, shift_mlp, scale_mlp, gate_mlp = self.
        adaLN_modulation(c).chunk(6, dim=1)

        # 通过注意力机制和自适应层归一化进行调制
        x = x + gate_msa.unsqueeze(1) * self.attn(modulate(self.norm1(x), shift_msa,
        scale_msa))
        x = x + gate_mlp.unsqueeze(1) * self.mlp(modulate(self.norm2(x), shift_mlp,
        scale_mlp))

        return x   # 返回处理后的特征
```

（5）FinalLayer 是 DiT 模型的最终层，负责将经过前面层处理的特征转换为输出。它使用自适应层归一化和调制技术，通过调整因子对输入进行处理，然后通过线性变换生成最终的特征输出。该层的主要功能是确保输出符合指定的补丁尺寸和通道数。

```
class FinalLayer(nn.Module):
    def __init__(self, hidden_size, patch_size, out_channels):
        super().__init__()
        self.norm_final = nn.LayerNorm(hidden_size, elementwise_affine=False, eps=1e-6)
        self.linear = nn.Linear(hidden_size, patch_size * patch_size * out_channels, bias
        =True)
```

```python
        self.adaLN_modulation = nn.Sequential(
            nn.SiLU(),
            nn.Linear(hidden_size, 2 * hidden_size, bias=True)
        )

    def forward(self, x, c):
        shift, scale = self.adaLN_modulation(c).chunk(2, dim=1)
        x = modulate(self.norm_final(x), shift, scale)
        x = self.linear(x)
        return x
```

（6）类 DiT 实现了一个深度生成模型，主要用于图像生成任务。该模型通过将输入图像分割为补丁、嵌入时间步和类标签，构建特征表示。

- __init__：类 DiT 的构造函数，初始化模型的各个参数，包括输入图像大小、补丁大小、通道数、隐藏层大小、层数、头数等。它还创建了嵌入层、多个 DiTBlock 层和最终层，并初始化权重。
- initialize_weights：该方法负责初始化模型中各个层的权重，这些层包括线性层、嵌入表和正弦余弦位置嵌入；还会将适应性层调制的权重和偏置设置为零，以便在训练开始时保持稳定。
- unpatchify：该方法将输入的补丁格式的张量恢复为完整图像格式。它通过重塑张量形状并使用爱因斯坦求和约定，重新组合补丁以生成最终的图像输出。
- ckpt_wrapper：该方法是一个包装函数，用于实现检查点功能。它接受一个模块作为输入，并返回一个前向传播函数，以便在训练时使用检查点节省内存。
- forward：forward 方法实现了 DiT 模型的前向传播，接受输入图像、时间步和类标签。它将这些输入传递给嵌入层和各个块，最终生成输出图像。
- forward_with_cfg：该方法扩展了 forward 的功能，同时处理无条件和有条件生成，以支持分类器无关引导。它通过合并条件和无条件的输出，以调整生成结果的质量。

```python
class DiT(nn.Module):
    def __init__(
        self,
        input_size=32,              # 输入图像的大小
        patch_size=2,               # 每个补丁的大小
        in_channels=4,              # 输入通道数
        hidden_size=1152,           # 隐藏层大小
        depth=28,                   # Transformer 块的深度
        num_heads=16,               # 自注意力机制中的头数
        mlp_ratio=4.0,              # MLP 的比率
        class_dropout_prob=0.1,     # 类标签的丢弃概率
        num_classes=1000,           # 类别数
        learn_sigma=True,           # 是否学习 sigma 值
    ):
        super().__init__()
        self.learn_sigma = learn_sigma
        self.in_channels = in_channels
        self.out_channels = in_channels * 2 if learn_sigma else in_channels
        self.patch_size = patch_size
        self.num_heads = num_heads

        # 初始化补丁嵌入层、时间步嵌入层和标签嵌入层
        self.x_embedder = PatchEmbed(input_size, patch_size, in_channels, hidden_size, bias=True)
        self.t_embedder = TimestepEmbedder(hidden_size)
        self.y_embedder = LabelEmbedder(num_classes, hidden_size, class_dropout_prob)
        num_patches = self.x_embedder.num_patches
```

```python
        # 使用固定的正弦余弦嵌入
        self.pos_embed = nn.Parameter(torch.zeros(1, num_patches, hidden_size),
        requires_grad=False)

        # 初始化多个 DiT 块
        self.blocks = nn.ModuleList([
            DiTBlock(hidden_size, num_heads, mlp_ratio=mlp_ratio) for _ in range(depth)
        ])
        self.final_layer = FinalLayer(hidden_size, patch_size, self.out_channels)
        self.initialize_weights()   # 初始化权重

    def initialize_weights(self):
        # 初始化 Transformer 层的权重
        def _basic_init(module):
            if isinstance(module, nn.Linear):
                torch.nn.init.xavier_uniform_(module.weight)    # 使用 Xavier 均匀分布初始化权重
                if module.bias is not None:
                    nn.init.constant_(module.bias, 0)   # 偏置初始化为零

        self.apply(_basic_init)   # 应用初始化函数

        # 初始化并冻结位置嵌入
        pos_embed = get_2d_sincos_pos_embed(self.pos_embed.shape[-1], int
        (self.x_embedder.num_patches ** 0.5))
        self.pos_embed.data.copy_(torch.from_numpy(pos_embed).float().unsqueeze(0))

        # 将补丁嵌入初始化为线性层
        w = self.x_embedder.proj.weight.data
        nn.init.xavier_uniform_(w.view([w.shape[0], -1]))
        nn.init.constant_(self.x_embedder.proj.bias, 0)

        # 初始化标签嵌入表
        nn.init.normal_(self.y_embedder.embedding_table.weight, std=0.02)

        # 初始化时间步嵌入 MLP
        nn.init.normal_(self.t_embedder.mlp[0].weight, std=0.02)
        nn.init.normal_(self.t_embedder.mlp[2].weight, std=0.02)

        # 将 DiT 块中的 adaLN 调制层的权重和偏置设为零
        for block in self.blocks:
            nn.init.constant_(block.adaLN_modulation[-1].weight, 0)
            nn.init.constant_(block.adaLN_modulation[-1].bias, 0)

        # 将输出层的权重和偏置设为零
        nn.init.constant_(self.final_layer.adaLN_modulation[-1].weight, 0)
        nn.init.constant_(self.final_layer.adaLN_modulation[-1].bias, 0)
        nn.init.constant_(self.final_layer.linear.weight, 0)
        nn.init.constant_(self.final_layer.linear.bias, 0)

    def unpatchify(self, x):
        """
        将补丁格式的输出转换为完整图像格式。
        x: (N, T, patch_size**2 * C) 输入补丁张量
        imgs: (N, H, W, C) 输出完整图像张量
        """
        c = self.out_channels    # 输出通道数
        p = self.x_embedder.patch_size[0]    # 补丁大小
        h = w = int(x.shape[1] ** 0.5)   # 计算输出图像的高和宽
        assert h * w == x.shape[1]   # 确保形状正确

        # 将补丁张量重塑为完整图像格式
        x = x.reshape(shape=(x.shape[0], h, w, p, p, c))
        x = torch.einsum('nhwpqc->nchpwq', x)    # 使用爱因斯坦求和重排列
```

```python
            imgs = x.reshape(shape=(x.shape[0], c, h * p, h * p))    # 重塑为输出图像形状
        return imgs

    def ckpt_wrapper(self, module):
        """包装函数，便于实现检查点功能"""
        def ckpt_forward(*inputs):
            outputs = module(*inputs)
            return outputs
        return ckpt_forward

    def forward(self, x, t, y):
        """
        DiT 的前向传播
        x: (N, C, H, W) 空间输入的张量（图像或图像的潜在表示）
        t: (N,) 扩散时间步的张量
        y: (N,) 类标签的张量
        """
        # 嵌入空间输入和位置嵌入
        x = self.x_embedder(x) + self.pos_embed    # (N, T, D)
        t = self.t_embedder(t)                      # (N, D)
        y = self.y_embedder(y, self.training)       # (N, D)
        c = t + y                                   # (N, D)

        # 通过 DiT 块进行前向传播
        for block in self.blocks:
            x = torch.utils.checkpoint.checkpoint(self.ckpt_wrapper(block), x, c)
            # (N, T, D)

        # 最终层处理和解补丁
        x = self.final_layer(x, c)      # (N, T, patch_size ** 2 * out_channels)
        x = self.unpatchify(x)           # (N, out_channels, H, W)
        return x

    def forward_with_cfg(self, x, t, y, cfg_scale):
        """
        带有无条件前向传播的 DiT 前向传播，用于无分类器引导
        """
        half = x[: len(x) // 2]    # 获取前半部分
        combined = torch.cat([half, half], dim=0)    # 复制前半部分以进行无条件前向传播
        model_out = self.forward(combined, t, y)

        # 使用分类器自由引导，默认情况下仅对三个通道应用
        eps, rest = model_out[:, :3], model_out[:, 3:]    # 获取前 3 个通道的输出
        cond_eps, uncond_eps = torch.split(eps, len(eps) // 2, dim=0)    # 拆分条件和无条件输出
        half_eps = uncond_eps + cfg_scale * (cond_eps - uncond_eps)    # 应用CFG缩放
        eps = torch.cat([half_eps, half_eps], dim=0)    # 拼接输出
        return torch.cat([eps, rest], dim=1)    # 返回最终输出
```

（7）下面代码实现了二维正弦/余弦位置嵌入的计算功能，用于为神经网络模型中的输入位置提供位置信息。它主要通过生成正弦和余弦函数的嵌入来表示每个位置的特征。此位置嵌入能够帮助模型理解输入序列中元素的相对位置。

```python
def get_2d_sincos_pos_embed(embed_dim, grid_size, cls_token=False, extra_tokens=0):
    """
    生成二维正弦/余弦位置嵌入
    grid_size: 网格的高度和宽度的整数
    返回：
    pos_embed: [grid_size*grid_size, embed_dim] 或 [1+grid_size*grid_size, embed_dim]（根据是否有 cls_token）
    """
    # 创建高度和宽度的网格
    grid_h = np.arange(grid_size, dtype=np.float32)
```

```python
    grid_w = np.arange(grid_size, dtype=np.float32)
    grid = np.meshgrid(grid_w, grid_h)  # 先生成宽度网格
    grid = np.stack(grid, axis=0)   # 将两个网格堆叠

    grid = grid.reshape([2, 1, grid_size, grid_size])  # 重塑网格形状
    pos_embed = get_2d_sincos_pos_embed_from_grid(embed_dim, grid)  # 获取二维位置嵌入
    if cls_token and extra_tokens > 0:
        pos_embed = np.concatenate([np.zeros([extra_tokens, embed_dim]), pos_embed],
            axis=0)  # 添加额外的零嵌入
    return pos_embed


def get_2d_sincos_pos_embed_from_grid(embed_dim, grid):
    """
    从网格获取二维正弦/余弦位置嵌入
    """
    assert embed_dim % 2 == 0  # 确保嵌入维度是偶数

    # 使用一半的维度来编码高度网格
    emb_h = get_1d_sincos_pos_embed_from_grid(embed_dim // 2, grid[0])  # (H*W, D/2)
    emb_w = get_1d_sincos_pos_embed_from_grid(embed_dim // 2, grid[1])  # (H*W, D/2)

    emb = np.concatenate([emb_h, emb_w], axis=1)  # (H*W, D)
    return emb


def get_1d_sincos_pos_embed_from_grid(embed_dim, pos):
    """
    从网格获取一维正弦/余弦位置嵌入
    embed_dim: 每个位置的输出维度
    pos: 要编码的位置列表,大小为 (M,)
    out: (M, D)
    """
    assert embed_dim % 2 == 0  # 确保嵌入维度是偶数
    omega = np.arange(embed_dim // 2, dtype=np.float64)  # 创建频率
    omega /= embed_dim / 2.  # 归一化频率
    omega = 1. / 10000**omega  # (D/2,)

    pos = pos.reshape(-1)  # 重塑为一维数组 (M,)
    out = np.einsum('m,d->md', pos, omega)  # 计算外积, (M, D/2)

    emb_sin = np.sin(out)  # 计算正弦嵌入 (M, D/2)
    emb_cos = np.cos(out)  # 计算余弦嵌入 (M, D/2)

    emb = np.concatenate([emb_sin, emb_cos], axis=1)  # 合并正弦和余弦嵌入 (M, D)
    return emb
```

(8)下面的代码定义了一系列配置函数,用于创建不同规模的 DiT 模型。这些函数通过指定不同的参数(如深度、隐藏层大小、补丁大小和头数)来生成不同版本的 DiT 模型。

```python
#################################################################################
#                                   DiT Configs                                  #
#################################################################################

def DiT_XL_2(**kwargs):
    # 创建一个 XL 规模的 DiT 模型,补丁大小为 2
    return DiT(depth=28, hidden_size=1152, patch_size=2, num_heads=16, **kwargs)

def DiT_XL_4(**kwargs):
    # 创建一个 XL 规模的 DiT 模型,补丁大小为 4
    return DiT(depth=28, hidden_size=1152, patch_size=4, num_heads=16, **kwargs)
```

```python
def DiT_XL_8(**kwargs):
    # 创建一个 XL 规模的 DiT 模型，补丁大小为 8
    return DiT(depth=28, hidden_size=1152, patch_size=8, num_heads=16, **kwargs)

def DiT_L_2(**kwargs):
    # 创建一个 L 规模的 DiT 模型，补丁大小为 2
    return DiT(depth=24, hidden_size=1024, patch_size=2, num_heads=16, **kwargs)

def DiT_L_4(**kwargs):
    # 创建一个 L 规模的 DiT 模型，补丁大小为 4
    return DiT(depth=24, hidden_size=1024, patch_size=4, num_heads=16, **kwargs)

def DiT_L_8(**kwargs):
    # 创建一个 L 规模的 DiT 模型，补丁大小为 8
    return DiT(depth=24, hidden_size=1024, patch_size=8, num_heads=16, **kwargs)

def DiT_B_2(**kwargs):
    # 创建一个 B 规模的 DiT 模型，补丁大小为 2
    return DiT(depth=12, hidden_size=768, patch_size=2, num_heads=12, **kwargs)

def DiT_B_4(**kwargs):
    # 创建一个 B 规模的 DiT 模型，补丁大小为 4
    return DiT(depth=12, hidden_size=768, patch_size=4, num_heads=12, **kwargs)

def DiT_B_8(**kwargs):
    # 创建一个 B 规模的 DiT 模型，补丁大小为 8
    return DiT(depth=12, hidden_size=768, patch_size=8, num_heads=12, **kwargs)

def DiT_S_2(**kwargs):
    # 创建一个 S 规模的 DiT 模型，补丁大小为 2
    return DiT(depth=12, hidden_size=384, patch_size=2, num_heads=6, **kwargs)

def DiT_S_4(**kwargs):
    # 创建一个 S 规模的 DiT 模型，补丁大小为 4
    return DiT(depth=12, hidden_size=384, patch_size=4, num_heads=6, **kwargs)

def DiT_S_8(**kwargs):
    # 创建一个 S 规模的 DiT 模型，补丁大小为 8
    return DiT(depth=12, hidden_size=384, patch_size=8, num_heads=6, **kwargs)
```

（9）创建字典 DiT_models，将不同规模的 DiT 模型映射到相应的配置函数。每个字典条目的键是模型的名称（例如'DiT-XL/2'），而值是调用该模型所需的函数。这样可以方便地根据模型名称动态访问和创建相应的 DiT 模型。

```python
DiT_models = {
    # 映射 DiT XL 规模模型的配置函数
    'DiT-XL/2':  DiT_XL_2,
    'DiT-XL/4':  DiT_XL_4,
    'DiT-XL/8':  DiT_XL_8,

    # 映射 DiT L 规模模型的配置函数
    'DiT-L/2':  DiT_L_2,
    'DiT-L/4':  DiT_L_4,
    'DiT-L/8':  DiT_L_8,

    # 映射 DiT B 规模模型的配置函数
    'DiT-B/2':  DiT_B_2,
    'DiT-B/4':  DiT_B_4,
    'DiT-B/8':  DiT_B_8,

    # 映射 DiT S 规模模型的配置函数
    'DiT-S/2':  DiT_S_2,
```

```
    'DiT-S/4':   DiT_S_4,
    'DiT-S/8':   DiT_S_8,
}
```

10.5.2 最小训练脚本

文件 extract_features.py 实现了一个最小的训练脚本，用于通过 PyTorch 的分布式数据并行训练 DiT 模型。它主要完成以下功能：加载图像数据集，使用预训练的 VAE 模型将输入图像编码为潜在空间并进行归一化，然后将提取的特征和标签保存为 NumPy 文件。这一训练过程支持多 GPU 训练，并且可以有效地处理大规模数据集。

```python
@torch.no_grad()
def update_ema(ema_model, model, decay=0.9999):
    """
    使 EMA 模型朝当前模型更新。
    """
    ema_params = OrderedDict(ema_model.named_parameters())
    model_params = OrderedDict(model.named_parameters())

    for name, param in model_params.items():
        # TODO: 考虑只对需要梯度的参数应用，以避免 pos_embed 的小数值变化
        ema_params[name].mul_(decay).add_(param.data, alpha=1 - decay)


def requires_grad(model, flag=True):
    """
    设置模型中所有参数的 requires_grad 标志。
    """
    for p in model.parameters():
        p.requires_grad = flag


def cleanup():
    """
    结束 DDP 训练。
    """
    dist.destroy_process_group()


def create_logger(logging_dir):
    """
    创建一个记录器，用于将日志写入日志文件和标准输出。
    """
    if dist.get_rank() == 0:  # 实际记录器
        logging.basicConfig(
            level=logging.INFO,
            format='[\033[34m%(asctime)s\033[0m] %(message)s',
            datefmt='%Y-%m-%d %H:%M:%S',
            handlers=[logging.StreamHandler(), logging.FileHandler(f"{logging_dir}/log.txt")]
        )
        logger = logging.getLogger(__name__)
    else:  # 虚拟记录器（不执行任何操作）
        logger = logging.getLogger(__name__)
        logger.addHandler(logging.NullHandler())
    return logger


def center_crop_arr(pil_image, image_size):
    """
    从 ADM 实现中心裁剪。
```

```python
    """
    while min(*pil_image.size) >= 2 * image_size:
        pil_image = pil_image.resize(
            tuple(x // 2 for x in pil_image.size), resample=Image.BOX
        )

    scale = image_size / min(*pil_image.size)
    pil_image = pil_image.resize(
        tuple(round(x * scale) for x in pil_image.size), resample=Image.BICUBIC
    )

    arr = np.array(pil_image)
    crop_y = (arr.shape[0] - image_size) // 2
    crop_x = (arr.shape[1] - image_size) // 2
    return Image.fromarray(arr[crop_y: crop_y + image_size, crop_x: crop_x + image_size])

def main(args):
    """
    训练新的 DiT 模型。
    """
    assert torch.cuda.is_available(), "当前训练需要至少一个 GPU。"

    # 设置 DDP：
    dist.init_process_group("nccl")
    assert args.global_batch_size % dist.get_world_size() == 0, f"批次大小必须能被世界大小整除。"
    rank = dist.get_rank()
    device = rank % torch.cuda.device_count()
    seed = args.global_seed * dist.get_world_size() + rank
    torch.manual_seed(seed)
    torch.cuda.set_device(device)
    print(f"开始训练，rank={rank}, seed={seed}, world_size={dist.get_world_size()}.")

    # 设置特征保存文件夹：
    if rank == 0:
        os.makedirs(args.features_path, exist_ok=True)
        os.makedirs(os.path.join(args.features_path, 'imagenet256_features'), exist_ok=True)
        os.makedirs(os.path.join(args.features_path, 'imagenet256_labels'), exist_ok=True)

    # 创建模型：
    assert args.image_size % 8 == 0, "图像大小必须能被 8 整除（用于 VAE 编码器）。"
    latent_size = args.image_size // 8
    vae = AutoencoderKL.from_pretrained(f"stabilityai/sd-vae-ft-{args.vae}").to(device)

    # 设置数据：
    transform = transforms.Compose([
        transforms.Lambda(lambda pil_image: center_crop_arr(pil_image, args.image_size)),
        transforms.RandomHorizontalFlip(),
        transforms.ToTensor(),
        transforms.Normalize(mean=[0.5, 0.5, 0.5], std=[0.5, 0.5, 0.5], inplace=True)
    ])
    dataset = ImageFolder(args.data_path, transform=transform)
    sampler = DistributedSampler(
        dataset,
        num_replicas=dist.get_world_size(),
        rank=rank,
        shuffle=False,
        seed=args.global_seed
    )
    loader = DataLoader(
        dataset,
        batch_size=1,
        shuffle=False,
```

```python
            sampler=sampler,
            num_workers=args.num_workers,
            pin_memory=True,
            drop_last=True
        )

        train_steps = 0
        for x, y in loader:
            x = x.to(device)
            y = y.to(device)
            with torch.no_grad():
                # 将输入图像映射到潜在空间并归一化潜在值:
                x = vae.encode(x).latent_dist.sample().mul_(0.18215)

            x = x.detach().cpu().numpy()    # (1, 4, 32, 32)
            np.save(f'{args.features_path}/imagenet256_features/{train_steps}.npy', x)

            y = y.detach().cpu().numpy()    # (1,)
            np.save(f'{args.features_path}/imagenet256_labels/{train_steps}.npy', y)

            train_steps += 1
            print(train_steps)

if __name__ == "__main__":
    # 默认参数将在这里训练 DiT-XL/2,使用本书中的超参数(除训练迭代外)
    parser = argparse.ArgumentParser()
    parser.add_argument("--data-path", type=str, required=True)  # 数据路径
    parser.add_argument("--features-path", type=str, default="features")  # 特征保存路径
    parser.add_argument("--results-dir", type=str, default="results")  # 结果保存目录
    parser.add_argument("--model", type=str, choices=list(DiT_models.keys()), default="DiT-XL/2")  # 模型选择
    parser.add_argument("--image-size", type=int, choices=[256, 512], default=256)
    # 图像大小
    parser.add_argument("--num-classes", type=int, default=1000)  # 类别数量
    parser.add_argument("--epochs", type=int, default=1400)  # 训练轮数
    parser.add_argument("--global-batch-size", type=int, default=256)  # 全局批次大小
    parser.add_argument("--global-seed", type=int, default=0)  # 随机种子
    parser.add_argument("--vae", type=str, choices=["ema", "mse"], default="ema")
    # VAE 选择(对训练无影响)
    parser.add_argument("--num-workers", type=int, default=4)  # 工作线程数
    parser.add_argument("--log-every", type=int, default=100)  # 每多少步记录一次日志
    parser.add_argument("--ckpt-every", type=int, default=50_000)  # 每多少步保存一次检查点
    args = parser.parse_args()
    main(args)
```

10.5.3 实现 DiT 模型

文件 models_original.py 实现了一个基于 Transformer 的扩散模型 DiT,用于处理图像生成任务。主要功能如下。

- 时序和标签嵌入:通过 TimestepEmbedder 和 LabelEmbedder 类将时间步和类别标签嵌入向量表示中,便于模型学习。
- DiT 模块:使用 DiTBlock 类构建模型的主要组件,包含自注意力机制和多层感知机,并引入自适应层归一化的调制。
- 最终层处理:FinalLayer 类用于输出图像的最终处理,结合了归一化和线性变换。
- 模型配置:提供多种规模的 DiT 模型(如 DiT-XL、DiT-L、DiT-B、DiT-S)配置选项,允许根据不同的深度、隐藏层大小和补丁大小创建模型。

- 位置嵌入：通过正弦和余弦函数生成位置嵌入，增强模型对空间信息的感知。

整体上，文件 models_original.py 为扩散模型的训练和推理提供了完整的结构和功能，适用于生成图像或处理与图像相关的任务。

10.5.4　DiT 模型的标准训练

文件 train_amp.py 是一个用于训练 DiT 模型的最小训练脚本，使用 PyTorch 和加速库来支持分布式训练，首先设置了日志记录和实验目录，然后创建和初始化 DiT 模型、优化器和数据加载器。脚本通过循环进行多个训练轮次，计算损失并更新模型参数，同时使用 EMA 来改进模型稳定性。定期记录训练损失并保存模型检查点。训练完成后，模型会进入评估模式，准备进行进一步的推断或评估。

```python
@torch.no_grad()
def update_ema(ema_model, model, decay=0.9999):
    """
    将 EMA 模型更新为当前模型。
    """
    ema_params = OrderedDict(ema_model.named_parameters())
    model_params = OrderedDict(model.named_parameters())

    for name, param in model_params.items():
        name = name.replace("module.", "")
        # TODO: 考虑仅应用于需要梯度的参数，以避免 pos_embed 的小数值变化
        ema_params[name].mul_(decay).add_(param.data, alpha=1 - decay)


def requires_grad(model, flag=True):
    """
    设置模型中所有参数的 requires_grad 标志。
    """
    for p in model.parameters():
        p.requires_grad = flag


def create_logger(logging_dir):
    """
    创建一个记录器，将日志写入日志文件和标准输出。
    """
    logging.basicConfig(
        level=logging.INFO,
        format='[\033[34m%(asctime)s\033[0m] %(message)s',
        datefmt='%Y-%m-%d %H:%M:%S',
        handlers=[logging.StreamHandler(), logging.FileHandler(f"{logging_dir}/log.txt")]
    )
    logger = logging.getLogger(__name__)
    return logger


def center_crop_arr(pil_image, image_size):
    """
    从 ADM 实现中心裁剪。
    https://github.com/openai/guided-diffusion/blob/8fb3ad9197f16bbc40620447b2742e13458d2831/guided_diffusion/image_datasets.py#L126
    """
    while min(*pil_image.size) >= 2 * image_size:
        pil_image = pil_image.resize(
            tuple(x // 2 for x in pil_image.size), resample=Image.BOX
        )
```

```python
        scale = image_size / min(*pil_image.size)
        pil_image = pil_image.resize(
            tuple(round(x * scale) for x in pil_image.size), resample=Image.BICUBIC
        )

        arr = np.array(pil_image)
        crop_y = (arr.shape[0] - image_size) // 2
        crop_x = (arr.shape[1] - image_size) // 2
        return Image.fromarray(arr[crop_y: crop_y + image_size, crop_x: crop_x + image_size])


def main(args):
    """
    训练新的 DiT 模型。
    """
    assert torch.cuda.is_available(), "训练当前需要至少一个 GPU。"

    # 设置加速器：
    accelerator = Accelerator()
    device = accelerator.device

    # 设置实验文件夹：
    if accelerator.is_main_process:
        os.makedirs(args.results_dir, exist_ok=True)  # 创建结果文件夹（保存所有实验子文件夹）
        experiment_index = len(glob(f"{args.results_dir}/*"))
        model_string_name = args.model.replace("/", "-")  # e.g., DiT-XL/2 --> DiT-XL-2
        （用于命名文件夹）
        experiment_dir = f"{args.results_dir}/{experiment_index:03d}-{model_string_name}"
"        # 创建实验文件夹
        checkpoint_dir = f"{experiment_dir}/checkpoints"  # 存储保存的模型检查点
        os.makedirs(checkpoint_dir, exist_ok=True)
        logger = create_logger(experiment_dir)
        logger.info(f"实验目录已创建于{experiment_dir}")
    #else:
    #    logger = create_logger(None)

    # 创建模型：
    assert args.image_size % 8 == 0, "图像大小必须是 8 的倍数（用于 VAE 编码器）。"
    latent_size = args.image_size // 8
    model = DiT_models[args.model](
        input_size=latent_size,
        num_classes=args.num_classes
    )
    # 注意参数初始化在 DiT 构造函数中完成
    model = model.to(device)
    ema = deepcopy(model).to(device)  # 创建模型的 EMA，用于训练后使用
    requires_grad(ema, False)
    diffusion = create_diffusion(timestep_respacing="")  # 默认：1000 步，线性噪声调度
    vae = AutoencoderKL.from_pretrained(f"stabilityai/sd-vae-ft-{args.vae}").to(device)
    if accelerator.is_main_process:
        logger.info(f"DiT 参数总数: {sum(p.numel() for p in model.parameters()):,}")

    # 设置优化器（我们在论文中使用了默认的 Adam betas=(0.9, 0.999)和恒定学习率 1e-4）：
    opt = torch.optim.AdamW(model.parameters(), lr=1e-4, weight_decay=0)

    # 设置数据：
    transform = transforms.Compose([
        transforms.Lambda(lambda pil_image: center_crop_arr(pil_image, args.image_size)),
        transforms.RandomHorizontalFlip(),
        transforms.ToTensor(),
        transforms.Normalize(mean=[0.5, 0.5, 0.5], std=[0.5, 0.5, 0.5], inplace=True)
    ])
    dataset = ImageFolder(args.data_path, transform=transform)
```

```python
loader = DataLoader(
    dataset,
    batch_size=int(args.global_batch_size // accelerator.num_processes),
    shuffle=True,
    num_workers=args.num_workers,
    pin_memory=True,
    drop_last=True
)
if accelerator.is_main_process:
    logger.info(f"数据集中包含{len(dataset):,}张图像({args.data_path})")

# 准备模型进行训练:
update_ema(ema, model, decay=0)  # 确保EMA使用同步权重进行初始化
model.train()  # 重要! 这使得嵌入丢弃用于无分类器指导
ema.eval()  # EMA模型应该始终处于评估模式
model, opt, loader = accelerator.prepare(model, opt, loader)

# 监控/记录的变量:
train_steps = 0
log_steps = 0
running_loss = 0
start_time = time()

if accelerator.is_main_process:
    logger.info(f"训练进行{args.epochs}个周期...")
for epoch in range(args.epochs):
    if accelerator.is_main_process:
        logger.info(f"开始第{epoch}个周期...")
    for x, y in loader:
        x = x.to(device)
        y = y.to(device)
        with torch.no_grad():
            # 将输入图像映射到潜在空间+归一化潜在值:
            x = vae.encode(x).latent_dist.sample().mul_(0.18215)
        t = torch.randint(0, diffusion.num_timesteps, (x.shape[0],), device=device)
        model_kwargs = dict(y=y)
        loss_dict = diffusion.training_losses(model, x, t, model_kwargs)
        loss = loss_dict["loss"].mean()
        opt.zero_grad()
        accelerator.backward(loss)
        opt.step()
        update_ema(ema, model)

        # 记录损失值:
        running_loss += loss.item()
        log_steps += 1
        train_steps += 1
        if train_steps % args.log_every == 0:
            # 测量训练速度:
            torch.cuda.synchronize()
            end_time = time()
            steps_per_sec = log_steps / (end_time - start_time)
            # 在所有进程中减少损失历史:
            avg_loss = torch.tensor(running_loss / log_steps, device=device)
            avg_loss = avg_loss.item() / accelerator.num_processes
            if accelerator.is_main_process:
                logger.info(f"(step={train_steps:07d}) 训练损失: {avg_loss:.4f}, 训练步数/秒: {steps_per_sec:.2f}")
            # 重置监控变量:
            running_loss = 0
            log_steps = 0
            start_time = time()
```

```python
            # 保存DiT检查点:
            if train_steps % args.ckpt_every == 0 and train_steps > 0:
                if accelerator.is_main_process:
                    checkpoint = {
                        "model": model.module.state_dict(),
                        "ema": ema.state_dict(),
                        "opt": opt.state_dict(),
                        "args": args
                    }
                    checkpoint_path = f"{checkpoint_dir}/{train_steps:07d}.pt"
                    torch.save(checkpoint, checkpoint_path)
                    logger.info(f"检查点已保存至{checkpoint_path}")

    model.eval()  # 重要! 这禁用随机化的嵌入丢弃
    # 进行任何采样/FID计算等操作时, 使ema (或模型) 处于评估模式...

    if accelerator.is_main_process:
        logger.info("完成! ")


if __name__ == "__main__":
    # 默认参数将训练DiT-XL/2, 使用我们论文中使用的超参数 (除了训练迭代)。
    parser = argparse.ArgumentParser()
    parser.add_argument("--data-path", type=str, required=True)
    parser.add_argument("--results-dir", type=str, default="results")
    parser.add_argument("--model", type=str, choices=list(DiT_models.keys()), default=
"DiT-XL/2")
    parser.add_argument("--image-size", type=int, choices=[256, 512], default=256)
    parser.add_argument("--num-classes", type=int, default=1000)
    parser.add_argument("--epochs", type=int, default=1400)
    parser.add_argument("--global-batch-size", type=int, default=256)
    parser.add_argument("--global-seed", type=int, default=0)
    parser.add_argument("--vae", type=str, choices=["ema", "mse"], default="ema")
    # 选择不影响训练
    parser.add_argument("--num-workers", type=int, default=4)
    parser.add_argument("--log-every", type=int, default=100)
    parser.add_argument("--ckpt-every", type=int, default=50_000)
    args = parser.parse_args()
    main(args)
```

对上面代码中各个函数的具体说明如下。

- 函数 update_ema: 功能是将EMA模型更新为当前模型的参数。该函数通过在当前模型参数上应用衰减因子, 使EMA模型逐步接近当前模型, 从而平滑模型训练过程中的变化。
- 函数 requires_grad: 功能是设置模型中所有参数的 requires_grad 标志。通过传入布尔值 flag, 用户可以控制模型参数是否计算梯度, 这对于冻结某些层的训练非常有用。
- 函数 create_logger: 功能是创建一个记录器, 用于将日志信息写入日志文件和标准输出。用户可以指定日志目录, 记录器将格式化输出并在运行时提供有关实验的详细信息。
- 函数 center_crop_arr: 功能是实现中心裁剪, 主要用于调整图像大小以适应指定的图像尺寸。该函数通过缩放和裁剪图像的中心区域, 确保输出图像的尺寸符合预期要求。
- 函数 main: 功能是执行模型的训练过程。该函数设置训练环境, 包括模型和数据加载器的创建, 训练循环的控制, 以及日志记录和模型检查点的保存, 确保训练过程的顺利进行。

10.5.5　DiT 模型的全精度训练

文件 train_baseline.py 的功能是训练一个基于 DiT 模型的图像生成系统, 它使用 DDP 进行全精度训练。通过深度卷积网络和变分自编码器 (Variational Autoencoder, VAE) 将输入图像映射到潜在空间，并在该空间中进行扩散过程。代码设置了必要的训练环境，进行了数据加载、模型初始化和优化器配置，并在训练过程中更新模型参数以及 EMA 模型。训练过程包括损失计算、日志记录、模型检查点保存等，最终完成模型训练。

（1）函数 update_ema 的功能是更新 EMA 模型的参数，使其逐渐靠近当前训练模型的参数，通过设定的 decay 值来控制 EMA 模型参数更新的速率。

```
def update_ema(ema_model, model, decay=0.9999):
    ema_params = OrderedDict(ema_model.named_parameters())
    model_params = OrderedDict(model.named_parameters())

    for name, param in model_params.items():
        ema_params[name].mul_(decay).add_(param.data, alpha=1 - decay)
```

（2）函数 requires_grad 的功能是设置模型中所有参数的 requires_grad 标志。此标志控制是否在反向传播过程中计算该参数的梯度，通常用于冻结或解冻模型的特定部分。

```
def requires_grad(model, flag=True):
    for p in model.parameters():
        p.requires_grad = flag
```

（3）函数 cleanup 的功能是在 DDP 训练结束后，销毁进程组，释放相关资源。这是确保训练进程正确退出的关键步骤。

```
def cleanup():
    dist.destroy_process_group()
```

（4）函数 create_logger 的功能是创建一个日志记录器，用于将训练过程中的信息输出到日志文件和标准输出（stdout）。如果不是主进程，则创建一个不执行任何操作的空日志记录器。

```
def create_logger(logging_dir):
    if dist.get_rank() == 0:  # real logger
        logging.basicConfig(
            level=logging.INFO,
            format='[\033[34m%(asctime)s\033[0m] %(message)s',
            datefmt='%Y-%m-%d %H:%M:%S',
            handlers=[logging.StreamHandler(), logging.FileHandler(f"{logging_dir}/log.txt")]
        )
        logger = logging.getLogger(__name__)
    else:  # dummy logger (does nothing)
        logger = logging.getLogger(__name__)
        logger.addHandler(logging.NullHandler())
    return logger
```

（5）函数 center_crop_arr 的功能是执行图像的中心裁剪操作，确保图像的尺寸符合要求。该函数从图像的中心裁剪出指定大小的区域，并返回裁剪后的图像。

```
def center_crop_arr(pil_image, image_size):
    while min(*pil_image.size) >= 2 * image_size:
        pil_image = pil_image.resize(
            tuple(x // 2 for x in pil_image.size), resample=Image.BOX
        )

    scale = image_size / min(*pil_image.size)
    pil_image = pil_image.resize(
        tuple(round(x * scale) for x in pil_image.size), resample=Image.BICUBIC
    )

    arr = np.array(pil_image)
```

```python
    crop_y = (arr.shape[0] - image_size) // 2
    crop_x = (arr.shape[1] - image_size) // 2
    return Image.fromarray(arr[crop_y: crop_y + image_size, crop_x: crop_x + image_size])
)
```

（6）函数 main 是训练过程的主函数，包含了模型的初始化、数据加载、优化器设置、训练循环、日志记录和检查点保存等步骤。通过调用此函数，可以执行完整的模型训练流程。

```python
def main(args):
    # 确保当前环境支持 GPU 训练
    assert torch.cuda.is_available(), "Training currently requires at least one GPU."

    # 设置 DDP:
    dist.init_process_group("nccl")   # 初始化进程组，使用 NCCL 后端进行 GPU 之间的通信
    assert args.global_batch_size % dist.get_world_size() == 0, f"Batch size must be divisible by world size."
    rank = dist.get_rank()   # 获取当前进程的全局排名
    device = rank % torch.cuda.device_count()   # 为当前进程选择一个 CUDA 设备
    seed = args.global_seed * dist.get_world_size() + rank   # 为当前进程设置种子
    torch.manual_seed(seed)   # 设置随机种子
    torch.cuda.set_device(device)   # 设置当前进程的 CUDA 设备
    print(f"Starting rank={rank}, seed={seed}, world_size={dist.get_world_size()}.")

    # 设置实验文件夹:
    if rank == 0:   # 只有主进程（rank 0）负责创建目录
        os.makedirs(args.results_dir, exist_ok=True)   # 创建结果文件夹（包含所有实验子文件夹）
        experiment_index = len(glob(f"{args.results_dir}/*"))   # 获取现有实验的数量
        model_string_name = args.model.replace("/", "-")
        # 将模型名称中的"/"替换为"-"，以便用于文件夹命名
        experiment_dir = f"{args.results_dir}/{experiment_index:03d}-{model_string_name}"   # 创建实验文件夹
        checkpoint_dir = f"{experiment_dir}/checkpoints"   # 存储模型检查点的文件夹
        os.makedirs(checkpoint_dir, exist_ok=True)
        logger = create_logger(experiment_dir)   # 创建日志记录器
        logger.info(f"Experiment directory created at {experiment_dir}")
    else:
        logger = create_logger(None)   # 其他进程创建空日志记录器

    # 创建模型:
    assert args.image_size % 8 == 0, "Image size must be divisible by 8 (for the VAE encoder)."
    latent_size = args.image_size // 8   # 计算潜在空间的大小
    model = DiT_models[args.model](
        input_size=latent_size,
        num_classes=args.num_classes
    )
    # 参数初始化在 DiT 构造函数中完成
    ema = deepcopy(model).to(device)   # 创建模型的 EMA 副本，用于训练后使用
    requires_grad(ema, False)   # 冻结 EMA 模型的参数，不更新其梯度
    model = DDP(model.to(device), device_ids=[rank])
    # 将模型包装为 DDP 模型，并将其分配到指定设备上
    diffusion = create_diffusion(timestep_respacing="")
    # 创建扩散模型，默认使用 1000 步线性噪声调度
    vae = AutoencoderKL.from_pretrained(f"stabilityai/sd-vae-ft-{args.vae}").to(device)
    # 加载预训练的 VAE 并分配到设备上
    logger.info(f"DiT Parameters: {sum(p.numel() for p in model.parameters()):,}")
    # 记录模型参数的总数

    # 设置优化器（使用 AdamW 优化器，默认 betas=(0.9, 0.999)，学习率 1e-4）:
    opt = torch.optim.AdamW(model.parameters(), lr=1e-4, weight_decay=0)

    # 设置数据:
```

```python
    transform = transforms.Compose([
        transforms.Lambda(lambda pil_image: center_crop_arr(pil_image, args.image_size)),
        # 中心裁剪图像
        transforms.RandomHorizontalFlip(),  # 随机水平翻转
        transforms.ToTensor(),  # 将图像转换为Tensor
        transforms.Normalize(mean=[0.5, 0.5, 0.5], std=[0.5, 0.5, 0.5], inplace=True)
        # 归一化图像
    ])
    dataset = ImageFolder(args.data_path, transform=transform)  # 创建图像数据集
    sampler = DistributedSampler(
        dataset,
        num_replicas=dist.get_world_size(),  # 设置副本数为世界大小
        rank=rank,  # 当前进程的rank
        shuffle=True,  # 是否随机打乱数据
        seed=args.global_seed  # 随机种子
    )
    loader = DataLoader(
        dataset,
        batch_size=int(args.global_batch_size // dist.get_world_size()),
        # 每个进程的数据批量大小
        shuffle=False,  # 不在DataLoader中进行随机打乱,采样器已处理
        sampler=sampler,
        num_workers=args.num_workers,  # 使用的工作线程数量
        pin_memory=True,  # 是否将数据加载到固定内存中
        drop_last=True  # 丢弃最后一个不完整的批次
    )
    logger.info(f"Dataset contains {len(dataset):,} images ({args.data_path})")
    # 记录数据集信息

    # 准备模型进行训练:
    update_ema(ema, model.module, decay=0)  # 确保EMA模型的参数与当前模型同步
    model.train()  # 将模型设置为训练模式(启用嵌入层的dropout)
    ema.eval()  # EMA模型应始终处于评估模式

    # 定义用于监控和日志记录的变量:
    train_steps = 0  # 训练步数
    log_steps = 0  # 日志记录步数
    running_loss = 0  # 累积损失
    start_time = time()  # 开始时间

    logger.info(f"Training for {args.epochs} epochs...")
    for epoch in range(args.epochs):  # 按照指定的epoch数进行训练
        sampler.set_epoch(epoch)  # 设置采样器的epoch(用于分布式训练中的数据重排)
        logger.info(f"Beginning epoch {epoch}...")
        for x, y in loader:
            x = x.to(device)  # 将数据加载到设备
            y = y.to(device)
            with torch.no_grad():
                # 将输入图像映射到潜在空间并归一化潜在向量:
                x = vae.encode(x).latent_dist.sample().mul_(0.18215)
            t = torch.randint(0, diffusion.num_timesteps, (x.shape[0],), device=device)
            # 随机生成时间步数
            model_kwargs = dict(y=y)  # 模型的额外输入(如标签)
            loss_dict = diffusion.training_losses(model, x, t, model_kwargs)  # 计算损失
            loss = loss_dict["loss"].mean()  # 平均化损失
            opt.zero_grad()  # 清除优化器的梯度
            loss.backward()  # 反向传播计算梯度
            opt.step()  # 更新模型参数
            update_ema(ema, model.module)  # 更新EMA模型的参数

            # 记录损失值:
            running_loss += loss.item()  # 累积损失
            log_steps += 1  # 增加日志记录步数
```

```python
                    train_steps += 1  # 增加训练步数
                    if train_steps % args.log_every == 0:  # 每隔一定步数记录日志
                        # 计算训练速度:
                        torch.cuda.synchronize()
                        end_time = time()
                        steps_per_sec = log_steps / (end_time - start_time)  # 每秒步数
                        # 在所有进程中减少损失历史:
                        avg_loss = torch.tensor(running_loss / log_steps, device=device)
                        dist.all_reduce(avg_loss, op=dist.ReduceOp.SUM)  # 在所有进程中求和
                        avg_loss = avg_loss.item() / dist.get_world_size()  # 计算平均损失
                        logger.info(f"(step={train_steps:07d}) Train Loss: {avg_loss:.4f}, Train Steps/Sec: {steps_per_sec:.2f}")
                        # 重置监控变量:
                        running_loss = 0
                        log_steps = 0
                        start_time = time()

                    # 保存模型检查点:
                    if train_steps % args.ckpt_every == 0 and train_steps > 0:
                    # 每隔一定步数保存检查点
                        if rank == 0:  # 只有主进程保存检查点
                            checkpoint = {
                                "model": model.module.state_dict(),  # 保存模型状态字典
                                "ema": ema.state_dict(),  # 保存 EMA 模型状态字典
                                "opt": opt.state_dict(),  # 保存优化器状态字典
                                "args": args  # 保存训练参数
                            }
                            checkpoint_path = f"{checkpoint_dir}/{train_steps:07d}.pt"
                            # 生成检查点路径
                            torch.save(checkpoint, checkpoint_path)  # 保存检查点
                            logger.info(f"Saved checkpoint to {checkpoint_path}")
                        dist.barrier()  # 同步所有进程

    model.eval()  # 将模型设置为评估模式 ( 禁用随机嵌入 dropout )
    # 使用 EMA 模型或当前模型进行采样、FID 计算等操作...

    logger.info("Done!")  # 训练完成
    cleanup()  # 清理分布式进程组
```

（7）函数 __main__ 的功能是解析命令行参数，并调用 main 函数启动训练。它允许用户通过命令行参数自定义训练的配置，如数据路径、模型类型、图像大小、批量大小等。

```python
if __name__ == "__main__":
    # Default args here will train DiT-XL/2 with the hyperparameters we used in our paper (except training iters).
    parser = argparse.ArgumentParser()
    parser.add_argument("--data-path", type=str, required=True)
    parser.add_argument("--results-dir", type=str, default="results")
    parser.add_argument("--model", type=str, choices=list(DiT_models.keys()), default="DiT-XL/2")
    parser.add_argument("--image-size", type=int, choices=[256, 512], default=256)
    parser.add_argument("--num-classes", type=int, default=1000)
    parser.add_argument("--epochs", type=int, default=1400)
    parser.add_argument("--global-batch-size", type=int, default=256)
    parser.add_argument("--global-seed", type=int, default=0)
    parser.add_argument("--vae", type=str, choices=["ema", "mse"], default="ema")
    # Choice doesn't affect training
    parser.add_argument("--num-workers", type=int, default=4)
    parser.add_argument("--log-every", type=int, default=100)
    parser.add_argument("--ckpt-every", type=int, default=50_000)
    args = parser.parse_args()
    main(args)
```

10.5.6 DiT 模型的特征预训练

文件 train_features.py 是一个专注于特征提取和训练过程优化的脚本,旨在使用 PyTorch DDP 对 DiT 模型进行训练,同时确保在特征提取阶段高效地利用数据和模型结构。文件 train_features.py 中各个函数的具体说明如下。

- update_ema(ema_model, model, decay=0.9999):更新 EMA 模型的参数,使其逐步接近当前训练模型的参数。通过控制衰减因子,可以调整更新的平滑度。
- requires_grad(model, flag=True):设置模型中所有参数的 requires_grad 标志。可以用来冻结模型的某些部分,避免计算梯度,从而减少内存占用和加速训练。
- cleanup():结束 DDP 训练,清理进程组,确保训练过程中所有的资源都被正确释放。
- create_logger(logging_dir):创建一个日志记录器,将训练过程中的信息写入日志文件和标准输出。确保只有主进程会生成实际的日志,而其他进程则使用空日志处理器。
- center_crop_arr(pil_image, image_size):对输入的图像进行中心裁剪,以满足特定的输入大小要求。该函数实现了一种自适应的裁剪方式,确保图像能够被缩放到目标尺寸。
- main(args):主训练循环,包括数据加载、模型初始化、训练过程的实现和日志记录。负责设置训练参数、管理训练进度、计算损失、优化模型权重,并定期保存检查点。

10.5.7 DiT 模型的原始训练

文件 train_original.py 是一个训练 DiT 模型的基础脚本,使用 PyTorch DDP 进行分布式训练,主要关注模型的训练过程、损失计算及检查点保存。文件 train_original.py 中各个函数的具体说明如下。

- update_ema(ema_model, model, decay=0.9999):更新 EMA 模型的参数,使其逐步接近当前训练模型的参数,通过控制衰减因子来调整更新的平滑程度。
- requires_grad(model, flag=True):设置模型中所有参数的 requires_grad 标志,以便冻结部分参数,从而减少内存占用和加速训练。
- cleanup():结束 DDP 训练,清理进程组,以确保所有资源都被正确释放。
- create_logger(logging_dir):创建一个日志记录器,将训练信息写入日志文件和标准输出,确保主进程生成实际日志,而其他进程则使用空日志处理器。
- center_crop_arr(pil_image, image_size):对输入图像进行中心裁剪,以满足特定的输入尺寸要求,确保图像适配模型的输入格式。
- main(args):主训练循环,包括数据加载、模型初始化、训练过程、损失计算和日志记录,负责设置训练参数、管理训练进度、计算损失、优化模型权重,并定期保存检查点。

整体而言,文件 train_original.py 作为训练 DiT 模型的基本实现,提供了一套完整的训练流程,专注于模型训练的各个环节,确保模型能够有效地学习和优化。

10.5.8 DiT 模型的禁用 TF32 模式训练

文件 train_tf32_disabled.py 是一个用于训练 DiT 模型的脚本,主要通过禁用 TensorFloat-32(TF32)模式来提升训练的稳定性,并使用 PyTorch 的 DDP 方法进行多 GPU 训练。文件 train_tf32_disabled.py 中各个函数的具体说明如下。

- update_ema(ema_model, model, decay=0.9999):更新 EMA 模型的参数,确保 EMA 模型逐渐接近当前训练模型的参数,通过设置衰减因子控制更新的速率。
- requires_grad(model, flag=True):设置模型中所有参数的 requires_grad 属性,控制哪些参

数在训练过程中需要计算梯度，以便于冻结不需要更新的参数。
- cleanup()：结束 DDP 训练过程，清理进程组，确保所有的资源和进程都被妥善释放。
- create_logger(logging_dir)：创建一个日志记录器，将训练信息同时写入日志文件和标准输出，以便于监控训练进度和调试。
- center_crop_arr(pil_image, image_size)：对输入图像进行中心裁剪，以保证图像的尺寸符合模型要求，通过调整图像的大小来适配模型输入。
- main(args)：执行训练的主循环，包括初始化训练参数、加载数据、创建模型、优化过程、计算损失并进行日志记录，控制整个训练流程，并定期保存模型检查点。

整体而言，文件 train_tf32_disabled.py 提供了一个稳定的训练框架，特别是在多 GPU 环境中，确保 DiT 模型能够有效地进行训练，同时通过日志记录和检查点保存来跟踪训练进度和结果。

10.6 生成图像

本项目使用 DDP 在多个 GPU 上进行训练，以生成高质量的图像。通过对输入数据进行处理和增强，将图像输入深度学习模型中，使用自适应动量估计（EMA）来优化模型参数，并最终将生成的图像保存到指定路径。这一流程支持大规模数据集，旨在提升训练效率和图像生成的效果。

10.6.1 预训练生成

文件 sample.py 的功能是利用训练好的 DiT 模型生成图像样本，该文件允许用户设置模型、图像尺寸、类别数量和采样参数，加载相应的模型和变分自编码器，并通过生成噪声和条件标签进行采样。最终，生成的图像会被保存为"sample.png"文件。通过设置参数，用户可以灵活控制生成的样本质量和样式。

```python
import torch
# 允许使用 TensorFloat-32（TF32）加速计算
torch.backends.cuda.matmul.allow_tf32 = True
torch.backends.cudnn.allow_tf32 = True
from torchvision.utils import save_image
from diffusion import create_diffusion
from diffusers.models import AutoencoderKL
from download import find_model
from models import DiT_models
import argparse


def main(args):
    # 设置 PyTorch 随机种子和禁用梯度计算
    torch.manual_seed(args.seed)
    torch.set_grad_enabled(False)
    device = "cuda" if torch.cuda.is_available() else "cpu"

    # 检查是否提供了检查点路径
    if args.ckpt is None:
        assert args.model == "DiT-XL/2", "只有 DiT-XL/2 模型可自动下载。"
        assert args.image_size in [256, 512]
        assert args.num_classes == 1000

    # 加载模型
    latent_size = args.image_size // 8
    model = DiT_models[args.model](
        input_size=latent_size,
        num_classes=args.num_classes
```

```python
    ).to(device)

    # 自动下载预训练模型或加载自定义的 DiT 检查点
    ckpt_path = args.ckpt or f"DiT-XL-2-{args.image_size}x{args.image_size}.pt"
    state_dict = find_model(ckpt_path)
    model.load_state_dict(state_dict)
    model.eval()   # 将模型设置为评估模式
    diffusion = create_diffusion(str(args.num_sampling_steps))
    vae = AutoencoderKL.from_pretrained(f"stabilityai/sd-vae-ft-{args.vae}").to(device)

    # 设置用于条件生成模型的标签（可以根据需要更改）:
    class_labels = [207, 360, 387, 974, 88, 979, 417, 279]

    # 创建采样噪声:
    n = len(class_labels)
    z = torch.randn(n, 4, latent_size, latent_size, device=device)
    y = torch.tensor(class_labels, device=device)

    # 设置无分类器引导:
    z = torch.cat([z, z], 0)
    y_null = torch.tensor([1000] * n, device=device)
    y = torch.cat([y, y_null], 0)
    model_kwargs = dict(y=y, cfg_scale=args.cfg_scale)

    # 进行图像采样:
    samples = diffusion.p_sample_loop(
        model.forward_with_cfg, z.shape, z, clip_denoised=False, model_kwargs=
        model_kwargs, progress=True, device=device
    )
    samples, _ = samples.chunk(2, dim=0)   # 移除空类样本
    samples = vae.decode(samples / 0.18215).sample

    # 保存和显示图像:
    save_image(samples, "sample.png", nrow=4, normalize=True, value_range=(-1, 1))


if __name__ == "__main__":
    # 定义命令行参数
    parser = argparse.ArgumentParser()
    parser.add_argument("--model", type=str, choices=list(DiT_models.keys()), default="DiT-XL/2")
    parser.add_argument("--vae", type=str, choices=["ema", "mse"], default="mse")
    parser.add_argument("--image-size", type=int, choices=[256, 512], default=256)
    parser.add_argument("--num-classes", type=int, default=1000)
    parser.add_argument("--cfg-scale", type=float, default=4.0)
    parser.add_argument("--num-sampling-steps", type=int, default=250)
    parser.add_argument("--seed", type=int, default=0)
    parser.add_argument("--ckpt", type=str, default=None,
                        help="可选的 DiT 检查点路径（默认：自动下载预训练的 DiT-XL/2 模型）。")
    args = parser.parse_args()
    main(args)
```

10.6.2 基于 DDP 的图像生成

文件 sample_ddp.py 实现了基于 DDP 的图像生成，加载训练好的 DiT 模型并利用扩散模型进行采样。首先，它设置了分布式环境并初始化随机种子，确保每个 GPU 在采样时能够生成不同的随机数。接着，它加载预训练模型和相应的变换器，创建保存采样结果的文件夹，并计算每个 GPU 需要生成的样本数量。代码通过循环进行图像采样，使用分类无关引导（CFG）可选地增强生成效果，最后将生成的图像保存为 .png 文件，并在所有进程完成后将图像汇总为一个 .npz 文件，方

便后续使用和分析。

```python
def create_npz_from_sample_folder(sample_dir, num=50_000):
    """
    从文件夹中的 .png 样本构建一个单一的 .npz 文件。
    """
    samples = []
    for i in tqdm(range(num), desc="Building .npz file from samples"):
        sample_pil = Image.open(f"{sample_dir}/{i:06d}.png")  # 打开样本图像
        sample_np = np.asarray(sample_pil).astype(np.uint8)  # 将图像转换为 numpy 数组
        samples.append(sample_np)
    samples = np.stack(samples)  # 将所有样本堆叠成一个数组
    assert samples.shape == (num, samples.shape[1], samples.shape[2], 3)  # 确保样本形状正确
    npz_path = f"{sample_dir}.npz"  # .npz 文件路径
    np.savez(npz_path, arr_0=samples)  # 保存为 .npz 文件
    print(f"Saved .npz file to {npz_path} [shape={samples.shape}].")
    return npz_path


def main(args):
    """
    运行采样。
    """
    torch.backends.cuda.matmul.allow_tf32 = args.tf32
    # 允许使用 TF32,提升性能,但可能导致小的数值差异
    assert torch.cuda.is_available(), "Sampling with DDP requires at least one GPU. sample.py supports CPU-only usage"
    torch.set_grad_enabled(False)  # 禁用梯度计算

    # 设置 DDP:
    dist.init_process_group("nccl")  # 初始化进程组
    rank = dist.get_rank()  # 获取当前进程的排名
    device = rank % torch.cuda.device_count()  # 根据排名选择设备
    seed = args.global_seed * dist.get_world_size() + rank  # 计算随机种子
    torch.manual_seed(seed)  # 设置随机种子
    torch.cuda.set_device(device)  # 设置当前设备
    print(f"Starting rank={rank}, seed={seed}, world_size={dist.get_world_size()}.")

    # 检查模型和参数
    if args.ckpt is None:
        assert args.model == "DiT-XL/2", "Only DiT-XL/2 models are available for auto-download."
        assert args.image_size in [256, 512]
        assert args.num_classes == 1000

    # 加载模型:
    latent_size = args.image_size // 8  # 计算潜在空间大小
    model = DiT_models[args.model](
        input_size=latent_size,
        num_classes=args.num_classes
    ).to(device)  # 将模型移动到设备上

    # 自动下载预训练模型或加载自定义检查点:
    ckpt_path = args.ckpt or f"DiT-XL-2-{args.image_size}x{args.image_size}.pt"
    state_dict = find_model(ckpt_path)  # 查找模型权重
    model.load_state_dict(state_dict)  # 加载权重
    model.eval()  # 切换到评估模式
    diffusion = create_diffusion(str(args.num_sampling_steps))  # 创建扩散模型
    vae = AutoencoderKL.from_pretrained(f"stabilityai/sd-vae-ft-{args.vae}").to(device)
    # 加载 VAE
    assert args.cfg_scale >= 1.0, "In almost all cases, cfg_scale must be >= 1.0"
    using_cfg = args.cfg_scale > 1.0  # 判断是否使用分类器自由引导
```

```python
# 创建保存样本的文件夹:
model_string_name = args.model.replace("/", "-")  # 替换模型字符串
ckpt_string_name = os.path.basename(args.ckpt).replace(".pt", "") if args.ckpt else
"pretrained"  # 检查点名称
folder_name = f"{model_string_name}-{ckpt_string_name}-size-{args.image_size}-vae-{args.vae}-" \
              f"cfg-{args.cfg_scale}-seed-{args.global_seed}"  # 生成文件夹名称
sample_folder_dir = f"{args.sample_dir}/{folder_name}"  # 保存样本的文件夹路径
if rank == 0:
    os.makedirs(sample_folder_dir, exist_ok=True)  # 创建文件夹
    print(f"Saving .png samples at {sample_folder_dir}")
dist.barrier()  # 所有进程同步

# 计算每个 GPU 需要生成的样本数量及需要运行的迭代次数:
n = args.per_proc_batch_size  # 每个进程的批处理大小
global_batch_size = n * dist.get_world_size()  # 全局批处理大小
# 为了使数量可被整除,我们将多采样一点,然后丢弃多余的样本:
total_samples = int(math.ceil(args.num_fid_samples / global_batch_size) *
global_batch_size)
if rank == 0:
    print(f"Total number of images that will be sampled: {total_samples}")
assert total_samples % dist.get_world_size() == 0, "total_samples must be divisible
by world_size"
samples_needed_this_gpu = int(total_samples // dist.get_world_size())
# 每个 GPU 需要的样本数量
assert samples_needed_this_gpu % n == 0, "samples_needed_this_gpu must be divisible
by the per-GPU batch size"
iterations = int(samples_needed_this_gpu // n)  # 迭代次数
pbar = range(iterations)
pbar = tqdm(pbar) if rank == 0 else pbar  # 进度条
total = 0
for _ in pbar:
    # 生成样本输入:
    z = torch.randn(n, model.in_channels, latent_size, latent_size, device=device)
    # 随机噪声
    y = torch.randint(0, args.num_classes, (n,), device=device)  # 随机类别标签

    # 设置分类器自由引导:
    if using_cfg:
        z = torch.cat([z, z], 0)  # 复制噪声
        y_null = torch.tensor([1000] * n, device=device)  # 无效类别标签
        y = torch.cat([y, y_null], 0)  # 合并标签
        model_kwargs = dict(y=y, cfg_scale=args.cfg_scale)  # 模型参数
        sample_fn = model.forward_with_cfg  # 使用带有配置的前向方法
    else:
        model_kwargs = dict(y=y)  # 模型参数
        sample_fn = model.forward  # 使用普通的前向方法

    # 采样图像:
    samples = diffusion.p_sample_loop(
        sample_fn, z.shape, z, clip_denoised=False, model_kwargs=model_kwargs,
        progress=False, device=device
    )
    if using_cfg:
        samples, _ = samples.chunk(2, dim=0)  # 移除无效类别样本

    samples = vae.decode(samples / 0.18215).sample  # 解码样本
    samples = torch.clamp(127.5 * samples + 128.0, 0, 255).permute(0, 2, 3, 1).to
("cpu", dtype=torch.uint8).numpy()  # 转换为图像格式

    # 将样本保存为单个.png 文件
    for i, sample in enumerate(samples):
        index = i * dist.get_world_size() + rank + total  # 计算保存的索引
```

```python
            Image.fromarray(sample).save(f"{sample_folder_dir}/{index:06d}.png")
            # 保存图像
        total += global_batch_size   # 更新总样本数量

    # 确保所有进程在尝试转换为.npz 文件之前已成功保存其样本
    dist.barrier()
    if rank == 0:
        create_npz_from_sample_folder(sample_folder_dir, args.num_fid_samples)
        # 转换为.npz 文件
        print("Done.")
    dist.barrier()
    dist.destroy_process_group()   # 销毁进程组


if __name__ == "__main__":
    parser = argparse.ArgumentParser()
    parser.add_argument("--model", type=str, choices=list(DiT_models.keys()), default=
    "DiT-XL/2")
    parser.add_argument("--vae",  type=str, choices=["ema", "mse"], default="ema")
    parser.add_argument("--sample-dir", type=str, default="samples")
    parser.add_argument("--per-proc-batch-size", type=int, default=32)
    parser.add_argument("--num-fid-samples", type=int, default=50_000)
    parser.add_argument("--image-size", type=int, choices=[256, 512], default=256)
    parser.add_argument("--num-classes", type=int, default=1000)
    parser.add_argument("--cfg-scale",  type=float, default=1.5)
    parser.add_argument("--num-sampling-steps", type=int, default=250)
    parser.add_argument("--global-seed", type=int, default=0)
    parser.add_argument("--tf32", action=argparse.BooleanOptionalAction, default=True,
                        help="默认使用TF32 矩阵乘法。这大幅加速了在Ampere GPU 上的采样。")
    parser.add_argument("--ckpt", type=str, default=None,
                        help="可选的DiT 检查点路径（默认：自动下载预训练的DiT-XL/2 模型）。")
    args = parser.parse_args()
    main(args)
```

10.7 调试运行

1. 设置环境

（1）创建 Conda 环境：项目中提供了一个 environment.yml 文件，用于创建 Conda 环境。如果您只想在 CPU 上运行预训练模型，可以从该文件中移除 cudatoolkit 和 pytorch-cuda 的依赖项。使用以下命令创建环境：

```
conda env create -f environment.yml
```

（2）激活环境：创建完成后，通过如下命令激活新的 Conda 环境：

```
conda activate DiT
```

通过上述步骤，即可成功设置好运行该项目所需的环境。接下来，可以开始使用预训练模型或进行模型训练。

2. 训练前的准备

要在一个节点的单个 GPU 上提取 ImageNet 特征，请运行以下命令：

```
torchrun --nnodes=1 --nproc_per_node=1 extract_features.py --model DiT-XL/2 --data-path /path/to/imagenet/train --features-path /path/to/store/features
```

- torchrun：用于运行分布式训练的命令。

- --nnodes=1：指定使用一个节点。
- --nproc_per_node=1：指定每个节点使用一个进程（GPU）。
- extract_features.py：提取特征的脚本。
- --model DiT-XL/2：指定要使用的模型。
- --data-path /path/to/imagenet/train：指定包含 ImageNet 训练数据的路径。
- --features-path /path/to/store/features：指定存储提取特征的路径。
- 通过上述命令，即可成功提取 ImageNet 数据集的特征，为后续的模型训练做好准备。

3. 训练 DiT

本项目提供了多个训练 DiT 模型的脚本文件，其中默认的脚本文件是 train.py。该文件用于训练基于类别条件的 DiT 模型，但也可以轻松修改以支持其他类型的条件。

（1）使用单个 GPU 在一个节点上启动 DiT-XL/2 (256×256) 训练：

```
accelerate launch --mixed_precision fp16 train.py --model DiT-XL/2 --features-path /path/to/store/features
```

（2）使用多个 GPU 在一个节点上启动 DiT-XL/2 (256×256) 训练：

```
accelerate launch --multi_gpu --num_processes N --mixed_precision fp16 train.py --model DiT-XL/2 --features-path /path/to/store/features
```

- accelerate launch：用于启动训练的命令。
- --mixed_precision fp16：启用混合精度训练以提高性能。
- --multi_gpu：指示使用多个 GPU 进行训练。
- --num_processes N：指定使用的 GPU 数量。
- --features-path /path/to/store/features：指定存储提取特征的路径。

此外，还可以使用 "train_options" 文件夹中的训练脚本，通过里面的脚本文件，可以根据需要灵活地进行训练。

4. 生成图像

（1）可以使用文件 sample.py 从提供的预训练 DiT 模型进行采样，会根据使用的模型自动下载预训练 DiT 模型的权重。该脚本具有多种参数，可以在 256×256 和 512×512 模型之间切换、调整采样步骤、改变无分类器引导尺度等。例如，要从 512×512 DiT-XL/2 模型进行采样，可以使用以下命令实现：

```
python sample.py --image-size 512 --seed 1
```

（2）自定义 DiT 检查点。

如果使用文件 train.py 训练了新的 DiT 模型，可以添加 --ckpt 参数来使用自己的检查点。例如，要从自定义的 256×256 DiT-L/4 模型的 EMA 权重中进行采样，可以运行以下命令实现：

```
python sample.py --model DiT-L/4 --image-size 256 --ckpt /path/to/model.pt
```

通过这些选项，可以灵活地使用预训练模型或自定义模型进行图像采样。

5. 性能评估

使用文件 sample_ddp.py 脚本，可以并行从 DiT 模型中采样大量图像，并生成一个样本文件夹和一个 .npz 文件。这个 .npz 文件可以直接用于 ADM 的 TensorFlow 评估工具，计算 FID（Fréchet Inception Distance）、Inception Score 和其他评估指标。具体地，可以通过指定 GPU 数量和要生成的样本数量来执行此操作，以便获取模型的性能评估数据。例如运行如下命令在分布式环境中并行生成图像并进行性能评估。

```
torchrun --nnodes=1 --nproc_per_node=N sample_ddp.py --model DiT-XL/2 --num-fid-samples 50000
```

运行上述命令后,将在一个节点上并行启动多个进程,从 DiT-XL/2 模型中生成 50000 张图像,并为后续的 FID 计算做好准备。

最终,通过使用预训练的 DiT 模型,用户可以根据文本输入生成高质量的图像,如图 10-1 所示。项目中的 sample.py 脚本支持使用训练好的模型进行图像生成,用户可以调整参数以满足不同的需求。整体上,项目提供了从特征提取到训练和图像生成的完整流程。

图 10-1　生成的图像